Self-Build
Manual

First published as *Build Your Own House* manual in 2010
Re-issued and updated as *Self-Build Manual* in January 2015
Reprinted April 2019

British Library Cataloguing in Publication Data:
A catalogue record for this book is available
from the British Library

ISBN 978 0 85733 803 7

Published by Haynes Publishing,
Sparkford, Yeovil, Somerset BA22 7JJ, UK
Tel: 01963 440635
Int. tel: +44 1963 440635
Website: www.haynes.com

Haynes North America Inc.
859 Lawrence Drive, Newbury Park, California 91320, USA

Printed in Malaysia

Acknowledgements

SPECIAL THANKS TO:
Basil Parylo MBEng Leeds City Council LABC and Luke Allin for
SAP advice.
Naomi Handford-Jones at The National Self Build & Renovation Centre.
Nigel Griffiths, author of Haynes Eco House Manual

PHOTOGRAPHY
Basil Parylo and the Author, except where credited

Thanks also to:
Charles Tidmarsh of Paddick Engineering, F&P Architectural Services.
Buildstore's NSBRC Swindon – especially Debbie Buggins and Jaclyn
Thorburn for assistance with photos and diagrams and David Snell for
general inspiration. Thanks also to architects Mark Kingsley, Peter Kent
and Piers Taylor and all who kindly submitted photos.

Self-Build
Manual

HOW TO PLAN, MANAGE AND BUILD THE HOME OF YOUR DREAMS

Ian Alistair Rock MRICS

CONTENTS

Self-Build Manual

1 INTRODUCTION

Photo: oakwrights.co.uk

Photo: Buildstore.co.uk

This book is about designing and building your own dream home – a stylish new property that's affordable and realistic to construct. Today, more and more people are doing just this – taking the plunge and creating the new home they really want, rather than settling for an overpriced developer-built house.

But before we start, let's nail one of the big myths about self-build. Despite the name, you don't have to physically construct the house yourself. In fact, to count as a 'self-builder' you needn't carry out any hands-on building work at all. What you must do is keep control of the overall process. Of course, those blessed with advanced building skills may relish tackling multiple tasks on site. But most self-builders find the best way to make an effective contribution is by helping organise their project, perhaps providing some DIY input along the way.

Regardless of the extent of personal involvement, most self-built houses not only cost less than equivalent second-hand properties but are of considerably better quality. Which is why more than 10,000 people in the UK make the decision each year to build their own home. This is more

than most corporate housing developers, accounting for around 5% to 10% of all new homes built in a typical year. Given the chance many more of us would take this route – an incredible 1 in 7 of the UK population say they want to build their own home.

Despite the impressive numbers, in this country we are still largely reliant on mass-produced housing, and the self-build sector is a fraction of the size of those in Australia, the US, and Japan. In other European countries the self-build market typically accounts for around half of all house production. In Germany it's common to commission one-off homes from local builders.

The main obstacle facing self-builders in Britain is, of course, finding the right plot. But times are changing. Under 'Right to Build' legislation local councils are required to release plots of land for sale at market value to self-builders. Even some mainstream housing developers are making sites of serviced plots available for large-scale 'custom-build' projects.

Photo: oakwrights.co.uk

Photo: oakwrights.co.uk

Building by numbers

- Average project cost of a self-build including the land: £ 260,000
- Average UK new build house size: 97m²
- % of total cost excluding land spent on materials: 56% (44% on labour)

Why self-build?

There are several good reasons for building your own home. Above all it's a way of acquiring the home you always wanted. Do it right, and you should end up with a house of better quality and value than if you'd simply bought a bog-standard, ready-made property. You will also have a valuable asset to raise funds against in the future.

Quality

One of the big attractions of self-building is that it affords you an excellent opportunity to create a truly impressive home, by venturing beyond the usual basic design, fittings, and materials.

Housing developers build to minimum standards at maximum speed, in order to increase profits. As a result buyers of new 'off the peg' houses commonly complain of paper-thin walls with poor sound insulation, squeaky floors, cracked plasterwork, and poor joinery and finishing, as well as inconveniently positioned sockets, switches, and radiators. In fact seven out of ten buyers of new homes notice faults after moving in. As the old saying goes, 'if you want a job done well, do it yourself'.

Whilst location and price remain the principal reasons behind housing choices, green design for low energy bills has become increasingly important. In Scandinavia, Holland, and Germany no one would dream of building a house without incorporating sophisticated energy-saving measures.

Price

There is a general belief that as a self-builder you will save large sums of money. But can you realistically expect to build a better quality home and still walk away with plenty of change? Obviously, the amount you save will depend to a large extent on how much time and work you're able to personally dedicate to the project. It also very much depends on how well things are managed.

Hard-nosed housing developers operate on a very approximate ratio of about 1/3 of the cost of a new home accounted for by the land, 1/3 construction costs (split about 40% materials and 60% labour) and up to 1/3 profit. Take for example a new house priced at say £300,000. The developer might have made as much as £100,000 profit, having spent about £60k on labour and £40k on materials. Potentially, as a self-builder, you could save all the labour costs as well as walking away with 100% of the profit!

Achievement

To make something that you can be proud of is a fundamental desire and a satisfying achievement. But in post-industrial Britain, modern working life affords few opportunities to create something worthwhile. The fact is, self-build offers greater rewards than simply getting a bargain house. You

Photo: oakwrights.co.uk

Photo: Piers Taylor Architects

should end up with a major accomplishment that you can proudly tell your children about.

Team spirit

Once bitten by the self-build bug, you somehow become part of a 'band of brothers', the community of like-minded people who've survived building site adversity and lived to tell the tale. This natural sense of camaraderie can extend even further with community self-build schemes. An alternative to a stand-alone build, such projects involve several individuals pooling their resources and working together, perhaps on a shared plot. Being involved with a joint programme as part of a small team means each person can trade building skills and jointly arrange funding, benefiting from economies of scale.

Before you start

Before you rush out to start digging foundation trenches, there are some sobering questions that need to be answered. After all, this project may end up being the biggest single investment of your life.

Can I afford it?

Simple. If you can afford a detached house in the area you want to build, you can afford to build your own home.

Photo: oakwrights.co.uk

If self-build is so brilliant, why isn't everyone doing it?

Building your own house is clearly a major commitment. For some it turns out to be a life-changing experience. Either way it's not something you want to enter into lightly. Not everyone has the necessary time or resources.

But there's one major disincentive: building plots can be hard to find. The planning system tends to conspire against individual self-builders, because when it comes to property development, local Councils 'think big' in terms of grandiose plans and statements. It's often easier to get permission for a huge new estate in the green belt than for a one-off home.

But there's another reason why fewer Brits take the plunge. Unlike Canadians or Scandinavians, we're culturally brought up to accept what's available. Which of course perfectly suits large developers, planners, and politicians.

Who exactly is a 'self-builder?'

Anyone who organises the design and construction of their own home counts as a 'self-builder'. These are homes commissioned by the people who plan to live in them.

If you're not sure whether you're the 'right sort of person' to self-build, remember that people from all walks of life have successfully built their own homes. You can get expert help at almost every stage.

However, there are two personal qualities that you will definitely need. Above all you need to be enthusiastic. Trades are likely to work harder and faster if they sense your enthusiasm. You also need generous reserves of patience, because the reality on site is that you're always waiting for someone or something – a particular trade, an urgent delivery, or just some dry weather. Even the best-laid plans can go wrong, and unexpected obstacles seem to arise out of nowhere.

Tommy Walsh – celebrity self-builder

Photo: Potton

How much should you do yourself?

If you can boast competent DIY skills, then you may be able to make a major contribution on site. If you've cut your teeth building a home extension, you'll be especially well prepared. But it's important to be realistic and recognise where professional assistance is likely to be needed.

If this is a team effort with your partner, one or both of you may have other skills to offer, such as being a good organiser, communicator, or workforce motivator. Many couples divide responsibilities between them, perhaps with one doing design work and the other financial management. Accounts and IT skills are useful for project management, or you may have interior design flair, the ability to draw plans, or an aptitude for things like joinery, plumbing, tiling, or decorating.

But there's one thing above all that you will need to personally contribute: it's essential that you're able to devote enough *time*, which will mean sacrificing other commitments such as work and family.

How long will it take?

The time from finding your building plot through to moving in to the finished house could be as short as 6 to 12 months if you employ a professional builder. But if you decide to do a lot of the work yourself, the build can take much longer. A typical self-build project can take a couple of years. It will depend on many factors, not least the time needed to find and buy a suitable site.

Probably the most frustrating stage is when you've made the decision to do a self-build and organised funding, but

finding a suitable site lies somewhere in the future. However, some of this time can be profitably and enjoyably spent developing your design, so that it's ready to activate directly your plot materialises.

Preparation

If you think you can contribute the time, and muster up the necessary skills and enthusiasm, the first step is to work out what you can afford. The crunch may come when you compare your budget with what you actually want from a house, and discover what you can realistically expect to build for your money. This is all part of the process of developing a picture of how your home could look.

Preparing a budget

Before embarking on a major development you need to put together a budget. At this early stage you won't know how much it will cost to build your house, so the figures will only be best guesses.

You probably won't know how much the plot's going to cost, or the level of specification, or whether construction costs are going to increase over the next year or so. For this reason your costings need to be regularly updated as you progress.

TYPICAL BUILD SCHEDULE

1 CLEAR AND PEG SITE MONTH 1
Get surveys to work out levels and peg out site as per drawings.

2 FOUNDATIONS MONTH 1
Dig trenches and begin to build up the blockwork footings.

3 FLOOR STRUCTURE MONTH 1
Finish footings and install ground floor slab or beam and block.

4 SUPERSTRUCTURE MONTH 2
Blocks or frame used to build external walls to first floor height.

5 SECOND LIFT MONTH 3
First floor structure installed, internal walls continue to roof.

6 ROOF MONTH 3
Walls finished, roof structure installed and covered with slates or tiles.

7 WEATHERTIGHT MONTH 4
Windows, doors and soffits fitted to make house weathertight.

8 FIRST FIX MONTH 5
Electrical and plumbing runs put in place – the stuff you won't see.

What comes first?

In an ideal world, before getting too committed you would first appoint a designer to draw up detailed plans of your dream home. Then you would find a stunning location to build it, and after buying the plot obtain planning permission – because land without consent costs only a fraction of the price. Finally, you would construct the dwelling.

Unfortunately, in reality it's impossible to do things in this order because the potential for disaster is far too high. Suppose, having paid for land and paid the architect, you found you couldn't get planning permission?

So most self-builders pursue a well-trodden path, where compromise is usually required to make your ideal design fit the realities of the plot and the budget:

- Organise your finances and draw up a budget.
- Identify a suitable plot of land with outline planning permission.
- Buy the plot, having checked ground conditions, legal conditions, and access.

Photo: oakwrights.co.uk

- Draw up detailed house plans to fit the plot and budget.
- Apply for detailed planning consent.
- Prepare the site and begin construction once full planning is granted.

Living arrangements

There is one major expense that's directly within your control – where you live during the build. Of course, the ideal arrangement would be to stay in the comfort of your existing home until the new place is ready. But for most people, raising funds to pay for the build involves liquidating their most valuable asset before starting on site. Clearly, the more you can save on accommodation, the more you can spend on the build, helping you afford a better home. Hence the need for a temporarily frugal existence. So if you can't live at home, and can't charge it to MP's expenses, what are the alternatives?

LIVING ON SITE

Slumming it in an old caravan on site for the duration of the build has become part of the culture of self-build. Living on site is a real plus when it comes to security and deliveries. It can even help develop a sense of shared adventure. And some hardy souls claim that enduring such hardships can actually enhance the ultimate feeling of pleasure when you finally move in.

GOING INTO RENTED

If the idea of living a hippy-traveller lifestyle doesn't float your boat, then rented accommodation may be a useful compromise. But if the rent costs about the same as your mortgage, you're not going to save much in terms of monthly outgoings.

Whichever form of temporary accommodation you plump for, there are four key points to consider:

- The impact on your family – as you're going to need their support.
- The distance you and your family have to travel to work and to school etc.
- Travelling time from accommodation to site.
- Any other major life-events that are likely to occur – babies, career change, or divorce!

9 PLASTERING	MONTHS 6–8
Usually a skim coat on top of plasterboard, after first fix complete.	
10 SECOND FIX	MONTHS 6–8
Connecting up the electrics and heating systems to include sockets.	
11 FIXTURES	MONTH 9
Kitchens and bathrooms in, flooring down, joinery complete.	
12 SNAGGING	MONTH 10
Builders back to deal with any issues, final exterior finishes.	

Website

For further technical information, plus sample documents and details of specialist firms, visit our website at www.Selfbuild-homes.com.

2 RAISING THE MONEY

Photo: oakwrights.co.uk

There's no point drooling over stunning drawings of beautiful dream homes unless you can actually afford to build one. So before going too far down the design route, one of your first tasks is to take a long hard look to see much money you can scrape together.

Where does the money come from?

A typical self-builder's budget is made up from:

- Mortgage funding
- Equity in your existing home
- Cash that you can raise

This is really no different from anyone looking to just move house, except that you'll need a special self-build mortgage (see page 16). But unlike people who are simply moving house, you will be responsible for a complex building project that depends for its success on smart budgeting.

Setting the budget

At this stage there's a great temptation to rush ahead and start sketching out designs and hunting for plots. But the fact is, you'll pretty soon get into trouble unless you budget properly. Without the right funding in place you could easily miss out on a great plot, or end up with a disastrous project that you can't afford to finish. However, at this early stage, without either a plot or a house design, it's obviously difficult to accurately predict costs. But you have to start somewhere, so some fairly broad assumptions need to be made about the likely cost of land and the size of house you hope to build.

It's important not to be over-optimistic or you may run into difficulty if the actual costs turn out significantly higher. Self-builders tend to be optimists by nature and are sometimes tempted to put the lowest possible costs into the budget. But there are very few true bargains in the development world – you tend to get what you pay for, so it's best to treat low figures with a degree of healthy suspicion. Ignore sales people fanning the flames of enthusiasm and architects who may disregard irritating budget constraints that stand in the way of their masterpiece. TV programmes and property magazines rarely highlight the failures – projects that crash in debt and divorce. With a bit of luck you may do extremely well financially – but treat this as a bonus.

Fortunately, the single biggest influence on the cost of your home is very much under your control. And that's how you run the project. Once you've set a realistic target budget you need to aim to bring the project in at that figure, regularly checking costs and updating your initial budget assumptions.

Where will all the money go?

The bulk of you budget is going to be spent on the three major costs of any building project – the land and those twin construction costs, materials and labour. But there are also a whole lot of additional 'hidden' costs that you'll need money to pay for, and must be included in your budget:

- Professional fees – architects, engineers, surveyors
- Local Authority planning fees
- Local Authority Building Control fees
- Insurances
- Warranties

- Finance and mortgage fees
- VAT paid out (before being reclaimed at the end)
- Payments for new service supplies for water/drainage/electric/gas etc
- Contingency fund
- Moving home fees and furnishings
- Site purchase costs and conveyancing

Land and construction

As a rule of thumb, your plot is likely to cost somewhere between a third and a half of your total budget. Anyone selling land is going to charge as much as they can possibly get, and in land-starved areas, supply and demand can dictate higher prices. Only once you've found your plot, and know exactly what it's going to cost, will you know how much money's left for the building budget.

The next big question is how much the building work will cost. Without a detailed design, all you can do at this stage is to estimate overall costs in terms of pounds per square metre. There are many factors that affect construction costs. Prices in London and major UK cities can be as much as 50% higher than average, whereas in some rural areas prices can be significantly cheaper. This is largely down to differences in local labour prices – the cost of materials is pretty similar wherever you build. So imagine how much you could save if you did all the building work yourself!

Photo: buildstore.co.uk

Estimating your build costs

Perhaps the most common complaint about modern mass-produced estate houses is their compact size. One of the attractions of designing your own home is to have a larger floor area.

This begs the question how exactly do you calculate 'floor area'? Just to make life confusing, different people measure it differently. Normally, for construction purposes it's the 'gross external floor area', which means the total area of all

Typical floor areas

	Gross external floor area
Small two-bedroom bungalow	75m²
Modern three-bedroom semi-detached house	100m²
Modern four-bedroom detached house	135m²
'Executive' four or five-bedroom family house or bungalow (study, utility, and en-suite to master bedroom)	200m²
Single garage	15m²
Double garage	28m²

What affects the cost of building your house?

Photo: Starline pools

HOUSE SIZE VERSUS QUALITY

The bigger your grand design, the more cash it will guzzle. So if you're building to super-generous proportions, you may need to trade off size against quality. For the same money you could build a smaller, higher quality house, or instead plump for a larger design with standard quality fittings to make it affordable. Such vows of austerity are, of course, easier said than adhered to, because there's always a strong temptation to upgrade on quality as the project progresses, taking you way over budget.

It's surprising how many people get carried away specifying top-of-the-range designer fittings, and are then genuinely shocked to discover that the project is running horribly over budget.

Once you know the size of plot and planning restrictions, the limits on size should become more apparent. But bear in mind that, in the long run, cheap materials tend to cost more in maintenance and can ultimately be a false economy.

Photo: opus.eu

BUILDING FOR PROFIT OR LIFESTYLE?

Most self-builders are 'building the dream', and the dream doesn't include bits dropping off after a few months. Investing in a quality lifestyle for yourself over the long term is a very different project from building speculatively, where to maximise profit you only spend money where it will help secure a higher sale price.

WHERE TO BUILD

It's no surprise that the dearest place to build in the UK is Central London, followed by South-East England. Construction in less densely populated parts of the country, such as rural Wales or North Yorkshire, can cost as much as 25% less. If you're developing in a posh postcode area, builders may be tempted to load their prices accordingly. Another major potential drain on funds is due to site limitations. Sites with problems, such as poor ground conditions and limited access, can very quickly run up big extra bills, reducing the benefit of cheaper labour costs in some more remote areas. Unless you deducted these from the price of the plot, they can seriously eat into your budget.

DIY INPUT

One of the big attractions of self-building is the prospect of saving money by doing some of the work yourself, thereby affording a

larger house built to a higher specification. But to be realistic, you need to account for the cost of your own time. It's pointless to slave away all day killing yourself doing hard labour when you could be earning money elsewhere to pay a professional builder.

How well you manage your project will also be crucial in keeping it on budget. To manage professionally, it's essential to have a plan with target dates for completing each phase.

When the going gets tough, there's a simple psychological trick than can help keep you going. Make a list of the three most important reasons why you're doing a self-build. Stick the list on the door of the site hut or mobile home so you see it every day. It just helps.

KEEPING IT ON BUDGET

As we shall see in Chapter 10, effective management of your project

is the key to bringing it in on budget. This means paying special attention to the following:

Preparation – Thoroughly investigating your plot is a major part of going in with your eyes wide open. The same goes for the design, because the more that's worked out in advance, the fewer expensive extras there will be. Design isn't just about making things look

good: it should design-out problems long before they can cause trouble – saving money and hassle.

Cost control – The reason costs sometimes escalate alarmingly is not always down to blowing big money on one or two luxury items. A whole bunch of small decisions made over a period of time often collectively result in tenders coming in way over budget. Time is money,

Photo: Roca baths

so keeping the project on schedule will keep costs down. Keeping accurate records is vital for effective project management and avoiding disputes, and is essential when it comes to claiming back every penny of VAT at the end of the build. When it comes to paying people, always pay promptly, and wherever possible pay in stages. Another golden rule is to never pay in advance, before a product or service has been provided.

Pick the right team – The team you choose should be strong in those areas where you personally are not. This means being aware of your strengths and weakness. A good builder or designer will usually pay for themselves in money saved, quality achieved, and stress avoided. A more expensive builder who's reliable, honest, and efficient will normally work out cheaper in the long run than one who quotes a very low initial price and then stings you with extortionate charges for extras and generally messes you about. And if there's only one supplier for something you want, it's still worth creating the impression that you're getting other prices.

Don't keep changing your mind – You can't be expected to decide on every last nut and bolt at the design stage. As we shall see, there are ways around this, such as leaving out particular fittings from tenders and instead using 'Prime cost sums'. But the fact is, changing your mind costs money. If you change something after you've received competitive prices, it will cost you more. But if you change your mind during the build it will cost you *considerably* more.

Photo: oakwrights.co.uk

Deeper foundations drink cash.

habitable floors (*ie* the ground and first floors in a two-storey house). Surveyors measure properties externally to include all the main and internal walls. This is done as a matter of course every time you change your mortgage, to calculate building insurance (the cost of rebuilding if it burned down). But estate agents normally describe 'floor area' as the smaller internal area of all the room sizes added together (*ie* just the usable internal floor area excluding all walls).

Price per square metre

The next step is to come up with a realistic price per square metre for the materials and labour in your area. Estimating build costs by '£ per sq m' can often be surprisingly accurate for an average developer-built home. But standard prices only allow for an average quantity and quality of windows, doors, fittings, and materials. It's the level of specification that's the biggest single factor that determines the ultimate cost of the build, and self-builders usually want better quality fittings and materials, so you may need to budget for a higher than average cost.

Building cost tables can be found in some of the monthly homebuilding magazines. These show different current prices depending on your chosen method of building. The most expensive method is to employ a main building contractor.

Alternatively, you could save money by employing your own direct labour force and doing some of the building work yourself. The difference between the two can be as much as 20%, depending on how much DIY input you contribute and how well the project is managed. Online calculators such as homebuilding.co.uk/buildcosts are worth exploring.

Adjustments

The problem with 'guesstimating' costs is that simple calculations can sometimes be misleading. Suppose, for example, that you've got a total budget of £300,000. If you budget the 'rule of thumb' one-third of this to buy the plot, it would leave you £200,000 to pay for the build. Assuming building costs of £1,500 per square metre you'd be able to afford to build about 135m² size of house. However, one danger with this approach is that you're ignoring all those 'invisible' costs described earlier.

But there's another potential trap. Suppose that you later need to adjust the size of the house you'd originally planned to build? Lopping a few square metres off your design won't necessarily translate into big savings. A 15m² room will normally have the same number of doors, windows, radiators, electrics etc as a 20m² room, so the only saving will be a bit less bricklaying. And even then, the reduction in materials may not make much difference since the surplus may be absorbed from having to order full loads to prevent wastage.

Contingency fund

On any build, there will always be unexpected costs. So it's essential to have some money in reserve.

Contrary to the impression sometimes conveyed by TV property shows, contingency funds are rarely used up by one or two major disasters; they tend to be gradually absorbed by the cumulative effect of a lot of smaller cost overruns throughout the project, as the harsh realities of building on site deflate earlier optimistic assumptions that everything would be plain sailing. On an average site you need to allow an extra 10% of your overall build budget. For more complex designs and challenging sites it's best to allow 20%. If, by some miracle, the sum remains unspent as the project reaches completion, you'll be able to afford a nicer kitchen.

Getting the best mortgage

- ■ Do your homework. Research mortgage lenders or make enquiries with specialist brokers before you draw up your budget or look for sites – shop around.
- ■ Prepare a detailed budget to demonstrate that your project is professionally planned.
- ■ Do a cash flow forecast and try to reduce the amount of cash you need to borrow in the early stages.
- ■ Compare how much cash different lenders will release at what stages.
- ■ Never borrow up to the max – it's wise to leave yourself a financial cushion to cover an emergency.

Mortgages

Before starting any serious plot hunting, its important to have the funds organised to pay for it. Otherwise there will be plenty of others with ready cash waiting to shove you aside and snap it up.

To finance a self-build project you can't just arrange an ordinary mortgage. For a start there's no house yet in existence for the bank to secure it against! So unless you're counting on inheriting a small fortune some time soon, you're likely to need a self-build mortgage.

For many self-builders the processes of buying the land and raising finance happen in parallel. Selling your existing home can provide enough capital to buy the plot, and may also cover the cost of preparing the land and getting the build started. The major difficulty is getting the timing right. You don't want to be roughing it in a pokey old caravan for years on end whilst searching for a site, so the best option is normally to remortgage your existing home to raise ready cash to buy the plot. Then you can take your time organising your self-build mortgage.

Do you qualify?

Self-build loans differ in certain key ways from conventional mortgages, and to qualify for one there are some extra requirements you'll have to satisfy. Firstly, the house you're about to build must be your own private dwelling, not for commercial use or as part of a property development business. Then, once you've secured your plot, you will need to provide a detailed cost analysis, along with drawings and evidence of current planning consent for the site, thus demonstrating that your project is properly planned and not just some wild pipedream.

The bank will normally want to instruct their own independent valuation of the plot, together with a professional estimate of how much the house will be worth once built. This will be done by a qualified chartered surveyor – but guess who's charged for the valuation fee (which could be as much as £500).

Self-build loans typically offer 50% to 75% of the land value and 75% to 80% of the completed property value. The maximum you can borrow, as with conventional mortgages, is typically around 3 to 4 times single income or 1.5 x joint. It may also be based on 'affordability' (taking into account your monthly outgoings, and the cost of monthly repayments in relation to your income). It's best to deal with either a building society or bank that specialises in self-build mortgages, or a trusted intermediary. The downside is that these tend to be smaller lenders who may impose more conditions and restrictions. For example, some insist that you appoint an architect working for an NHBC-registered builder as your project manager – which could add thousands to the cost.

Stage payments

Unlike buying a house with a standard mortgage, a building project incurs costs progressively over time, so you don't need all the money up front in one go. Typically, you need a large dollop of money at the start for the land purchase, followed by a fairly long delay while you get organised before regular payments to builders need to be made.

Self-build mortgages pay out tranches of cash at key stages of the build, usually in arrears. But the amount of money released by the bank in stage payments doesn't relate to how much you've actually spent. They may base it on their surveyor's valuation, and lend a percentage of the estimated value of the project to date. So you need a reliable cash flow forecast to be sure that you'll be able to cover the bills as they arise.

GREEN CHOICE

Green funding

Non-conventional designs can scare away mainstream mortgage lenders. However, a few – such as the Co-operative Bank and Ecology Building Society – actively support eco housing with development loans that can later be converted into mortgages.

One way to seriously boost your cash flow is to arrange an 'accelerator mortgage' that makes stage payments *in advance* of the key stages, rather than in arrears. This helps prevent the stop-start syndrome, the curse of so many self-build projects.

TYPICAL STAGES FOR PAYMENT
- Purchase of plot
- Foundation or DPC level
- Wall plate level (or, for timber frame houses, when the kit is erected)
- Building watertight (i.e. roof and windows in place)
- First fix
- Final completion

Keeping finance costs down
The less you spend on the mortgage, the more you can invest in the property. To minimise payments you should:

- Buy the land using personal funds if possible, so the loan is only needed to pay for the construction. This will reduce the amount borrowed, significantly cutting your interest payments.
- Arrange a 'stage payment' self-build loan, where you only borrow money for each phase of construction (up to a pre-agreed total amount).
- Arrange adequate site insurance, to cover yourself for damage to the building and theft of materials etc.
- Check lending conditions, for example:-

What to check in a self-build mortgage
- What's the maximum Loan To Value ratio (LTV)? What percentage of the plot value do they lend, if any, and how much for construction?
- What's the interest rate? What other upfront charges are there – arrangement fees, legal fees, valuation fees etc? Are there any hidden charges?
- What certification do they want – *eg* NHBC or LABC warranty, Building Control completion certificate, and/or architects certificate?
- Do they insist on you employing a main contractor, or can you employ individual trades and do some work yourself?
- Are you allowed to self-manage the construction? Or do

they insist you employ a professional project manager or architect?
- Do they want full/detailed planning permission before releasing any money, or is outline planning permission sufficient?
- Do they restrict you to using traditional 'brick and block' construction methods?
- What site valuation inspections do they carry out, and will they charge you for these?
- At what stages do they release money, and in what amounts? Are staged payments made in advance or in arrears of each stage? If paid in arrears, you will probably need to borrow more to cover labour and materials.
- Are there any termination/redemption fees when it comes to paying off the mortgage later?
- Are there any penalties for delays that cause the build to go beyond agreed timescales?
- Can you take out an interest-only mortgage (cheaper than a repayment mortgage)?

Protection: insurance and warranties

Take cover
Death, disaster, and gruesome injuries aren't part of the self-build dream. But no matter how much of an organisational genius you are, with a professionally run site, life can still throw you a 'black swan' – an unexpected event. For example, it's not unknown for intruders who've injured themselves whilst stealing from sites to successfully submit legal claims against site owners. So it's essential to insure yourself against all possible claims for personal injury, death, and loss. Be sure to include your family in the cover, as well as site visitors and the people you employ.

The fact is, as soon as you own a plot of land you're immediately responsible for pretty much everything that happens on it. Security fencing can be purchased, or hired for the duration of the build, and depending on the location you may even consider installing security lighting and an alarm system. If you're living on site, perhaps in a mobile home, you'll also need to guard your personal possessions.

Specifically, you need to arrange insurance cover for:

■ Employers' liability (a legal requirement where you employ subbies)
■ Public liability
■ Contract works
■ Plant, tools, and machinery
■ Your own on-site living accommodation
■ Services connection (in case you damage a main drain etc)
■ Legal expenses (eg if you're sued for damages)
■ Damage to buildings (during and after construction)

This boils down to three types of insurance:

■ Contractors' All Risk (CAR)
■ Employers' Liability (EL)
■ Public Liability (PL)

Insurance or warranties?

If a building collapses during the construction period, the cost of rebuilding should be covered by CAR insurance. If a subcontractor is injured as a result of the collapse, this would be covered by EL. But any consequential damage to a neighbour's property (or to a neighbour) would be covered by PL. If your house falls down once it's complete, your household buildings insurance policy should cover it.

Warranties provide cover for structural defects caused by faulty design, workmanship, or materials. So should the building happen to fall down at any time within the warranty period (normally ten years) you should be covered for the cost of rebuilding it, although not for any injury (EL) or damage to others' property (PL).

The best-known providers are NHBC (*Solo* for self-built properties) and LABC (*New Home Warranty*). Warranties are a common condition with self-build mortgages. But even where you don't need a

mortgage, if you think you might want to sell the house within ten years, your purchaser's lender will probably demand one. The same applies if you remortgage in the first few years, which most people do in order to switch to a cheaper deal.

Where an architect (or another suitably insured professional, such as a chartered surveyor) is employed to inspect the building work at various stages, they normally issue a CML Professional Consultants' Certificate at completion. Although not actually a warranty, some lenders accept them in lieu. But any liability the architect has is not passed to subsequent owners. See Chapter 10.

Tax

The taxman is uncharacteristically generous to self-builders. Indeed, one of the great attractions of self-building is the mouth-watering prospect of claiming back all that VAT from the government.

But to make the system work for you, it's essential to play by the rules.

VAT

If you're building a private house in the UK, you can claim back the VAT you've paid out on most of the materials purchased for construction. This is recovered at the end of the project. New build is zero rated, which means you should not pay VAT on the labour. If the builder is zero-rated for his work, then so will the materials he supplies. If you're mistakenly charged VAT by your builder, or you inadvertently pay them VAT for labour, it could prove an expensive mistake, since the taxman may not accept your claim. You just have to ask nicely and hope that the builder gives it back!

If you're converting a non-residential building such as an old school into a dwelling, you can similarly recover the full rate of VAT that you've paid on the materials at the end of the project. However, you will have to pay a reduced rate of 5% VAT on the labour.

Photo: Kermi

Get organised

It's essential to be properly organised right from the start. First of all, get into the habit of keeping accurate records. List every item you order, and make a note of

who supplied them and how they were paid for – by cash, cheque, card, or on account. Above all, retain receipts for everything! Always ask for a VAT receipt and check that the supplier's VAT number is visible. Each receipt must clearly show:

- A description of the goods
- The date items the items were purchased
- The price paid
- The supplier's name
- Their invoice/reference number
- Your name and address on all receipts for more than £100

Do I qualify?

To qualify for a VAT refund, the property you're building must be for use as a residential dwelling (but you can still work from home using a room or two). You don't have to do any of the building work yourself. If you employ builders, you can still claim a VAT refund for any materials that you've purchased and provided them with. Even if you bought a ready-built shell from a builder, you can claim for finishing the 'fitting out'.

But you can't claim for extra work you might do to an already completed building, such as adding an extension, garage, or conservatory.

What's *not* included?

You can get a refund of VAT that you've paid on 'building materials' used to construct your house and its site. However, HMRC have some slightly odd rules about what's included and what isn't (see boxout).

When you buy building materials, most suppliers charge VAT on them at the standard rate. For example, if you buy stuff at B&Q the VAT will be included in the price. But most of the materials are likely to be acquired on your behalf by your builders. It will then depend on their VAT rating whether they then pass on the VAT charge to you, and therefore whether you will in turn need to claim it back from the VATman.

Generally you're not allowed to claim for anything that doesn't form part of the structure of the dwelling, for example 'movable items' like carpets and curtains. You also can't claim for items used outside the site.

What's the deadline?

You only get one pop at claiming back your VAT. You're allowed to submit a single claim, so it's essential to get it right. There's also a strict time limit – you have just three months from the completion date in which to submit your claim. 'Completion' is generally understood to mean the date the Building Control completion certificate was issued. So that you can claim as much as possible, you ideally want to wait until all the building work is complete. But if you apply too late, or use the wrong forms, you could get nothing – an expensive mistake.

If you have good reason to anticipate a delay, contact HMRC, who have a certain amount of flexibility to extend the deadline in valid cases. Payment is normally made to you within 30 working days of your claim being successfully submitted.

Things you *can't* reclaim VAT on

- Professional services, eg project management, architects, surveyors, designers, consultants.
- Electric or gas appliances, such as washing machines, fridges, doorbells, waste-disposal units etc.
- Fitted furniture (but you can claim for fitted kitchen units).
- Carpets and underlay – but you can claim for fixed flooring such as vinyl, ceramic tiles, parquet, and other wooden/laminate systems, and even curtain rails and blinds.
- Garden sheds, greenhouses, ponds, and ornaments.
- The cost of hiring or buying tools and equipment, plant, and machinery, such as scaffolding, skips, ladders etc.
- Consumables not incorporated into the dwelling – such as sandpaper, paintbrushes, white spirit etc.
- The plot of land (on which VAT shouldn't be charged anyway).

FILLING IN THE FORM

It's amazing how interesting a form can suddenly become when you know it's worth thousands of pounds to you! The HMRC website has a 'Self-Build Claim Pack' which includes all the relevant forms and notices. The actual form you need for a VAT refund is known as 'Notice 719 – VAT refunds for do-it-yourself builders and converters'.

Other taxes

STAMP DUTY

Stamp duty is payable on the purchase of land just as it is when buying a house, where the agreed price is above the minimum tax threshold. It is paid by the purchaser on the cost of the plot, but is not payable on building costs, and is usually added to the solicitor's bill.

Because buying a plot of undeveloped land is considerably cheaper than buying a house, one of the attractions of self-building is that you pay less stamp duty. A lot of plots cost less than the minimum threshold, so depending on what part of the country you're in there's a good chance of avoiding this major expense altogether.

Even if your plot costs more than the threshold, at least you won't have to pay it on the full value of the house, as you would when moving home conventionally. At the present rate of 1%, a £200,000 plot would set you back £2,000. More expensive plots will incur 3% to 5% of the full price paid – see website for current rates.

Council Tax

Your new home should not be liable for Council Tax until it's ready for occupation. Normally, if you don't yet live in the house, and it's not furnished, you shouldn't have to pay for six months. However, where a property is yet to be decorated it can still be habitable in the eyes of the council. So where Building Control have issued the completion certificate, confirming that it's structurally sound and physically safe, the property could be considered 'chargeable'.

There are discounts and exemptions, but they're not automatic – you have to apply for them:

Caravans and mobile homes are not exempt, so living on site won't let you off the hook. But if your main home is elsewhere and the caravan is effectively a furnished site office, you can apply for it to be empty-rated for six months. Once your new house is complete, Council Tax can be minimised on an unoccupied home by keeping it unfurnished. It will then be exempt for six months, after which it's subject to a 50% empty charge.

Students in full-time education are completely exempt from Council Tax,

Keeping it unfurnished cuts Council Tax .

so completion could be a good time for you to enrol as a mature student!

CAPITAL GAINS TAX

CGT is not payable on your main home. But if you still own your old home when you occupy your newly built house as your main residence, the old place will become CGT chargeable when you come to sell it (unless you're an MP, obviously). But the gain on any profit made is only taxable for the period that it was no longer occupied as your main home. In most cases your annual exempt amount should easily cover this.

Things look even better when you claim back the VAT!

3 PLOT HUNTING

Many self-builders will tell you that they just 'stumbled across' their plot. This may be perfectly true, but to 'get lucky' you need to be able to recognise an opportunity when you see it and be ready to act quickly to secure it.

The truth is, finding building land in our overcrowded island can be the hardest part of the whole self-build business. You'll be up against stiff competition not only from other self-builders, but also from small local builders and professional land-finders. You may even find yourself pitched against the mighty new homes development industry. But developers need a quick turnaround and an easy profit, and are not generally interested in 'difficult' sites with unusual ground conditions or a lack of mains services because it doesn't make economic sense to devote the necessary time to overcome such challenges just for a one-off house. Consequently, such plots may stay on the market for ages, with the lack of appeal reflected in the price. For self-builders this can be a golden opportunity, and you may even bag a bargain.

There's no point starting to look for sites until you've worked out a budget and have a fairly clear idea of the size and quality of house you want to build. This will tell you roughly how much change you'll have left to afford for the plot to build it on.

When you do finally discover the perfect plot, you need to be ready to spring into action with your finances already in place. Vendors won't want to wait while you sell your house, and there are plenty of other plot hunters out there eager to beat you to it.

The reality for most of us is that you need to allow a considerable time to find the right plot – perhaps a year or more. There are always plots around, as you can see from land-finding agencies' websites. But the more rigid your location requirements, the more plots you'll reject and the longer the process will take.

The main reason for rejecting sites is because, as self-builders, we don't want to compromise. Perhaps we'll accept nothing less than our dream plot – which may not actually exist. So before looking, think carefully about what you really need, and where you'd be prepared to compromise.

Prime location

You probably already have a pretty good idea of your ideal location. For many the self-build dream is set in a rural environment, amidst open countryside or within a charming village. Others may prefer urban living, or a compromise location in the semi-rural outskirts of town. Either way, it's essential to narrow your options and focus on specific areas before going plot hunting. Also consider whether your requirements are too rigid. For example, if you want a classic village location but insist on a cutting-edge steel and glass box design, your chances of getting planning permission will be pretty slim.

The scarcity of virgin plots means that, increasingly, self-builders are willing to consider more than just lush, green fields. Many find their dream plot comes in the guise of a small decaying bungalow occupying a larger plot, ripe for demolition and rebuilding.

Land without planning

After many months of fruitless searching, plot hunters sometimes become a little desperate. There are firms who seek to take advantage by offering what appear to be prime potential sites at bargain prices, marketed as 'land with prospects'. The catch, of course, is there's no planning consent. Land without permission to build isn't a plot, it's a field. Carefully worded sales blurb may strongly suggest that, in time, the land will most likely get planning approval. In truth, however, if there was any realistic prospect of getting consent the owners would have applied for it themselves, and reaped the massive price increase.

For larger plots with potential for two or more units, big developers with knowledge of long-term local planning policies are sometimes willing to take the risk to acquire future development rights. So if anyone offers a bargain plot, unless they're a relative or trusted friend, forget it. Never buy a plot without planning consent, unless you're fully prepared for the risk of not getting it, and can afford to gamble and lose money.

The only exception to this rule, as we shall see later in this chapter, is where you agree to pay only once consent has been granted. This is normally done by taking out a legal agreement known as an 'option'. So it can still be well worth tracking down potential plots without planning.

Building in the garden

Amidst all the excitement of getting ready to go plot hunting, there's one thing that is sometimes overlooked. The perfect plot may have been staring you in the face. It might be possible to build at the bottom of your own garden if there's space to create a driveway to the side of your house. Perhaps there's a decrepit old garage or outbuilding that could be redeveloped. Corner plots often have bags of potential, and even an end terrace house may have scope to extend to the side with a new dwelling.

Plot hunting

Finding a suitable plot requires a mix of good luck, careful planning, and shedloads of perseverance. Unlike conventionally moving house, simply putting your name on an estate agent's list usually isn't the best way to get results. So before you hit the streets, it's essential to draw up a plan of action.

Research

A lot of wasted mileage and unnecessary footslogging can be saved by some thorough research in advance. To plan your search effectively, you first need to carefully select your target areas. Trying to cover too broad a territory will spread your resources thinly. It's best to focus only on a few key areas, such as two or three villages or suburbs, so that you can cover them thoroughly. Even if you're sticking to a neighbourhood you know well, some research will still be required in order to help spot opportunities with fresh eyes.

To familiarise yourself with the detail of a specific area Ordnance Survey (OS) maps are an excellent tool. OS maps at 1:1250 scale show streets and the outline of buildings. This can help pinpoint potential plots by making it easy to spot low-density areas where buildings are spaced well apart. This can be combined with computer images from 'Google Earth' and 'Street View', although these can be some years out of date and resolution is fairly low.

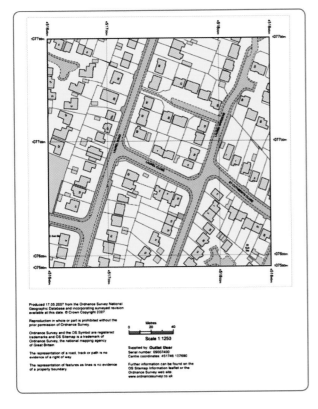

Scouting for land

Armed with a plan that narrows down the areas you're
going to focus on, it's time to get out there and start

scouting around. Initially driving around your chosen areas
can help pick up useful clues, but driving a car limits your
concentration and you can easily miss less obvious potential
sites. Getting out and exploring on foot is the best way to
become familiar with a specific area. It also provides an
opportunity to ask around and have a chat with local
residents in pubs and shops. If there's a post office in the
vicinity, stick a card in the shop window. It's always
possible that somebody will know someone else who's
got suitable land.

Once you're in site-spotting mode it's surprising how
many potential plots seem to pop up in virtually every street.
Although very occasionally a plot will jump out at you in the
form of a private sale board announcing 'Residential Building
Land For Sale', in most cases success comes about from a
mix of footslogging, sleuthing, and sheer persistence.

Having identified a potential plot, be sure to note its
precise location and size, and mark it on a map. Take plenty
of photos and draw sketch plans. If it looks promising the
next step is to track down the owner (see below).

Pre-emptive strikes: land that's not on the market

The smart way to find a plot is to go direct to landowners to
register your interest in advance. This should put you one
step ahead of the competition. But to really be ahead of the
game, you need to identify sites with potential for planning
consent that the owners haven't yet considered selling. But
where should you start looking?

Key characteristics

There are certain essential attributes that land should have in order to
qualify as a potential plot:

- **Space** – Clearly there needs to be enough room to accommodate a
  new house with gardens and parking, and without obviously
  overlooking the neighbours. Sometimes, plots that appear fairly
  small are actually capable of supporting a decent-sized footprint,
  and shouldn't be automatically dismissed.
- **Surroundings** – It's best to look in areas where the existing houses
  are of a similar size and style to the one that you'd like to build.
  Otherwise, if you build your grand mansion in a street of old back-
  to-backs, the chances are you'll be 'over-developing' – ie spending
  far more money on a disproportionately large dwelling than you
  could ever sell it for. This is the opposite of 'under-developing',
  where you fail to make the most of a site's potential.
- **Vehicle access** – Plots without some kind of existing car access from
  the road can make it harder to get planning approval. For remote
  sites, you need to factor in any infrastructure costs such as building
  a new access road.
- **Frontage** – The width of the plot that fronts the road shouldn't be too
  narrow, unless the plot is L-shaped or is accessed by a driveway
  leading from the road.

THE PLANNING REGISTER

When anyone makes a planning application, the details are
published and made available to the public. This explains
why, shortly after submitting an application, you get
bombarded with letters from builders and suppliers. It also
explains how developers sometimes snap up land before it
even gets to the market – by gathering information on
current planning applications.

There's nothing to stop you adopting the same tactics.
What you're looking for is recent applications, preferably for
Outline Planning for one-off houses (ie those without
detailed drawings). With outline consent granted, the
owners may be hoping to sell the plot, since there's no point
them getting detailed plans which may later need to be
changed. However, in some cases the planners insist that full

applications are made, so make a note of these also. Ideally you want those that have just been submitted. This gets you ahead of the queue, as plots aren't normally advertised until permission's been granted, since the consent massively increases the plot value. If you spot these early enough you can make an approach before other potential buyers are even aware.

Most councils make their planning registers accessible online from their websites. You can search by local area and see a list of all recent applications and ones currently under consideration. When you spot a suitable application, note the applicant's name and address, since they're usually owners of the plot, particularly if they live next door to it (there may be a clue in the description, *eg* 'land adjoining 23 High Street'). You will also often find details of an agent who has submitted the application on behalf of the owners.

Having noted any likely-looking applications, the next step is to write to the applicants. Explain that you're hoping to build your own home, and if they want to talk to you about buying their land, you have funds ready and can pay a good price. You might hear nothing, or you could well get a call – a private sale cuts out the middleman and saves them thousands in agents' fees, and a quick sale may be welcome.

Offers from self-builders are often better than those from property developers (or via their agent chums). Developers commonly spin out the purchase hoping to get consent to squeeze an extra house on to the plot. The fact is, as a private buyer you have some aces up your sleeve – you're not only offering to buy the land, but to be their neighbour! If you can arrange to talk to them as potential neighbours, about the quiet life you want to lead, it may overcome their fears. And you can offer them the best possible price right now. See below – *Making contact.*

LOOKING BEYOND THE OBVIOUS
Most people's idea of a building plot is a rectangle of grass neatly marked with a boundary fence or hedge, tucked away along a quiet road. But most available sites with planning permission are not going to be lush country pastures with stunning views. Often reality only bites after you've viewed a succession of dismal plots. The fact is, potential plots often come in disguise. To identify a site you may need to use some imagination, and look beyond the obvious. Empty land can appear deceptively small, so it can be hard to visualise how the site might look with your grand design perched on it.

When people put their houses on the market, they usually go to some trouble to present them attractively, perhaps with the aroma of coffee and baking bread to create the desired effect. Not so with plots, which are often covered with rubbish, profuse undergrowth, and abandoned cars. Owners may even chuck old trash on them in the hope that the builders will get rid of it. Land that's overgrown or

has a derelict building already on it may not look very inspiring. So you have to try and imagine how it would look once an eyesore is removed, and with the finished house on it. In other words, look at the plot as it's going to be, not as it is now.

Infill sites: spotting gaps
It's often possible to get planning consent for small 'infill' developments between existing houses. If there's space beside or behind a house with access from the road, it's a potential plot. Often good infill sites can be found where a

Options – buying land without planning

Earlier we warned of the dangers of buying plots without planning consent. However, in some cases such plots can be profitably considered, if you agree to buy subject to receipt of satisfactory planning permission. You don't actually have to own a plot of land to make a planning application. All that's required is that you serve notice on the owners. But if you don't enter into a prior legal agreement, making planning applications on other people's

land could easily backfire. Such an agreement can either be in the form of an exchange of contracts 'subject to receipt of satisfactory planning consent', or in the form of a legal 'option to purchase' – a very useful legal tool. Without this, there'd be nothing to stop the vendor thanking you for going to the trouble of getting planning permission for them and then promptly selling their land to someone else!

Taking out an option on a site means you enter into a legal agreement with the owner of the land that they will not sell to anyone else for a fixed period of time. If at the end of this period you still haven't bought it, the owner can either keep it or sell to someone else.

Options are widely used by developers to secure the right to buy a piece of land at a future date. In return for payment, the landowner grants the developer an 'option' to buy the plot within an agreed period of years, if the developer can get planning permission. If successful, the developer buys the land at an agreed discount to cover the risk and cost of getting consent and the upfront payment.

However, vendors won't usually want to agree to a long delay unless they know the chance of getting planning is pretty slim. They'll know that getting consent themselves and then selling it will be more profitable.

garage occupies a generous side garden, something that's fairly common in 1930s to 1970s suburbs. A new detached house may only need to be 6m wide or less, and can be squeezed on to a fairly narrow site. Large back gardens where access can be provided to the side of the existing house sometimes lend

themselves to 'backland development' with a new house built at the bottom of the garden.

Start by looking for gaps in the street scene on OS maps. Look out for opportunities in streets where some infill development has already taken place but where there are still generous spaces. Some fairly new houses may have been built in the large back gardens of existing houses. But of

course, maps can be out of date, and there are sometimes physical reasons why a site can't be developed. If an area looks promising, you could go from door to door distributing a leaflet offering a generous cash reward, making it clear that you're a private individual looking to build your own home.

Piecing a plot together

It's possible to assemble a site by taking a bit of space from two or more existing gardens and stitching them together. A small amount of surplus space combined with a similar piece of adjoining garden can create the prefect plot. Assembling the jigsaw pieces of a plot from individual 'parcels of land' in

this way takes considerable patience, tact, and usually a fair amount of time. But when the homeowners realise how valuable a piece of their garden can be, they may suddenly become as keen as you are. Professional developers are wise to the potential of 'site assembly', knowing that it can reap big profits, but they won't be in the market for single-unit sites.

Look for houses with large back gardens and space to the side of the house for car access. Also keep an eye open for properties on corner plots where access could be gained from a side road directly into the garden.

Where a property has a generous road frontage, perhaps with a long fence adjoining the pavement, it could potentially lead to a viable new plot carved from a number of back gardens.

The problem with assembling sites is that you have to negotiate with more than one vendor. The correct approach is to devise the scheme and tie it all up with a legally binding

'option' agreement with each of the owners. This will allow you to purchase the land once planning consent has been achieved. But having a formal agreement in place first is essential, because once you've got planning one of the owners could change their mind and hold you to ransom.

Brownfield sites

Once they're cleared of clutter and buildings, unlikely sites such as old garages or scrapyards can make an excellent

New driveway down the side of the existing house makes it possible to develop rear part of the back garden

location for a small detached house. Such sites are easy to miss, because it can take a lot of imagination to visualise an old factory or petrol station reincarnated as a

beautiful home. Even then there will be issues on planning and possible contamination risk.

Who owns the land?

There's no point going to the trouble of tracing and contacting the owners of a potential plot if it's a non-starter with the planners. So your next step is to visit the planning office to ascertain what chance there would be of obtaining consent for the plot. The planners may be very negative, in which case it's probably not worth pursuing. Or they could simply be non-committal. But in the light of recent government pressure to utilise land more intensively, they may accept the potential to build. If so, ask if they would prefer an outline or full planning application.

If you come away feeling encouraged, you will need to find out who owns the plot. Tracking down the owners can sometimes be a problem, although in many cases you may only need to look as far as the neighbouring houses. The main source of information is the Land Registry (www. landregistry.gov.uk or www.ros.gov.uk). Provided you know the postal address, for a small fee you should be able to inspect maps of the plot online and trace the registered owner – unless, of course, the land isn't registered (see next chapter).

Making contact

This is the hard part – trying to convince the owner to sell it to you. The best approach when making contact with the landowner is to write a pleasant, friendly letter, and if possible arrange a meeting. There's an art to gently commencing negotiations. Above all avoid being too pushy.

It's important to stress that what you're proposing to build won't be detrimental to the future enjoyment of their home or to the value of the adjoining houses. The best approach is to be frank and present yourself as an individual looking to build your own home. Their first reaction might be to tell you to bugger off, but it's surprising how the tempting prospect of a big cash windfall can change minds over time!

When people get planning consent for part of their garden, they are initially delighted at the prospect of untold riches. Then, as time passes, they start to worry about loss of control over what is going to happen to the plot. So you can put their minds at ease by offering some involvement at the design stage. A charm offensive can help them realise that the alternative to your house could be much worse – a developer might cram a monstrous block of flats on to the site. Or you could throw in a sweetener by offering something they've long wanted – perhaps improved access to their garage.

Replacement and conversion

Demolition plots

A large proportion of self-build homes aren't built on greenfield land. They are 'one for one' replacements. There are plenty of older properties, some as recent as the 1970s, that are in need of costly renovation but are sitting on generous plots, so that it often makes economic sense to demolish and rebuild. Such properties are likely to come on to the market in the usual way via estate agents, who may not stress the development potential in their published particulars. Taken at face value such a property may not sell easily, perhaps hanging around for many months, so it's worth registering with local agents for details of properties in need of refurbishment, as well as building plots. But buying run-down properties with a large mortgage can prove difficult, as valuation surveyors are likely to find extensive defects that put off mortgage lenders, with long lists of essential repairs and large retentions.

However, compared to buying a virgin plot, replacement dwellings can offer real advantages. The downside, of course, is the cost of demolition, but this shouldn't make too

much of a dent in your overall budget (even allowing for the extra cost of disposal of any asbestos-based materials). The real bonus comes in the form of services and infrastructure, which can save a small fortune. Ready-connected sewers can provide a major saving, even where some of the old pipework has to be replaced. Then there are further potential savings from having existing driveways, gardens, and fencing.

The big question is always what size replacement the planners will accept. Councils normally limit the size of replacement dwellings to a percentage increase over the original building. You may even be able to negotiate this figure upwards with your knowledge of Permitted

Original bungalow provides temporary accommodation until new house completed.

Development Rights (PDRs) – see Chapter 7.

If the plot is large enough, there may be a further bonus: you may be able to utilise the old house as accommodation whilst building your new home in the garden. Once the new home is occupied the original building can then be demolished. However, if by some miracle the planners allow you to retain the existing property once the new house is complete, then you have really hit the jackpot. In such a scenario, both would probably be accessed via a new shared driveway and there would need to be sufficient distance between the two properties so that each had a reasonable sized garden.

Demolition is an activity that's extraordinarily dangerous in the wrong hands, and is therefore best carried out by specialists. Where there are adjoining buildings in close proximity, there's a further risk of damage to foundations and cellars. With period properties the cost can sometimes be reduced by salvaging any valuable architectural material,

Bungalow demolished, ready for new larger replacement.

such as fire-surrounds, period doors, handles, light fittings, stained glass, slates, roof tiles, and bricks.

Conversions, makeovers and renovations

Viewers of Kevin McCloud's *Grand Designs* will be familiar with the adventurous species of self-builder prepared to create a home from the most unlikely properties – such as redundant water towers, lighthouses, and windmills. Although not everyone's idea of a self-build, all kinds of structures can make suitable residential projects if the planners permit it – barns, mills, railway stations, schools, chapels, castles, wartime pill boxes, you name it. But there

Photo: potton.co.uk

are generally a lot more headaches and red tape with conversions compared to new build.

Of course, wacky buildings aren't the only source of successful conversions. Ugly ducklings can be remodelled into beautiful abodes. Some ghastly 1970s eyesores occupying superb plots have sound structures and perfectly good layouts. So why waste money demolishing and rebuilding? Even the dullest buildings can be transformed into contemporary palaces with a few tricks of the trade. New high-quality windows and doors can provide an instant facelift. Applying smooth white render over bland brickwork, set off by a spot of tasteful timber cladding, can work wonders (perhaps having first taken the opportunity to install external insulation panels). Kerb appeal can be transformed by a new highly glazed gabled front extension. Re-cladding shallow roofs with tasteful small plain clay tiles, adding small 'cottage dormers' to new loft rooms, or even rebuilding the roof to a traditional steeper pitch, can all add considerable charm and style to even the plainest buildings.

Photo: Piers Taylor Architects

Other sources of land

Estate agents

Not all estate agents deal in land. There are usually just one or two local firms in any area, often the older traditional ones. There are also specialist 'land agents' who deal in agricultural land and plots – with or without planning consent. The large corporate estate agency chains rarely have plots on their books, but occasionally their housing

developer clients instruct them to market surplus sites they've released in order to raise funds. To ascertain which local agents deal in building land, it's worth taking a trawl through websites such as Rightmove, searching for plots currently on the market. This is also an easy way to get a feel for current land prices. Having identified the relevant local agents, you need to visit them to register your interest and then keep in regular contact by phone and email.

The main problem with trying to source land via estate agents is that they tend to favour local developers and small builders with whom they have a business relationship. Selling the land to their regular contacts may earn them a 'finding fee' from the developer in addition to normal sales commission from the landowner client. Alternatively, the developer may agree to instruct them on the sale of the houses that ultimately get built on the plot, thus securing future sales commission.

Although as a self-builder you're up against stiff competition, there are still one or two killer cards you can play.

Firstly, unlike developers, most self-builders don't have to allow for a fat profit margin and can afford to pay a higher price, and are often in a position to pay cash with ready funds. Self-builders won't play the kind of dirty tricks that developers sometimes adopt, such as delaying the purchase whilst seeking revised planning consents for higher density developments, only to pull out months later if the application is refused. 'Serial self-builders' may even be able to offer repeat business in future. As if that wasn't enough, there's another very juicy carrot you could dangle in front of the agent to help swing things in your favour. If you need to sell your existing home to raise funds to finance the build, you could offer to sell it via whichever agent can sell you the plot (assuming you've already got sufficient cash to buy the plot).

If all else fails, there's nothing to stop you conveying the impression that as a 'private house builder' you may ultimately want to instruct the agent to sell the house you're planning to develop. Such minor deceptions are perfectly valid because, at the end of the day, the landowner will probably get a better deal from you as a purchaser, for the reasons stated above. The fact is, the landowner is not employing the agent to pass it to their developer chums at a lower price, in return for some sort of backhander. The agent's first duty should always be to their client, the owner of the plot, and it's the agent's job to get the best price for their client. Maybe this is why truly savvy plot owners are likely to cut out the middleman and advertise their land privately, thereby reaping a higher price with massively reduced selling fees.

Auctions

Land sold at auction often comes about as result of repossession or bankruptcy. The main drawback of buying at auction is that you have to pay for the cost of carrying out all site surveys, checks, and legal work in advance, because the highest bid is binding. This, of course, is money wasted if you're not the winning bidder. It's also easy to get carried away with auction fever and bid over the odds. The dates of forthcoming auctions are published in local papers and online.

Plot-finding services

The first stop for many self-builders is to subscribe to land-finding agencies such as Plotfinder and Buildstore's PlotSearch, both of which have several thousand plots with

planning consent on their books. These are gleaned from estate agents, private sellers, and large landowners such as the MOD. A subscription of around £50 a year buys you a membership so you can search by county and have 'Plot-alerts' emailed to you, although there may be some overlap between competing websites.

There are a number of smaller land-listing agencies that collect information on available plots from estate agents and then email lists to subscribers. But a lot of people subscribe to get this information, so any decent sites are likely to have already been snapped up. Some plots are marketed without planning consent, but where there's a history of failed applications it's unlikely that you would fare any better. In some cases information is not up to date, and upon contacting the seller you may discover that a plot was sold several months ago.

Professional land-finders

Large developers sometimes employ professional firms of land-finders. Although they're only really interested in sites with potential for multiple units, they occasionally come across plots with more limited potential, for single dwellings, so it's worth a call.

Friends and family

Spread the word. Make sure that everyone you know is aware you're on the lookout for a plot. Offer a sizeable 'finder's fee' reward. An example where this sometimes works is where someone knows of a potential plot in the large garden of a friend, relative or neighbour.

Media

Occasionally plot-owners place classified ads in local papers and websites under 'land for sale' in order to save on agent's fees. Several self-build magazines publish lists of building land, but because they go to print a couple of months before the magazines hit the shelves many plots will already have been sold. So you need to take out a subscription and register to check the current listings on magazine websites. It can also sometimes be worth placing 'wanted' ads in local parish magazines. If you see a plot advertised, you need to act promptly before the seller is bombarded by offers from estate agents (who are looking to act as selling agents or to buy for their developer clients).

Self-build clubs

As the old saying goes, 'it's who you know'. Networking with other like-minded plot hunters, perhaps by joining a club, means you may hear of a suitable plot through the

grapevine. The Association of Self-Builders is a club that keeps a record of land for sale, including Council-owned plots. Some builders' merchants also run self-build clubs, holding useful seminars and issuing newsletters.

By teaming up with other aspiring self-builders, you can immediately multiply your buying power, putting larger plots within your budget. And co-buying a site with two or three individual plots means you get to choose your neighbours!

Spare plots

Sometimes self-builders acquire larger sites that are too big for their needs and may have planning consent for more than one dwelling. They may plan to split it into multiple plots, and sell spare plots to another self-builder. Housing developers sometimes buy up land that is too small for an estate and instead of developing it themselves will sell individual plots to self-builders, having first put in the services. Buildstore run a 'Plotshare' service for people looking for a single plot on a multiple site.

But where development sites of more than four units require new highway access and sewers, local authorities commonly impose a legal requirement for a bond leading to formal adoption, which should have already been satisfied.

Photo: oakwrights.co.uk

Community self-build

Sometimes groups of aspiring self-builders join together to form 'self-builder collectives' to build their houses together, often with assistance form Housing Associations or Local Authorities. These tend to be low-cost schemes, part funded by grants, with different folk contributing different skills and labour or 'sweat equity'. The Community Self-Build Agency is worth investigating – see website.

Package deal kit firms

Photo: potton.co.uk

If you're planning to buy a 'kit house', such as a ready-designed timber frame package, the suppliers may be able to offer building plots as part of a package deal. Self-builder kit firms sometimes acquire large sites and split them into individual plots. By keeping in touch with these firms, an opportunity may arise to take over a plot where another client has dropped out, perhaps for personal or financial reasons. Local architects and surveyors may also have clients who pull out or have more land than they need.

Custom-build

Custom-build is a growing part of self-build where 'enabling developers' supply serviced plots on a relatively large scale. Custom-build developments are a good way for local authorities to meet the demand for plots from individuals. In some cases it may have been a condition of granting planning consent for a new large development that some plots are made available to self-builders.

As well as providing a serviced site in the first place, the developers can manage the construction work and even arrange the finance for you. This is more of a 'hands off'

approach with less risk. The downside is that plots tend to be bunched together on a shared development, and although the lower risk will be factored into the cost, the developers benefit from sizeable economies of scale.

Some custom-build developers provide a menu of options – e.g. they may just sell you a serviced building plot which you then take over and organise; or they might offer to build your home to a watertight stage so that you can then fit it out to your requirements. See www.nacsba.org.uk

Builders

Over the years builder-developers tend to accumulate land banks to provide them with a continuous supply of ready plots. But occasionally they may decide to sell off surplus plots to raise funds to ease cash-flow problems. Or they may want to dispose of a site where planning consent has been refused for four or five houses and they aren't interested in building just one or two. However, they may want to tie the sale to an agreement to appoint them to build the house to your design. But without a competitive tender their construction prices are likely to be on the high side.

Local authorities

A few enlightened Councils have policies to set aside development land for self-builders, so it's worth checking this with the planners. It can also be worth contacting the Estates Department of your local authority or County Council, as they sometimes have spare plots that have been acquired over time, perhaps in relation to larger projects, that would be suitable for a single dwelling.

Farmers

When you own hundreds of acres of land, you could probably sell off the odd half-acre here or there without living to regret it. Farmers are, of course, legendary for knowing how to charge, so it's rare to negotiate a bargain plot. But if you think you can get planning consent on a field, it may be worth entering into an option with the farmer.

Occasionally farmers get

planning consent to build a new home subject to agricultural ties, where a planning condition stipulates that at least one occupant must earn their living from farming activities, which significantly depresses the property's market value.

Utility companies

Privatised water, gas, telecoms, and electricity suppliers inherited vast amounts of under-utilised land, often in unusual locations or comprising odd shapes. In some cases advances in technology have made pumping stations or telephone exchanges the size of small bungalows

redundant. The planners may even be happy to see the street tidied up with a semi-derelict building replaced by a gleaming new house.

The downside of such plots is the potential for ground contamination, and possibly large amounts of equipment to be removed (although this may have some scrap value). So it's worth keeping an eye open for such opportunities or sending enquiries to utility companies.

Other major landowners

Network Rail is another huge organisation that owns large amounts of land, not all adjoining railway tracks. It has an estates department that disposes of surplus plots. Organisations such as universities and British Waterways also sell off parcels of land for development from time to time, so keep an eye on the local news.

English Partnerships are responsible for New Towns and sometimes sell fully serviced plots around the UK. Land is sometimes allocated for self-build purposes, for example in Milton Keynes, where large acreages were split into smaller individual plots for sale to self-builders, with developers specifically excluded. Additional planning controls may apply to such schemes to prevent a potential mishmash of clashing styles in the same road.

Plot hunting summary

- Look beyond the tumble-down shacks, prodigious undergrowth and dumped, rusting vehicles.
- Estate agents are not always good at preparing details for plots – a badly written description and lack of photo could hide a potential gem.
- Get your funding ready. Good plots don't hang around, so you need to show you can come up with the readies to beat rival buyers by offering a swift completion.
- It doesn't pay to be shy. Finding your own plot may mean knocking on a lot of front doors to ask if the owners want to sell.
- Confining yourself to one village or a small radius could mean your area is too restricted. Widening your horizons can boost your prospects.
- Be pro-active – let everyone know you're looking.
- Don't limit yourself to perfect greenfield sites. Many successful homes result from one for one replacements. The extra cost of demolition can be balanced by the benefit of existing services already connected.
- If the plots you like the look of are all above your budget, consider spending a little more on the plot and make savings on the build, by extending or improving later when funds permit.
- Don't be too fussy. You could wait forever for that generously proportioned village plot with dream views. Consider a less ideal plot, perhaps as a step up the ladder making money to finance your ultimate self-build home.

Solicitors

Occasionally solicitors help manage property portfolios and may also be appointed to dispose of land and property. Executors of a will may be attracted by the prospect of a swift private sale.

Building plots can often be created by hiving off part of the garden.

4 CHECKING AND BUYING YOUR PLOT

At last you've spotted a site that actually appears suitable for building your new home. But even the most perfect-looking plots can harbour hidden problems, and you don't want any disasters erupting once you've bought it. On the other hand, if you dither for too long before making your offer someone else could steal it from under your nose. So time is now your enemy. It won't be possible to wait for detailed checks, but you must at least narrow down the risk with some essential checks.

To save time, buying the site can usually take place in tandem with your detailed site investigations. The process usually proceeds as follows:

- ■ A swift initial appraisal and essential checks.
- ■ Negotiating your offer and commencing conveyancing.
- ■ Detailed site investigations.
- ■ Exchange of contracts and completion of the purchase.

Doing your homework is especially important for self-builders. This is because all the 'easy' plots tend to get snapped up quickly by developers, and the ones that hang around are often the more unusual sites, which are sloping, or oddly-shaped, or for some reason aren't so well suited to standard designs.

Keeping secrets

Once you've found the right plot, be careful about who you tell. It may sound paranoid, but at this early stage you can't afford for the news to seep out so that someone gets wind of it and proceeds to outbid you. To be on the safe side, only discuss it with professional advisers who you know can be trusted. It's best not to reveal the precise location to anyone except your conveyancer until you've completed the purchase. Even then, to emphasise the need for discretion, you may want them to sign an undertaking that the details are to be kept strictly confidential until the plot is secured.

The initial appraisal

You already have a gut feeling that the plot is suitable for the type of house you want to build, otherwise you wouldn't have picked it. At this early stage you should already be aware of fundamentals such as the plot shape and size, the planning situation, and the likely sale price. But now is the time to swap those rose-tinted specs for a microscope, because in a very short space of time you need to identify any significant site problems and decide how much it's likely to cost to solve them. These costs then need to be reflected in the price you're willing to pay for the site. There are two main sources of information – physically checking the site and desk research.

Go walkabout

To identify any problems, start by taking a slow, methodical walk across the site. Note the boundaries, access gates, and the location of any manholes. Do a rough sketch to record the position of key features. You can do a simple preliminary check yourself before making an offer to buy the plot, and a more detailed investigation once your offer has been accepted. Take lots of photos, as it's surprisingly easy to forget details.

A simple site survey should check

- The north point.
- Site access.
- Location and type of existing boundaries – walls, hedges, fences, ditches etc.
- Buildings on and adjacent to the site.
- Roads, footpaths, and pavements adjacent to the site.
- Trees and vegetation.
- Any manholes and ducts for services.
- Any features – trees, ponds, pits, streams etc.
- Ground levels, including those outside the plot, in relation to a fixed datum point.

Do your research

There's plenty of useful research that can be done online to flag up potential risks, such as flooding, noise, and radon – see website. But there's other key information that will not be accessible via the Internet. You don't want to belatedly discover that your house is built on the site of an old plague pit, so it's important to check out local history. Councils and local libraries can be a goldmine of information about historic land uses. Possible contamination risks, such as old leather tanneries, can be flagged up, along with long-forgotten graveyards. Local street names can reveal a lot. For example 'Willow Avenue' or 'Swan Lane' may suggest a high water table or a flood plain.

It's also worth talking to potential neighbours who may know the history of the site and reveal things that the vendor has kept secret. In the unlikely event that they're extremely hostile or offensive, you may decide you don't want to live next door to them and start looking for another site!

Japanese knotweed – highly invasive – see p45.

Questions to ask prior to purchase

Time is of the essence, so there are some key questions you need to get answers to pretty quickly. All are discussed in more depth later in this chapter.

Why hasn't it been built on already?

This always has to be your first question. Maybe it's just that no one thought of it until now, or it's been part of somebody's garden. It's possible that the planners would have refused such a development until quite recently, spurred on by government pressure.

Perhaps the plot has been used for some other purpose, such as access to a now defunct pumping station. Some obstacles, such as access problems, may be negotiable. Even where the plot was simply uneconomic to develop in the past, because of bad ground, it may now make sense thanks to higher property values.

Is this a suitable location?

In many ways, the choice of location will depend on the same criteria you'd apply when moving home conventionally – *ie* proximity to local amenities, such as schools, shops, and transport. But you're always going to judge a plot for your future dream home more harshly than one that's merely a stepping-stone to your next project. In other words, if you're developing property, what counts is minimising costs and maximising profits in the short term, rather than whether the plot meets your personal lifestyle aspirations and offers sufficient space for future extension. If you plan to put down roots, you also have to ask yourself whether the plot instinctively 'feels' right.

From a design perspective, the geographical position is a major consideration. With a plot at high altitude, more than 500 feet above sea level (the 'snow line'), there will be a risk of severe winters, something that's not obvious if viewed in the summer. Sites on, or near, flood plains obviously need to be researched for any history of flooding that might indicate a future problem.

Is there access?

To qualify as a plot, a piece of land needs a legal right of access. If you can't actually get to it, either legally or physically, it's obviously not much use.

Some sites on the market turn out to be 'land-locked', where you need to cross land owned by someone else to gain access. Purchasing land is always more complex where new legal rights of way are required. Because of the extra risk, and the costs for legal work and building a new driveway, you would expect this to be reflected in the price of the plot.

If there's a site access road, take a look at the condition of the road surface. Could it cope with convoys of trucks delivering concrete, unloading giant pallets of bricks, and manoeuvring JCBs? Is there scope for widening, or is the land either side owned by someone else?

What can you build?

What is it that transforms a field, or part of someone's garden, into a valuable building plot? The answer, of course, is 'planning permission'. Without this a plot is virtually worthless, unless you want to graze herds of llamas.

So the planning status of your plot is one of the first things to check before making your offer. Either outline permission (OPP) or full planning approval should already be in place, but either way, you would normally want to submit a revised application for your own house design. If it just has outline planning consent, you only have the assurance that you can build a new dwelling of a certain size and type. All the design detail has yet to be approved.

To ascertain what the planners are likely to accept, look for clues in local architectural styles and the types of properties close to the site, especially newer houses. To get a fuller picture, it's important to make contact with the planners at the earliest opportunity. If possible talk to the officer who handled the existing outline consent and take a look at the file containing the planning history. See Chapter 7.

Although it is possible to *agree* to buy land *before* it has planning consent, you must only *pay* for the site once permission's been granted. In other words, your offer must be conditional upon consent being granted. See 'Options' on page 43.

Future developments?

Imagine building your new home on a fabulous plot surrounded by glorious open countryside. Then just as you're about to move in, looking forward to enjoying the splendid views, a JCB starts tearing into the adjoining fields. It transpires that the land next door is being redeveloped into an industrial estate. After all, if you managed to get planning, perhaps the owners of the adjoining land can too.

Prior to purchase, your solicitor will carry out the usual

Photo: BECO

local searches, which should reveal any road-widening schemes and suchlike in the immediate vicinity. But surprisingly enough, local searches won't tell you about future developments on adjoining land that could drastically affect your property. So be sure to instruct your solicitor to make additional enquiries with the planners to confirm the 'use' for which the surrounding land has been designated.

It's also a simple matter to go online to your local authority's website and search for details of any planning applications in the locality.

Is it serviced?
Some sites are sold as 'serviced', but vendors and agents should always be asked to clarify details, as the cost of providing new services can be phenomenally expensive. Make a point of checking for manhole covers and access points on the ground, both within the plot and in adjoining streets, to identify any services in close proximity.

Is there any contamination?
If the grass glows in the dark, or you notice strange humming noises emanating from the plot, you may suspect problems with contamination. Alarm bells should ring where a site's had a 'brownfield' non-residential former use, and local authority records must be checked. Although very few sites are so badly damaged that they cannot be used at all, whether such a plot is viable for development or not ultimately comes down to money. And becoming liable for the massive expense of cleaning up serious pollution could bankrupt you. If a former industrial operation was illegal there may be no records, so if you have any suspicions at all it's essential to instruct a professional contamination report. If any remedial works are necessary, the purchase price will obviously need to be renegotiated.

Pollution comes in several guises other than chemical contamination, notably noise, smell, and vibration. In fact noise from traffic and aircraft are probably the two most common forms, but these can to some extent be designed out. It's surprising how few locations don't suffer from some degree of background road noise. So before buying the site, try to visit it at different times, especially during peak hours and at weekends. Listen also for noise from factories, flight-paths, and farms. Wind can carry sound for surprisingly long distances, although if it normally blows away from your site even nearby sources of noise may not be particularly intrusive, so check the prevailing wind direction.

Will you need a land survey?
Most building plots are overgrown and uneven. However sites with irregular shapes or significant changes in level will need a measured land survey. This is also advisable where there are major potential obstacles or natural features such as mature trees that are to be retained. Professional measured surveys are carried out by land surveyors using modern laser equipment to plot accurate computer plans.

Once you have an accurate survey drawing of the site,

you know exactly what you're buying. It also means that your planning drawings will be accurate. Get the size or shape wrong and expensive legal battles with neighbours could rage for years afterwards. Give a copy to your conveyancing solicitor to compare with land registry plans of the same plot.

What lies beyond?

Adjoining factory could detract from value.

The most common mistake people make when first visiting a site is not looking much beyond the site. So be sure to look beyond the immediate boundaries, to put it in the context of its surroundings. Start a mile or so away, and look for changes in the social profile of the area. The value of our homes depends to a large extent on those around us. If you're thinking of building an upmarket house, having to approach it via a large Council estate would detract from its value. Consider what sort of design would fit into the existing street scene. If you go against the trend for what's acceptable in the local market, the completed property is likely to be worth less than you think.

Buying the site

For any good plot there will be plenty of rivals eager to push you aside. Speed is of the essence, so having satisfied yourself with some brief initial checks that this is the site for you, the next step is to negotiate an offer. At this stage it's essential to have your funding in place, to demonstrate to the seller that you're deadly serious. Once your offer has been accepted, the objective is to exchange contracts as quickly as humanly possible. In order for the purchase process to proceed as rapidly as it can you'll need to work as a team, with your solicitors, effectively becoming their eyes and ears on site as they carry out the necessary searches and enquiries.

This nail-biting few weeks can be profitably utilised to carry out detailed investigations, to verify that there are no hidden nasties. Only once you've concluded these checks and exchanged contracts can you start to relax, with completion of the purchase taking place on an agreed date shortly after exchange (although legally it's possible to do both simultaneously). But before making your offer, there's one big question that needs to be answered: how much is the plot actually worth?

How much should you pay?

After devoting endless months or even years to plot-hunting, you'll probably already have a pretty shrewd idea whether a site is worth the price being asked. Even so, to be sure you're not paying over the odds it's advisable to do a few sums before opening your wallet.

The market value of any plot depends directly on how much the property built on it will ultimately be worth, which in turn is largely a function of planning permission. Despite the additional element of uncertainty, in most cases a site with Outline Planning will be worth pretty much the same as one with detailed consent. The land value can then be calculated by working backwards, starting with the value of the completed house, less all the costs involved in building, allowing for a profit margin.

Up until this stage, most self-builders operate according to the rule of thumb that you generally expect to pay between a third and a half of your total project cost for the land. So, for example, if you bought a plot for £100,000 and it cost £200,000 to build your house, the plot would obviously have cost a third of your combined land and construction costs. If the finished house then had a market value of £400,000 you'd be sitting on a 25% 'developer's profit'. By waiving this profit margin you could afford to pay more for the plot, or you might choose to invest it in better quality materials and fittings. Or a bit of both. Some people work backwards using the following rule: Land costs (A) + build costs (B) + a 20% to 30% margin = the finished market value of the house (C).

No matter how fierce the competition to buy, all plots have a ceiling value, a price beyond which it's just not economic for anyone to build. Developers will not go above this figure, as they need to leave a fairly generous profit margin after all the costs of land, labour, and materials. The fact is, the most anyone could sensibly pay for a plot is the market value of the finished house less the total build cost. This, of course, would leave no profit margin. And the only people who might be willing to sacrifice some or all of the profit would be self-builders!

Assessing a plot's true value

Asking prices are often set optimistically in the hope that a desperate buyer, perhaps someone who's half-crazed from losing out on a previous purchase, will pay over the odds. But like everything else, whether or not a price is correct will depend on market conditions – supply and demand. As everyone knows, house prices can move down as well as up, which has a direct affect on the economics of building plots. For example, the value of development land across the UK actually fell by an average of 50% during the course of 2008

according to Knight Frank, as a result of falling house prices and problems securing development finance.

The first shock you get when starting plot-hunting, is when you realise how incredibly expensive plots with planning permission are compared to those that can only boast 'potential'. The difference between the two can be as much as 1,000%, because plots without planning pose far greater risk, as the use to which the land can be put is uncertain.

There are two main valuation methods used by professional valuation surveyors – comparable evidence and development potential.

COMPARABLE EVIDENCE

When it comes to valuing houses, surveyors and estate agents simply look down the road to see how much similar

properties have sold for. Then, based on their experience, a judgement is made to reflect the relative pros and cons of the property's size, condition, and immediate location.

But plots are far harder to compare, like with like. Even in the same road, plots can vary quite dramatically in terms of planning status and size. Plus considerably fewer building plots are sold than houses, which means comparable evidence tends to be rather patchy and less reliable.

However, this method can sometimes be used successfully because, at the end of the day, plot values are a function of house prices. So a plot with consent for a 200m² four-bedroom detached house is likely to be worth roughly the same as another plot with a similar consent in the same local area. Also, you tend to instinctively develop a feel for the 'going rate' after months of trudging round plots, simply based on comparing the prices being asked.

Land is sometimes described in terms of pounds per acre,

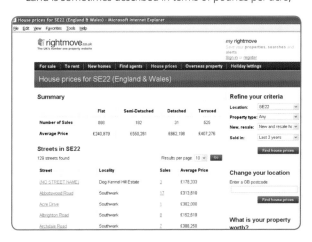

notably agricultural land. But estate agents adopt a similar approach when selling large, multi-unit residential building plots. Depending on what part of the country you're talking about, these can be priced in millions of pounds per acre. But when it comes to smaller single-unit plots of less than about half an acre, this method can be misleading.

Photo: buildstore.co.uk

DEVELOPMENT POTENTIAL

To accurately assess the value of a piece of land you need to consider its development potential. In order to determine this, there are two key questions to ask: does it have planning consent, and could more be built on the plot?

As we saw earlier, once you know this the plot value can be found by simply deducting (a) the cost of construction and (b) the profit, from what it would sell for when built:

What will it sell for?

To calculate the market value of your completed house, ask the advice of some good local estate agents. But agents can sometimes be a tad over-optimistic, so back this up by checking land registry figures online to see what similar properties have actually sold for. These prices can then be adjusted to reflect likely market increases or falls over the next year or so.

How much will it cost to construct?

As long as you know the size of the house you plan to build, it's a simple matter to apply an average price per square metre, excluding VAT. There are several online guides to estimating construction costs – see website.

Transaction costs

Factor in the site purchase costs, such as stamp duty and conveyancing fees, plus theoretical sale costs of the completed property, including agents' fees.

Contingency fund

Allow a margin of at least 10% for unexpected complications or unduly optimistic prices.

Developer's profit

The average profit margin most builders anticipate is around 25%.

But when it comes to valuing a plot, there's a potential problem for self-builders, something that's unlikely to bother developers: 'the goldmine trap'. What happens if the piece of land you're interested in buying is actually capable of being developed more intensively than just one house sitting in generous gardens? Answer: it pushes up the value. Imagine that a plot with consent for a single detached house could actually get planning for a block of ten flats, each with a market value of £150,000. This could be a potential goldmine for a developer, and will be reflected in the price.

But bigger is not always better. If the plot is in an area where it would best be developed as a four-bedroom house, and you decide to go to the extra expense of building a six-bedroom house, it might not be worth much more, despite spending all that extra money on building costs. The plot would have been 'overdeveloped'. On the other hand, someone who only built a two-bedroom bungalow would be 'underdeveloping' the plot, not realising its full potential.

Making the big decision

At the end of the day, you may conclude that the asking price is too high. Perhaps there are too many risks with the site, in which case it's better to pull out before you get stuck with a complete lemon. More likely your preliminary investigations won't have uncovered anything too scary and the decision can be made to go ahead. To secure the plot you will need to demonstrate to the seller that you're a serious buyer in a position to act swiftly, with your solicitor appointed and with sufficient money in place to proceed.

Your offer should be subject to the vendor withdrawing the plot from the market, to reduce the risk of gazumping, where rival bids are submitted before you've exchanged contracts. It should also be subject to the assumption that no major hidden defects, such as site contamination, will be revealed by your ongoing 'due diligence' investigations.

Negotiating the purchase

If a plot is being sold via an estate agent, as soon as you show interest it's not unknown for them to announce that there's a rival bidder waiting in the wings with their heart set on acquiring 'your' site. This could, of course, be perfectly true. Or it may be a crude attempt to get you to bump up your offer. You have to judge how genuine such tales of rival bidders are, and if necessary walk away.

Later in the process, to speed things along, they may remind you that 'whoever is ready to exchange first will get the land'. If you feel an agent isn't being entirely straight it can often pay to establish contact directly with the vendor, particularly if they live locally or next door to the plot. As a potential new neighbour they're unlikely to mess you about and may be happy to confirm whether anyone else has made a formal offer and is in the running.

Picking the right solicitor

Buying land for development is a specialist area. Although the conveyancing process is similar to buying a house, and

Photo: jnp.co.uk

may sometimes be simpler, it's important to pick a solicitor with experience of land purchases. The ability to recognise when an obscure problem can be solved, perhaps by arranging a simple indemnity policy, can make the difference between success and failure. If possible appoint a solicitor recommended by other self-builders.

■ A local solicitor should be familiar with the workings of local Councils. Also, being located within a reasonable proximity means you should be able to sign key documents at short notice, which can save valuable time later.
■ Speed and efficiency are important, so it's worth paying a little more for a faster service, to help beat others who are pursuing the same plot.
■ Any mining in the area, old or existing, must be checked by your solicitor, as it could affect the stability of the ground.

Most solicitors will not visit the site, so check that the shape of the plot on the legal plan is consistent with the actual boundaries on site. Provide your solicitor with plenty of photos and an Internet aerial photo.

Different methods of selling land

There are a number of different ways in which land is marketed.

Private treaty

This is the most common method for buying property and land. But whereas houses are normally marketed with an 'asking price' pitched slightly above the figure the sellers hope to achieve, land is frequently marketed with a 'guide price' above which 'offers are invited'. One disadvantage of buying land via the 'private treaty' system is that vendors and agents sometimes harbour unrealistically high expectations, which can make it a long, drawn-out process. Incidentally, plots of vacant land do not require a Home Information Pack; nor do properties sold for demolition.

But as anyone who's bought or sold a house will know, the big problem with this system is that between agreeing an offer and exchanging contracts, either party is free to pull out for any reason. Which means you could be gazumped if the seller decides to be 'unfaithful' and accepts a higher

Key stages in a private treaty purchase

■ The buyer makes an offer 'subject to contract', either via an estate agent or direct to the seller.
■ A price is agreed and both sides instruct their solicitors.
■ The buyer's solicitor prepares a draft contract and sends this, along with 'preliminary enquiries', to the seller's solicitor.
■ If the buyer needs a mortgage, their bank will instruct a mortgage valuation inspection.
■ The buyer's solicitor instructs a local authority search, and land registry searches.
■ The buyer's solicitor raises further enquiries and checks the terms in the contract.
■ When both sides are satisfied, the buyer's deposit (normally 5% or 10%) is paid and contracts can be exchanged (which in practice is done over the phone).
■ Completion takes place at an agreed time after exchange, often two or four weeks, at which point the balance of the purchase price is paid.
■ The conveyance document is prepared and the transaction recorded at the Land Registry.

For more on the conveyancing process and buying property, see the *Haynes Home Buying & Selling Handbook*.

Buying in Scotland

Buying land or property in Scotland is more akin to buying at auction. Once a sale is formally agreed, it is binding on both parties. This largely prevents the plague of gazundering and gazumping that goes on in the rest of the UK. Where there is competition to buy, offers are made by sealed bids above a set guide price, and submitted by a given date. But as with auctions, all your preparation and research must be done in advance.

offer from a rival buyer. On the other hand, this process allows you the luxury of carrying out your detailed site investigations prior to exchange, so if you belatedly discover that it's sitting on top of a massive mine shaft it won't be too late to pull out. The best advice is to push hard and get to exchange of contracts as quickly as possible.

Auction

Plots sold at auction tend to be those that are tricky to value, such as sites with odd shapes or with hidden problems that have caused previous sales to collapse. Buying land at auction isn't for the fainthearted. It tends to favour professional developers who are familiar with the process. After all, speculating with company capital is always going to be easier than for self-builders betting their life-savings.

The stated 'guide prices' are usually set on the low side to stimulate bidding, so what appears to be an amazing bargain will still have to reach its 'reserve price' (which is kept secret) before a bid is binding.

Watch out for spoof bids sometimes made at the beginning by staff or associates of the vendor to get things going, or to help push bidding above the reserve price. Be careful not to get carried away, but at the same time don't lose a plot for the sake of a few pounds if you sense other bidders are getting to their limit. Sometimes vendors will accept an early bid put in before the auction starts. Conversely, if bidding doesn't reach the reserve price the vendor may be willing to accept a lower offer after the auction.

The main drawback of buying at auction is that all your key site investigations, finance, and legal work must be completed in advance. This is because, when the

auctioneer's gavel falls, the highest bidder is legally committed to the purchase. So if you missed the quicksand and piles of toxic waste – tough. This means that unless you win, all the money you've spent on upfront investigations will have effectively been flushed away. But minimising the risk of abortive fees by only paying for the minimum upfront checks means you'll increase the chance of buying a site with hidden problems.

Sale by tender

Where there are several competing purchasers, estate agents often ask each party to submit their 'best and final offer' by a stated deadline in the form of 'sealed bids'. For fear of being outbid by a few pounds, bids tend to be submitted slightly above the nearest round figures, perhaps £150,509 rather than a round £150,000. Although the tender process has similarities to buying at auction, there are some key differences. For a start, you don't know how much others are bidding. And it's not always the highest bid that's accepted – being in a position to proceed, with funding readily available, can push a lower offer to the top. Sometimes further negotiations go on afterwards with the preferred bidders. Perhaps surprisingly, the winning bid is not binding, so even if you win you can't afford to relax until you've exchange contracts.

Options

As we saw earlier, taking an option on a site is a technique that's widely used by developers where a site doesn't have planning consent. The buyer enters into a legal agreement with the landowner that they won't sell to anyone else for a fixed period of time. If at the end of the period you have still not bought it, the owner can keep it or sell to someone else. But when acquiring a single plot the vendor will probably not want to agree a long delay. However, developers sometimes enter into long-term options over several years. This only makes sense where getting planning consent is a long shot, perhaps dependent on a scheduled new road being built. It only takes one such deal to succeed to more than compensate the developer for the cost of all the others.

Conditional contracts

Plot owners may find a conditional contract a more attractive proposition than an option. This commits the potential purchaser to buying, subject to a stated condition occurring, such as getting planning, or suitable access or another piece of land being acquired. This deal has benefits for both sides because the vendor knows that the purchaser is obliged to buy, and the buyer knows the land won't be sold to someone else after they've invested a lot of time and money.

When buying subject to planning consent being granted, it's necessary for conditional contracts to be exchanged *before* the date of the planning committee meeting. This is to safeguard against you getting permission and thus making the site much more valuable without any obligation for the vendor to sell to you at the agreed price!

Detailed investigations

With your purchase safely progressing you now have a second chance to explore the site in greater depth. Your initial appraisal should have identified any major concerns visible on site, but such information may only ring alarm bells when combined with results gleaned from desk research.

You are responsible for managing this process because although your solicitors will be making enquiries and delving through old files, they will be relying on you to flag up any concerns spotted on the ground. For example, you might notice an old caravan on site, which could suggest that someone has rights of occupation over the land.

So far you've not spent a great deal of money. But to be sure that you really can build the house you want without any major unforeseen financial disasters, there are five key areas that now need to be thoroughly checked prior to exchange.

- ■ Ground conditions
- ■ Planning considerations
- ■ Access considerations
- ■ Services considerations
- ■ Legal considerations

Ground conditions

The quality of the ground must be carefully checked, as it can drastically affect the cost of your build. Get this wrong, and the extra expense of special foundations could cause you to run out of money later in the project. But one patch of undergrowth can look pretty much like another, so how can you tell exactly you're looking at?

Most sites have a topsoil covering of earth (vegetable loam) on top of a deeper subsoil comprising a mixture of clays, sand, and stones. The foundations have to penetrate down through this surface layer until they reach load-bearing subsoil. Nature can often provide clues to the type of ground – for example, trees such as willows and alders are usually found in moist ground comprising wet loam, often close to watercourses. And ground with a high water table often needs special foundations.

If there are any suspicions of ground problems, a professional soil investigation must be carried out before contracts are signed. In any case the type of subsoil will need to be verified before detailed drawings can be produced. You probably need to budget around £500 to £1,000 for a soil investigation, but it's a worthwhile expenditure for most sites.

You can sometimes spot clues to ground problems as you approach the site. If there are any buildings under construction in the surrounding area, try to see if they're

using conventional trenches or special foundations. If possible peer into any open trenches, and note how deep they are and whether the sides are holding up without support.

Photo: HSS Hire

Before concluding that all is well below your feet, have a chat with Local Authority Building Control, as they're normally aware of local ground conditions, and can help alert you to problems in the area. A quick phone call can help rule out obvious nasties within the immediate locality.

Trial holes

The usual way to discover what's going on underground is to dig three or four trial holes around the site, just outside the area where you plan to build. Unless the ground is reasonably firm, these sometimes need to be dug to a

What are you building on?

Different types of ground offer varying levels of support, depending on their 'bearing capacity'. What you're really interested in is the deeper subsoil into which your foundations will be excavated. The design and cost of your foundations will depend on what this is made of. Although most subsoils comprise a mix of different elements, if it turns out that you've mainly got solid rock lurking down below then it's fair guess that your foundations would get better support than from a layer of boggy peat, for example.

ROCK

Building on top of solid rock – 'nature's concrete' – must surely give a house underlying strength of Biblical proportions. The only problem is that you can't just build straight on top of it. Special cutting machinery may be needed to form a level base in the rock before concrete can be poured. However, where you've got loose rock or shale it can be difficult to get the trench sides to remain stable without support.

CHALK

Chalk is generally considered to be the best subsoil on which you can build. It drains well, and can also be cut to fairly precise trench widths. But chalk frequently occurs alongside other, less robust soil types, so the foundations may need to be designed to prevent any risk of differential settlement.

CLAY

Much of the ground in central and southern England comprises clay. Of course, the problem with clay is that it's notorious for swelling up and expanding in wet weather and then shrinking in dry weather. Parched summer ground often opens up with dramatic ravine-style surface cracks. These seasonal forces can be incredibly powerful, affecting the top metre or so of the foundations. But some clays give a bumpier ride than others. Different types are graded from 'low shrinkability' to high. Fortunately, the solution in most clay areas is simply to dig deeper foundations to at least 1.2m (compared to the minimum depth for good load-bearing ground of only 600mm).

Ground conditions are made worse where you have large, thirsty trees growing in heavy clay. Trees absorb an enormous volume of water from the subsoil, and if removed or killed by building works the subsoil will expand and rise up or 'heave' with a pressure that's strong enough to crack foundations. See 'Trees and shrubs' on page 45.

SAND

Sand might sound like the last thing you'd want to build on, but it actually has good load-bearing strength. As anyone familiar with building seaside sandcastles will know, firm sand isn't too difficult to excavate, but when it becomes waterlogged digging can be almost impossible.

As the 'Great Escape' tunnellers discovered in World War Two, a sandy subsoil can assist successful excavation, but the resulting tunnels easily become unstable. For this reason, the sides of trenches excavated in dry sandy soil will require timber shoring.

GRAVEL

Gravel can provide an excellent load-bearing subsoil providing it's compact. It's not affected by water and very little by frost.

PEAT

Peat has very poor load-bearing capacity, making it unsuitable for the direct support of foundations.

considerable depth. Digging down 2 or 3m isn't a DIY job, being too deep for easy excavation by spade and with potential for danger, so you'd probably need to hire a local groundworker with a digger.

What you find will determine the foundation design, by confirming the type of subsoil, its load-bearing capacity, the presence of any substantial tree roots, and the likely depth of the water table.

In most cases, a two-storey house will only require simple trench foundations. But you could be unlucky. If you keep digging and there's still no firm ground, the chances are you'll need to use special foundations, such as concrete piles. See Chapter 11.

Try to arrange for the engineer or architect designing your foundations to take a look at the exposed subsoil. Best of all, local Building Control Surveyors (formerly known as 'Building Inspectors', 'District Surveyors', and 'Building Control Officers') are well practised in the dark art of looking down holes and reading the runes. But despite taking such prudent precautions it's not unknown for trial holes to expose good virgin ground by sheer fluke, and it later transpires that the rest of the site comprises dodgy filled ground. So it will be Building Control who have the final word on the ultimate design and depth of your foundations when the trenches are excavated.

Site levels

Sites with dramatic landscapes are generally rejected by developers, who are only interested in churning out bog-standard designs and don't want the extra work. Hilly or sloping sites are therefore more likely to remain available for self-builders. This presents a creative challenge. With a touch of flair, the slope can be used to generate a stylish design.

Where your plot is at a significantly higher level than adjoining plots, or where a site has a steep slope, construction of expensive retaining walls may be required. One clue that could indicate poor ground stability is where trees or garden walls are tilting severely, which may perhaps necessitate deeper, specially designed foundations.

To a geologist's eyes, changes in ground levels can indicate something interesting buried underground, whether hidden features or filled, made-up ground. More obviously, if

a site is anywhere in the vicinity of cliffs any history of coastal erosion should be checked. In April 2008 residents had to be evacuated from bungalows at Knipe Point, Scarborough, after their rear gardens disappeared down the cliff face. But such hazards weren't entirely without precedent. In the same area, 15 years earlier, the Victorian Holbeck Hall Hotel had dramatically collapsed down the cliff.

Trees and shrubs

In many cases mature trees can be retained to enhance a design. Indeed, the planners may insist on this, or they could

be protected with a Tree Preservation Order (TPO – see below). But in order to accommodate such natural assets, it may be necessary to modify the footprint of the building. Tree roots can penetrate deep into the ground to draw moisture, so even after they've been cut down they can still affect the stability of the ground. Consequently, when inspecting sites you need to look for stumps and any dips in the ground where one might have been removed. As we saw earlier, the expansion and contraction of clay is exacerbated by nearby trees. The worst-case scenario is an aggressive species on high shrinkability clay, which could necessitate going to the added expense of mini-pile foundations. See Chapter 11.

But trees aren't the only worry. Fast-growing Japanese knotweed is a highly invasive weed that has become widespread in recent years and is strong enough to damage foundations, concrete, and tarmac. Complete eradication is difficult because the roots can penetrate to depths of 3m, requiring a mix of weeding and pesticide. Flourishing in any soil however poor, knotweed-contaminated soil is classed as 'controlled waste' that has to be taken away by a licensed operator to a designated landfill site and buried to a depth of 5m.

Filled ground and demolition

Where a site shows signs of having been excavated at some point in the past, and subsequently filled back in, alarm bells should ring. One clue is where there's uneven ground with broken fragments of tiles and bricks. Once excavated, soil expands quite dramatically, and land that's been backfilled won't be compacted sufficiently to safely take much loading. Such ground will therefore settle under the weight of the new building, unless the foundations are taken right down to rest on undisturbed ground. This could mean having to dig a lot deeper or may require special foundations – at extra cost.

Where an existing building is to be demolished to make way for the new one, Building Control may not allow new foundations to be built onto or cut through the old ones, particularly where there are old cellars or where unstable ground may be concealed. However, a new suspended floor can bridge over any areas of backfill and rubble left below ground.

Mining

Any mining activity – past, current, or proposed – should be picked up by your solicitor's search. Both the Council planners and Building Control should have records of mining. Nonetheless, in mining areas a mining survey will be required, because you can never be 100% certain as some old seams were unrecorded. For example, old 'bell mines' found in parts of Cornwall are often unmapped and occasionally suffer spectacular collapses. As the name suggests, these have a narrow entrance on the surface that opens into a balloon-shaped cavity beneath – not something you really want to be living on top of. The solution is to fill them with a large volume of concrete – at huge expense.

In areas where there are localised weak spots in the ground, raft foundations can provide a viable solution. These comprise a large platform made from reinforced concrete that's designed to 'float over' any weak ground, spreading the load.

In traditional mining areas you occasionally come across small seams of coal within the foundation trenches. These must not be left in place, and thin seams can be dug out.

Flooding

A charming *Wind in the Willows* home nestling beside the lazy river is, for some, a self-build dream. But the effects of global warming combined with sprawling property development

blocking natural flood plains, mean that flooding and rising sea levels are now a serious prospect in many areas.

Sites in close proximity to the sea, or to rivers and canals, will need to be assessed for their flood-risk. What appears in summer to be a dry site could disappear under a metre of water as the water table rises in winter. Flood plains cover around 10% of land in England and Wales, and nearly six million people live on them. Fortunately, a quick check of the Environment Agency website can reveal whether your chosen plot is smack in the middle of one or is in the vicinity of a watercourse and potentially at risk.

But this doesn't mean you have to rule out buildings close to water, or those located in high water table areas. What it does mean is that your design will need to assume that flooding is a possibility. One way to do this is to design a raised ground floor level that's safely above the maximum known flood level, leaving the area under the building to facilitate the

Photo: Helifix

free passage of water. Such a house will be supported on a series of brick or concrete piers, in turn resting on piled foundations, using a suitable sulphate-resistant concrete. But it doesn't have to look like a precarious Vietnamese bamboo stilt dwelling. A conventional facade can disguise a sophisticated water-resistant structure, with concrete ground floors and continuous damp proofing, or the ground-floor could comprise a gated car parking area rather than conventional rooms with easily damaged timber joinery and carpets.

Another approach is to protect the property by building a physical flood barrier in the garden. Excavated soil can be recycled to form a 'bund' wall, a kind of dyke. A 'sump and pump' disposal system can be designed so that any water that gets in is immediately expelled back outside the barrier. This can be constructed using a series of perforated pipes set in the ground leading to a sump with a dual pump.

The biggest immediate problems caused by flooding are the mixing of sewage with floodwater and the backing up of the drains into the house. The risk of damage can be minimised by siting any private sewage systems well outside the bund, and fitting non-return valves to the foul drains.

One practical problem is buildings insurance, to the extent that massive premiums can make development uneconomic. In high-risk areas, such as floodplains with a recent history of flooding, buildings insurance may even be refused, effectively making a property unmortgageable, and consequently slashing its market value.

Radon

In certain parts UK, there is a naturally occurring radioactive gas that seeps up from the ground. Radon is an invisible, odourless gas that is present everywhere in the atmosphere, accounting for half of all natural background radiation. Normally it's quite harmless, but in modern highly sealed draught-free homes concentrations can build to levels where it's a potential health hazard. Indeed, radon is now recognised as a contributory factor to lung cancer.

How much you need to worry about this depends largely on where you're building. In some areas radon levels are relatively high, such as in parts of Cornwall, Devon, Derbyshire, Northants, Wales, the Yorkshire Dales, the Midlands, and North Oxfordshire. It is said to be especially evident in districts where the underlying rock is granite. To check your area visit www.hpa.org.uk.

The worry is that radon can build up unnoticed over time in poorly ventilated areas. So for new houses in locations with high radon levels Building Control will want to see some simple protective measures. Precautions normally involve making ground floor slabs gas tight, by incorporating a 'radon barrier' – a continuous sheet or membrane – and providing ways to disperse it into the atmosphere. Sometimes a sump is required in the ventilated space under the building to disperse any gas before it can accumulate to damaging levels. Fortunately, the work isn't difficult or expensive, but it has to be done unless you can prove that radon isn't a significant risk with a certificate from the British Geological Society.

Methane

When a bungalow in Loscoe, Derbyshire, suddenly exploded in 1986, the somewhat distressed occupants were surprised to discover the cause. Invisible methane gas had been seeping into their home over a number of years from a nearby refuse landfill site, its normally highly distinctive rotten smell having somehow gone unnoticed.

Methane, unlike radon, is largely a man-made threat that not only causes nausea and headaches, but is dangerously explosive when mixed with air. The policy of disposing of much domestic waste by burying it in landfill sites – giant holes in the ground – means that increasing numbers of properties are potentially at risk. The main problems occur where organic vegetable matter such as old food, garden refuse, and dead animals, rots over time, slowly decomposing to produce flammable gases that can percolate up to the surface. Which is why the bin men don't like it when you conceal garden waste in your black bin liners. Methane can also occur in the form of 'sewer gas' in old cesspits.

Sites most at risk are obviously those near where landfill has taken place (the official 'risk zone' is within 250m of a landfill site), since methane can travel over 200m from its point of origin. Local Authority searches should reveal the locations of existing and former landfill sites.

Methane can enter buildings through gaps around service pipes and cracks in floor slabs and walls below ground, as well as through wall cavities. The danger is that it tends to accumulate in confined spaces such as voids under floors, in drains, and in cupboards, making them potentially vulnerable to explosion.

The design solution is pretty much as described for radon, with plenty of air vents to disperse gas away from the sub-floor areas and gas-tight seals to services etc. Use of gravel-filled shallow pits known as 'French drains' around the outside of the main walls can also help disperse gases.

Contamination

If you're buying an old petrol station or the site of a disused factory, the potential risk of contamination is pretty obvious. But there are many other sources of pollution and some can affect adjoining plots. Sites in the immediate vicinity of busy roads, intensive farms, GM crops, and fields that are crop-sprayed may all be affected.

Pollution solutions

A plot with contamination issues may not necessarily need to be rejected. As long as the sale price reflects the extra cost of putting it right, and makes an allowance for the risk and hassle, it could be a bargain in disguise. So how exactly do you fix such a problem?

CONTAMINATION

Old filling stations are most likely to have leaking underground tanks (brick or concrete). All the contaminated soil will need to be dug out and carted to an approved dump. Including the cost of replacing it, this work could easily total more than £20,000.

However, there are alternative solutions when making contaminated land safe for residential development. One solution, used in the Greenwich peninsular in London (home of the O2 dome and site of a former gas works), is for the site to be capped with an impermeable layer, to keep the contamination safely at bay. Alternatively, contaminated soil can be left in place and treated using 'bio-remediation', where microbes break down the hydrocarbons and neutralise the pollution.

Not all hazards need treatment. If present in low enough concentrations, the risk may be deemed manageable. For example, no action may be required on a site with old deposits of arsenic where it can be proved that the amounts are unlikely to pose a threat.

NOISE AND VIBRATION

You may have found an idyllic site except for one thing – noise pollution, perhaps emanating from a motorway or factory. Fortunately, with some smart design the building itself can help to neutralise the problem – for example, the house should be oriented so it faces away from the noise source. Landscaping can also be used as a defensive weapon (see Chapters 5 and 14).

Plots near busy roads, railways, or airports may also suffer from vibration. In some cases the problem can be designed out by fitting floating floors with multiple layers of thick foam underlay (which also lessens the impact of earthquakes!).

To be classified as 'contaminated', land must be deemed to present 'significant hazard to people or to the environment'. Owners of polluted land are legally required to notify their local authority, and Councils have a statutory duty to go out and inspect any contaminated land in their area and to keep a register of all such sites. But some

landowners may not consider that their plot presents a 'significant hazard'. Or they simply may not be too keen to report it, perhaps preferring to pass the polluted parcel on before they get stuck with an enormous bill to sort out the mess. As a result, Council lists of contaminated sites are not always comprehensive.

It's possible that the person you're buying the site from may not choose to volunteer such negative information, so it's essential that your solicitor specifically asks about this in writing. If you can prove that a vendor wilfully failed to disclose known contamination, they could be open to litigation from a purchaser.

Where contamination is identified, the local authority has the power to enforce its clearing up by serving a 'remediation notice'. Usually the polluter pays, but if they can't be found the burden falls on the current owner. If ignored, the Council can do the work and charge the landowner. So any clean-up costs should be borne by the vendor. Although solicitors' searches sometimes reveal a risk of contamination, if there's the slightest doubt it's advisable to instruct an additional environmental search on the land.

Some plots are described as 'cleared sites' – which means remedial works have already been carried out to neutralise former polluting operations.

Planning considerations

If you buy a piece of land without planning consent all you've really got is a 'potential building plot'. Which is a pretty expensive lottery ticket. The plain fact is, no one's going to sell their land at a fraction of its true value if there's the slightest chance of getting planning consent. So if there's one thing above all that you need to get right it's the planning status of the site. At this stage you normally just need to make sure the site has at least got outline planning permission (OPP) for the type of house you want. Getting the detailed consent for your dream home is discussed in Chapter 7.

Where a piece of land is being sold by an agent they should provide details of the planning permission, including the planning reference number. But sometimes all that's shown is the front page of the letter granting permission. This doesn't tell you the full story. So first of all, check that you've actually been given the complete document, which should normally run to at least two pages. It's not unknown for several pages of onerous planning restrictions to have been 'accidentally' mislaid, so it's essential to check all the conditions listed because these impose restrictions on what you can build. Also check the precise wording of the approval. Addresses for plots can sometimes be a bit vague, such as 'the plot of land next to The Old Cottage', so make sure that this corresponds to the plot you're buying.

Renewal

It's essential to check the date on which permission was granted and when it's due to expire. Normally permission will expire if construction doesn't commence on site within three years of it being granted. So if you're buying land where work has already started – perhaps the foundations have been built – check with the planners that they accept the consent has been implemented.

There's a potential risk that if one of the planning conditions in the original consent hasn't been satisfied it could still be possible for the permission to be invalidated. For example, one common condition is where the planners need to approve the external materials in writing before commencement of work. This may have been overlooked, so any work that's been done could, strictly speaking, be deemed unlawful and in breach of the conditions. Fortunately, in most cases the planners will be quite relaxed about this. But in some cases Councils may decide to oppose renewal of an expired consent, particularly where it had been granted against their wishes on appeal after they'd initially turned it down. They may also be less accommodating where an original application had faced strong local opposition, or was granted in defiance of green belt policies.

Sometimes land comes on to the market with permission that's expired. Again, the assumption may be that having granted consent once the planners will agree to renew it.

Although in most cases the principle of development will have been clearly established, there's never a guarantee that they'll grant permission a second time, especially where there's been a change of policy in the meantime. Or they may only allow you to build a smaller property.

Land use

Residential land is sometimes referred to in planning terms as 'pink land', so it's also worth verifying the site's current planning designation or 'use class'. Delving back in time by viewing old maps and town plans can reveal previous uses, which may turn up something unexpected, or maybe provide ideas for an interesting house name once your new home is finished – perhaps 'The Old Gas Works' or 'Plague Pit View'?

On the edge of towns and villages, land that appears to be a large garden, with potential for development, may actually be defined in planning terms as 'paddock land'. To planners, the difference between garden and paddock is massive. Paddock land is regarded as nothing more than a meadow. It requires a separate planning application for 'change of use' to get it redefined as residential garden. So such a building plot may turn out to be far smaller than it might seem at first glance. Similarly, if you're buying an existing building such as an old barn, to convert into a dwelling, it's likely to need consent for 'change of use'.

Green belt

The green belt was established in the late 1940s to prevent uncontrolled urban sprawl scarring the countryside. Until recently, green belt land was widely regarded as sacrosanct. Today, however, it's no longer possible to confidently declare 'They can't build on that – it's green belt.' Times have

changed, and governments have declared that large swathes of the countryside need to be concreted over with sprawling 'eco'-towns. This may be music to the ears of corporate house-builders and landowners, but applications by self-builders to construct one-off homes on green-belt land are still likely to be strongly resisted.

The most likely scenario for development of one-off new homes on green-belt land is likely to involve an application for demolition and replacement, for which planning is normally possible, subject to size restrictions.

Listed Buildings

As a self-builder you're only likely to encounter Listed Buildings where you want to build 'within proximity' of one. Listed Buildings are those deemed to be of 'special architectural or historic interest'. In most cases they will be more than 100 years old. When a building is Listed, it means consent is required for alterations that would normally be permitted, both externally and internally. As you might expect, applications for outright demolition and replacement are normally rejected, with very few exceptions.

But the main problem for self-builders is that all the adjoining land that was in the same ownership as the main building at the time of listing is also included. Which means that all structures within the curtilage (ie the enclosed area of land or garden around a dwelling) are also listed. Initial discussions with planners should reveal any such restrictions.

However, things can sometimes get a bit silly, because the listing may well include decrepit old outbuildings as well as assorted boundary walls, gazebos, and statues. To alter one of these without consent is a criminal offence, with potentially hefty fines. Hopefully, in most cases the local Conservation Officer would have a fairly relaxed attitude to proposals to alter or remove such low-grade structures. But there's always the possibility that a 'strictly by the rule book' interpretation could be applied.

Proposals to build close to a Listed Building in England will attract the attentions of English Heritage; in Wales, Cadw; and in Scotland, Historic Scotland. There are three grades in England and Wales: Grade 1, Grade II star, and Grade II, which covers the vast majority. The equivalents in Scotland are A, B star, B1, and B2.

Conservation Areas

Conservation Areas are 'areas of special architectural or historic interest', and there are currently around 11,000 of them in the UK. Here the planners have a duty to ensure that any developments or visible alterations 'preserve or enhance' the character and appearance of the area. This means that designs for sites within Conservation Areas will come under very close scrutiny, including your precise choice of materials, which can substantially increase construction costs. Stringent controls will apply, so designs that are 'out of keeping' are likely to be deemed unacceptable, although this doesn't expressly rule out *well-designed* modern buildings.

As a result, obtaining planning consent in a Conservation Area is harder than normal. When buying a piece of land, your solicitors should pick this up in their local search, but you can check online at your Council's website. Specific restrictions include any significant demolition work, which will require planning consent. In addition, within the designated boundary no cutting, removal, wilful damage, or destruction is allowed of trees that are over 75mm in diameter at 1.5m above ground level without giving prior notification to the Council, who then have six weeks to decide whether the tree should be made the subject of a Tree Preservation Order. There also tends to be a presumption in favour of conservation when it comes to any significant boundary walls or hedges. Conservation Officers at local Councils are a useful source of information.

National Parks, AONBs, and SSSIs

As you might expect, any application to build within a National Park or an Area of Outstanding Natural Beauty (AONB) is likely to be vigorously resisted. The Planning Authority will oppose anything that they feel is not in accordance with stated policies for these special areas.

You may also come across Sites of Special Scientific Interest (SSSIs). These are areas designated as having flora, fauna, or geographical features of national importance, sometimes only comprising a single field. Here endangering or digging up certain rare plants is legally prohibited. As a self-builder it's possible that you might encounter an SSSI without ever setting foot in one. This could happen where consent has been granted for a new house to be built *near* such a site, and planning conditions may be applied to prevent your works from causing harm to it. The respective bodies to consult about such areas in Britain are English Nature, the Countryside Council for Wales, and Scottish Natural Heritage.

Wild things

You know it's not your lucky day when your plot turns out to be home to creatures with friends in high places. Certain species, such as bats, owls, badgers, and newts, are protected by law and it is an offence to disturb them. So if there happens to be a badger's sett just where you want to build, you could be in trouble. On such sites, the planners need to consult the appropriate body (*eg* English Nature). This is likely to cause considerable delays and will affect the time of year that work can be carried out. At worst it could even prevent development from taking place.

Tree Preservation Orders

If there are any trees on your site that are protected with Tree Preservation Orders (TPOs) it can seriously restrict where you can build, perhaps constraining the footprint of your design.

You are perfectly free to apply for planning consent to fell or prune protected trees, but any unauthorised removal, significant pruning or deliberate damage is an offence that carries a hefty fine of up to £20,000, plus a requirement to replace the tree – not the best way of spending your contingency budget.

Councils can place a TPO on any size or species of tree, or a group of trees, that they believe is worth protecting. This not only prevents felling, but also limits pruning to authorised work that won't harm the tree's appearance or health.

Unorthodox method of tree preservation!

However, there are a number of exceptions where such work is permitted. The obvious situation is where the existing planning consent for a new house already includes removal of a tree. Utility firms such as water and power companies are also permitted to cut down or prune trees where their supplies are adversely affected. And where a tree is dangerous, dead, or dying, such work may be permissible – assuming it's not part of a scam!

But no matter how much of a tree-hugger you are, one that's located smack in the middle of where you want to build will have to go. Even where a tree that you want to remove isn't protected, there's still scope for trouble where an aggrieved local resident, determined to throw a spanner in the works of your development, tries to get it protected. Or perhaps a neighbour wants the tree preserved as it affords a degree of privacy. Councils have the power to instantly slap on 'provisional TPOs' for a period six months if they believe a tree is in imminent danger. So where developers sense the risk of an emergency TPO being issued, they may want to do the dastardly deed as quickly as possible at the weekend, when Councils are less likely to be around to react.

As noted above, trees in Conservation Areas are similarly protected, and any work to fell, lop, or prune a tree requires six weeks' prior written notice to be given to the local authority.

Time team

The prospect of buried treasure may delight you, but other ancient remains can have serious cost implications and cause delays. From time to time, when trenches are being excavated on site, human remains are dug up. When massed bones are discovered in the course of building works, the local press tends to fly into a frenzy, publishing hysterical stories about pits from the Great Plague and unexplained hauntings. But such speculation is usually very wide of the mark. Until the 18th century it was common for urban areas to be sown with small cemeteries, such as those beside chapels, or pauper graveyards attached to workhouses. As cities expanded, increasing pressure on space meant it wasn't uncommon for graveyards to be sold off and built over. Most disappeared beneath builders' yards, workshops, and rows of small tenements, with their secrets locked away until redevelopment many years later. The only remaining clue may have been in street names, such as 'Deadman's Place' or 'Crossed Bones Yard'.

But it's not just mouldering skeletons that can give you the creeps. Councils can decide that an area has 'archaeological significance' for many reasons, even the suspected existence of old buried pottery. Once designated as such, it can restrict planning approval for such areas. So any planning consent is only likely to be granted with one of the following conditions:

Photo: Wessex Archaeology

- An archaeological site survey may be required before they'll grant approval. This is likely to involve carefully excavating holes and trenches – again costing you time and money, perhaps adding an additional £5,000 or more to your budget.
- Fortunately, in most cases planning requirements can normally be met where consent is granted subject to a 'watching brief', where you allow aspiring Tony Robinsons to peer into your foundation excavations as the work proceeds. They will check that there are no valuable historic artefacts or ancient remains. This shouldn't prove too restrictive, although it'll all be done at your expense.

In extreme cases, special foundations, such as concrete stilts, may be stipulated to facilitate future access to the site below the building.

Excavating next to foundations is best avoided!

Access considerations

Land without a legal right of access is virtually worthless. Most plots will therefore already have some kind of existing path or driveway leading from an adopted road, or will have planning consent that confirms a new right of access. But what if your access route turns out to be owned by someone else who may not be willing to let you cross it, or they want to charge a king's ransom for the privilege? Such 'ransom strips' typically take the form a small piece of land located somewhere between the public highway and the plot but perhaps only a few inches wide. It may have been sneakily put there by a previous owner in a bid to cash in on any increase in value from future development. It may even have been deliberately concealed, so that there are no visible clues to this deal-breaker on site – see 'Legal considerations' below.

In order that this legal time-bomb doesn't blow-up in your face, it's essential to compare all dimensions on site with those from the Land Registry plan (or in the title deeds). If you notice any discrepancy it might indicate the existence of a ransom strip. Checking this on site is up to you, as solicitors rarely venture beyond their offices. The next step is to track down the legal owner (the beneficiary) and agree to purchase the plot or a legal right over it, paid for by the person from whom you're buying the land. However, in some cases owners can't be traced, so the only solution may be to obtain an indemnity insurance policy, again paid for by the vendor. The best precaution is to have your solicitor make the purchase of the plot subject to a legal right of access. And where a plot abuts a private road, check that there's a legal right for you to use that road.

Some plots already have rights over someone else's land, such as a right of access over a shared driveway that actually belongs to a neighbour. If this arrangement isn't 100% legally watertight, an indemnity insurance policy may again be needed to guard against the risk of someone challenging your right of way in future.

Planners stipulate shared driveway.

Photo: Wooden Hill

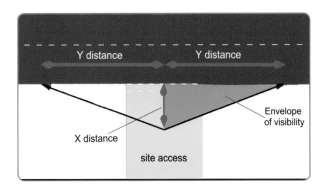

Visibility splays

Most sites need vehicular as well as pedestrian access. So before buying your plot, there's another potential problem that it's essential to spot. A common planning condition is that the driveway must meet the requirements of the local Highways Authority. For plots on fairly busy roads, this means there may be a requirement for 'sight lines' and 'visibility splays'. In other words, a car exiting the drive should have a clear unobstructed view of the road either side. The resulting triangles of land therefore need to be kept free of any obstructions.

If the land is within your ownership, there shouldn't be any major difficulty complying with this (unless a stray telegraph pole needs relocating).The real problem is where the sight lines cross neighbouring land. To satisfy the planning condition, you would have to enter into a legal agreement with the neighbours to keep this land free from obstruction in future. It's possible they might refuse to cooperate. Or, more likely, they may demand some sort of payment, which again should be made by the vendor of the plot, not by you.

New access

Construction of a new access route to a previously 'land-locked' piece of land can, of course, massively transform its value. It may even be the key to unlocking objections to getting planning consent. But many a hard battle has been fought over attempts to achieve access. Not only do you need a legal right of access over a piece of land, for example where the verges abutting a road are privately owned, you

also need consent from the Highways Authority to build a driveway, and, in some cases, confirmation from the planners that they have no objection. So never assume a new access driveway can be built just because it looks feasible. Bitter legal disputes have been fought all the way to the High Court over such claims, by Parish Councils concerned to protect a 'registered village green' from any kind of development.

However, it is possible to acquire a legal right of access where you can prove 'long-usage'. If you can demonstrate that a property has enjoyed unencumbered and uninterrupted access over land for a period of 20 years or more, a legal right can be established, known as a 'prescriptive easement'.

So, for example, the owner of a property who can prove they have driven over a piece of land for at least 20 years can apply to the landowner for a formal legal right of vehicular access. Naturally, this is subject to a number of caveats. The route must have been used 'continuously' for access, which means at least once every year. The right must have been obtained without 'force, secrecy or permission'. The right must also attach to a Freeholder, rather than tenant, and access must not be illegal – which rules out railway land or land owned by British Waterways. Finally, a payment must be made to the person whose land you're crossing, of up to 2% of the value of the house that the access route serves (to which it would, of course, add considerable value).

Services considerations

Providing new mains services to a site can eat up a huge chunk of your budget. So plots with services already available, or, more likely, available in the street, can be worth considerably more than an

isolated field somewhere out in the sticks. So before you buy, it's important to check which services are available on or near your site. This, of course, means contacting the relevant electricity, gas, and water/drainage companies. They should be able to provide an OS map (for a small fee) that clearly shows the location of supply pipes and cables in the vicinity.

The converse problem is where a site, rather than being devoid of services, has too many! It's not unknown for hapless plot owners to remain totally oblivious to the presence of a public sewer running through the middle, until the water authority intervenes at the last minute to stop the build as the site is about to be stripped. To prevent such calamities, it's essential to carry out thorough enquiries prior to purchase to flag up anyone else's pipes and supply cables that happen to be running through your plot. Although the accuracy of old maps isn't guaranteed, it's still better to know approximate positions in advance rather than cutting blindly into the ground with a JCB.

Such unwanted services may need to be diverted before you can start building, which could obviously cost a substantial sum and should be deducted from the sale price of the plot. It's important to make contact with the utility companies as soon as you're seriously interested in buying a site, as some are notoriously slow to respond.

There is, however, another way of checking what lies under the ground – if you're not put off by things new-agey. Dowsing for water is widely acknowledged as a traditional skill that's fairly easy to master. Most major oil and mineral exploration companies retain a dowser somewhere on the payroll as an aid to locating mineral deposits, and a skilled practitioner should be able to accurately pinpoint the path of underground pipes and cables using a pair of dowsing rods.

Where there are no existing service supplies, getting new ones connected can be shockingly expensive. So it's important to get estimates to include in your budget as soon as possible. You need to write to the relevant service providers including a site plan. Water and electricity are considered to be the two essential supplies, and laying new electricity cables or mains water supply pipes is generally less expensive than pipework for new mains drainage or a gas supply. But even in a typical suburban street, a new electricity

supply and connection can leave little change from £1,000. Connecting to existing water and sewage systems can add another £1,500. But of course, for isolated rural sites the costs can be totally uneconomic, so it's just as well there are often viable alternatives.

Fortunately, supplies for most plots can be brought in from the nearest road. Where the costs can seriously start to mount is if you have to cross someone else's land in order to connect to the main supply as they will probably want to charge you for the privilege. In some cases you may need to connect directly to neighbour's services, so you'd need to budget for additional connection costs as well as legal expenses.

Water

A mains water supply probably tops most people's lists of essential services as the appeal of congregating round a communal village pump can soon wear thin. Thankfully, unless your plot really is in the middle of nowhere, this shouldn't be a problem. Even where there isn't a mains supply in the road, if there are any other houses within spitting distance there should be mains water pipes to tap into. However, bear in mind that modern pressurised hot water systems are dependent on a consistently strong mains pressure.

If there really is no mains supply available, alternatives such as rainwater harvesting and grey water recycling are unlikely to provide a reliable substitute – see Chapter 6. Even if you're a hardened eco-warrior, prepared to squeeze drinking water from your garden cactus, there's still the problem that a house without mains water will seriously

deter buyers when you eventually come to sell. It could even condemn the property as unmortgageable.

If the cost of a new supply is prohibitive, an alternative is to drill a borehole, but this can cost more than £10,000 depending on depth, and is only really viable for larger housing developments where costs are shared. It's also necessary to satisfy your local authority Environmental Health Department that the well water is safe for drinking. To screen out pollutants, water treatment systems need to be installed.

Electricity
There are many otherwise perfect sites that have been rejected because there's no mains electricity supply in the locality. When you're talking in terms of kilometres, the cost of running a new supply in can be prohibitive. The fact is, a mains supply is an essential requirement for most people. Even if you're prepared to run the telly on bicycle power, when you come to sell your house most buyers will run a mile. Self-generated green energy is still largely dependent on a connection the national grid, and isn't yet capable of providing a fully independent power supply, even with banks of batteries.

Happily, with nine out of ten new developments you should be able to connect directly to the existing system via the street supply. However there's always the possibility that when you apply to extend the supply to your house, the system could turn out to be fully loaded. And if an underground electricity cable has to be upgraded, the cost can run into thousands of pounds.

In villages and rural areas it's not unusual for houses to be supplied by above-ground cables. Where such a high voltage cable is directly in the way of building works, it will be necessary to have it diverted, usually at significant cost. But where there are high-voltage power lines nearby, and your works (especially scaffolding) come within 3m, you can arrange for the electricity company to sleeve them, or erect safety screens. This should be done free of charge. Even where cables are buried underground, they may be nearer the surface than you'd imagine, posing a serious hazard. So always check before digging.

Gas
A gas supply is not essential. In rural areas oil is widely used for heating and electricity for cooking. If you insist on gas, LPG can be used for both (electricity is not normally cost-effective for heating). There are three ways of having LPG gas delivered to site – in an above-ground tank, a below-ground tank, or a cylinder. But the plot needs to be large enough to incorporate storage tanks, as there are minimum rules on how close tanks can be sited to a dwelling, and access will be needed for lorries to deliver refills. See Chapter 8.

Oil
Oil is a well-established and popular alternative to gas heating. Of course, oil prices can be extremely volatile, but then all fuels have suffered dramatic price-hikes in recent

years. As with gas tanks, there are strict rules about siting and construction of oil tanks. See Chapter 8.

Drainage
In most cases a public sewer will be available. However a site without mains drainage needn't be a problem. Where there's no existing connection to the plot, and no public sewer within striking distance, private drainage systems such as septic tanks can be installed. See Chapter 6. However, where there's a mains sewer closer than 30m to the site, even if you prefer the idea of a private drainage system, the drainage authority may insist that you to connect to it.

As part of your site inspection it's important to check all existing sewage routes, because if new pipes have to cross someone else's land it will mean having to negotiate and pay for a legal right to be established. Lift any visible inspection chamber covers and try to figure out the direction of flow and whether the drain runs are shared with adjoining houses. If there are none, look for manholes in the road.

Modern drainage systems are divided into surface water for relatively clean rainwater dispersal and foul waste from sinks, bathrooms, and WCs. Some older single drains may take both foul and surface water together, but 'combined drains' are no longer acceptable. This is because in storm conditions surface water can arrive in mighty, surging torrents, and where mixed, it all has to be treated as foul, potentially overwhelming sewage works. So even where there's a combined mains drainage system, such as in London, it's unlikely that you'll be permitted to connect your new surface water drainage to it.

Sterile zones
One reason why a site may not have been built on before could be because it's blighted by a utility company's pipes or cables running straight through it. Such legal rights of way for services are known as 'easements', and may prohibit any building being carried out within a certain distance of the pipes or cables. This creates a 'sterile zone', making it impossible to develop. But there is sometimes a way round the problem.

Although when it comes to water mains, public sewers or gas pipes there's often no realistic solution, services such as power lines are sometimes capable of being moved – at a price. Of course, the utility company may be distinctly unenthusiastic and you may have a long battle on your hands. But if you can at least can get a quote from them, it means they accept that the work can be done. Naturally, the quotation is likely to be eye-wateringly expensive, but if it makes the difference between a building plot and an unusable piece of land, the plot's value will skyrocket. It may even be possible to negotiate to do some of the work yourself to reduce the price.

But it's not all bad news. Where a *private* sewer crosses part of your site, it may even be of some benefit, since you can usually connect to it. A private sewer is one that takes waste from two or more properties up to the junction with the main public sewer, usually running under private gardens. Even if it has to be diverted to avoid the new house, it should still be cheaper than paying for a connection to a public sewer in the highway.

EMFs and mobile masts

There is some concern, but as yet no conclusive evidence, that continued exposure to magnetic fields could be harmful. Whenever electricity is used an electromagnetic field (EMF) is produced. Principle sources are:

- High-voltage transmission lines on pylons, which give off high-power electric and magnetic fields. It is believed that people living within 200m may be affected.
- Low-power substations are fairly common in built-up areas. Screening with thick concrete walls can minimise risk.

There is similar debate about mobile phone masts. It is believed that high masts (typically 15m tall) are safer than lamp-post-type masts of a mere 6.5m or so, which could potentially direct radiation into homes.

Legal considerations

Normally, when you buy a house, checking who legally owns it couldn't be simpler. This is because whenever land or property is sold or mortgaged it has to be registered with the Land Registry. So anyone buying it in future only has to do a quick check on title to get guaranteed accurate information. Registration of land and property sales has applied to most of the country for many decades (although it only became compulsory everywhere by 1990). This is obviously good news when you're moving house. But the problem with buying a plot of vacant land is that it may not have been sold in living memory, and may still be unregistered.

Proving ownership

If the plot you're buying turns out to be unregistered, proving who legally owns it can be a much more long-winded process than with registered land. Your solicitor will need to trawl through lots of dusty old documents. Problems sometimes arise where deeds have been mislaid, perhaps mouldering away in a forgotten bank vault. The true owner may be long dead or might have emigrated years ago. So you might need to turn detective and ask around the locality, checking with neighbours or inspecting the electoral roll. However, placing ads in local shop windows or papers could potentially alert rival plot hunters, so that by the time the legal owner finally surfaces there may be a queue of buyers waving their chequebooks.

Defective title

Even where it's fairly obvious whose land it is – for example, where the landowner lives close by – absolute proof of legal ownership, known as 'Title Absolute', might not be available. They may only have a weaker form of legal ownership known as 'Possessory Title'. This sometimes applies where the original deeds have been lost, or where a claim to ownership is only supported by 'adverse possession' (see below).

In such cases, the solution may involve your solicitor obtaining *statutory declarations* from the vendors, and from other related parties such as neighbours. Armed with these, it should then be possible to arrange insurance (a defective title indemnity policy) to cover you against the risk of someone popping up and claiming ownership of your land. It should be possible to arrange this for a one-off payment of no more than £500.

Adverse possession

Tracing who owns a piece of land can be made more challenging thanks to 'squatter's rights' – a noble tradition of pinching someone else's land that dates back several centuries in England and Wales.

When someone has occupied a piece of unregistered land without the consent of the true owner, they may be able to acquire the title by 'adverse possession'. To 'occupy' could mean living in a caravan on the site, or cultivating the land, or even fencing part of it off. Where a piece of unregistered land has been occupied for at least 12 years, and this has gone unchallenged, the occupier (or 'squatter') may legally register 'possessory title'. After another 12 years they can apply to convert this to 'absolute title'.

Unregistered land with an untraceable owner is more likely to be the subject of such a claim. Where land has been registered, it is much harder to establish adverse possession, because the Land Registry notifies the registered owner directly someone attempts to claim part of their land.

However, cases of adverse possession rarely arise because of the deliberate actions of squatters. It's more likely to come about by mistake, where the boundaries between two properties have been unclear for many years.

Fencing it off

In practice, such legal curiosities are only likely to affect self-builders buying a site where part of the land turns out to have a different form of ownership to the main part. It may be that in years gone by, the person who's now selling you the land decided to enlarge the plot by moving their fence over and 'borrowing' some adjoining undergrowth or unused land. In which case the main plot will then be offered with *absolute title*, and the acquired bit with *possessory title*. So if you're offered a plot which includes any such 'grabbed land', where parts of the title are not absolute, you will need to take out an indemnity

Photo: Grange Fencing

policy to cover the slim possibility that an absent owner could one day turn up to dispute your occupation. The premium for this should be paid by the vendor.

Of course, fences can just as easily be moved in the other direction. It's possible that a neighbour may at some point in the past have cunningly 'liberated' a few metres of your site in a bid to make their garden bigger, assuming that no one would miss a bit of overgrown land. Either way, what you see on the ground may be different from what you're actually buying as shown on the legal plan.

Boundaries

Although your solicitor will know the approximate dimensions of the land from the deeds or Land Registry plans, these are usually drawn to a minuscule 1:1250 scale

that shows very little detail. And where land is unregistered, old maps found with conveyancing documents may not even be drawn to a reliable scale. So it's pretty much down to you to make sure the land you're buying is the same size and shape on the ground as it appears on your solicitor's title plan. But pinpointing the precise borders of an overgrown site can sometimes require a bit of guesswork, for example where boundaries comprise ancient hedges and ditches.

Boundary disputes are a common cause of disagreements between neighbours, and can equally be an issue when

Photo: HSS Hire

buying a plot. Over the years neighbours may have planted hedges or built walls that intrude on to your plot, confusing the identification of the correct boundaries. The origin of old boundary changes may now be lost in

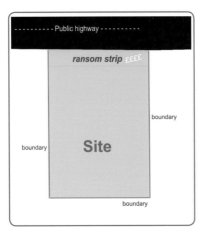

the mists of time, so depending on how long they've been there, it could prove legally difficult to move them. Where there's any doubt, it's advisable to instruct an independent measured survey, so that the existing boundaries can be clearly identified and agreed with the vendor. If there's any discrepancy with the legal site plan, you can then mark it with a red line showing where you and your solicitor believe the correct boundary should be. Probably the most important part of the plot to check is the access route from the street, to spot any 'ransom strips' owned by someone else, as discussed earlier.

If you're confident that there are clear signs of recent encroachment where a neighbour has 'borrowed' some land, there's nothing to stop you taking direct action to recover the land. With some wooden stakes, clearly mark out the correct legal boundaries. Then have your solicitor send a letter by recorded delivery to the neighbour confirming that you have established your legal boundaries (including a copy of the deed plan). The neighbour then has to actively prove any claim they have to the land. But if enough years have passed to allow them to claim adverse possession, you may need to renegotiate with the vendor if the plot you're actually buying turns out to be significantly smaller than shown in the deeds.

Compulsory purchase

It's possible that a plot, or part of it, has remained undeveloped because some years ago it was earmarked for compulsory purchase by the Council, for a now defunct scheme. The most common reason for compulsory purchase is where land is taken for road building or widening. Many Councils acquired large amounts of land in the 1960s and '70s for grandiose road schemes that were then postponed and eventually abandoned. So it may now be possible to negotiate to buy it back.

Any such land acquisitions should be recorded on the deeds and registered at the Land Registry. If it turns out that part of your plot is owned by the Highways Authority, you can't just ignore it, perhaps in the hope of establishing ownership by adverse possession over time. Highways land cannot be legally acquired through 'squatting'.

Footpaths and rights of way

If there's a public right of way running straight through the middle of your site, there may be few if any visible clues to this on the ground. Yet this could put the kybosh on any hopes of developing the plot, so it's essential to carefully check the deeds and legal plans. Where footpaths cross fields the exact routes might not be recorded, but they normally follow the shortest route between the gates or stiles at either end.

Footpaths are protected by law, and jealously guarded by local Councils, backed up by bodies such as the Ramblers' Association. So although it is technically possible to have footpaths or public rights of way moved, it's a potential legal nightmare – a major undertaking that can take several years. Any proposed diversion must be proved to benefit those using the footpath. The most realistic approach is to assume that the route will remain unaltered, and try to plan your development to accommodate it. Footpaths near the boundaries of a plot can normally be screened from the house with hedges.

Covenants

It's possible that a piece of land hasn't been developed because, tucked away in the deeds, there's a legal prohibition against doing so. Such 'restrictive covenants' are legal nasties that date back to when a property was originally built, or the land changed hands. Binding on all future owners, they place restrictions on what you can do with the land or property, regardless of whether or not you've got planning consent. Over time these can become anachronisms that make no sense. For example, there may be a requirement that a stretch of land should remain undeveloped, perhaps with the original intention of protecting the view from a Manor house that's long been demolished.

It is sometimes possible to have restrictive covenants removed, or legally 'overturned', but this can be time-consuming and expensive. The usual solution is to pay for an indemnity insurance policy to cover you in the unlikely event that anyone should try to enforce the restriction.

One of the more usual covenants that affects land is one that requires the builder of any new house to formally seek the approval of the vendors of the land for the proposed design. However, there is also normally a stipulation that the vendors 'must not unreasonably withhold their consent'. So you could argue that the granting of planning permission where the adjoining owners were formally consulted but chose not to object, means that to oppose your plans now would be unreasonable.

Section 106 agreements

Where you buy part of a larger plot, it could still be subject to a piece of planning law known as a 'Section 106 agreement'. These are often applied by Councils when consent is granted for large developments, as a way of enforcing payments towards infrastructure costs like schools and roads or affordable housing, known as 'planning gain'. These attach to the deeds and are legally binding on future owners, so it's possible you could inherit a share of the obligation to contribute funding – something your solicitor should spot.

'Ag Tags'

If consent is granted for a new home in a rural area where there's normally a planning presumption against development, it may have been granted subject to an 'agricultural tie'. This means planning comes with a strict condition that anyone occupying the property must be 'wholly or mainly engaged in agriculture, last engaged in agriculture, or the widow or widower thereof'.

This 'encumbrance' has the effect of dramatically reducing the property's market value, and in some cases rendering it unmortgageable.

Such conditions are often legally enforced via Section 106 agreements. This binds the owners of the property and successors in title – including anyone to whom they sell. However, it's sometimes possible to have an 'Ag Tag' overturned. A planning application has to be made for its removal, and you must demonstrate that the need for the tie has lapsed.

Easements and wayleaves

Ever wondered how the British countryside came to be scarred with thousands of ugly pylons carrying high voltage cables? To cross land with electricity cables, gas pipes, sewers and the like, service providers can impose or negotiate 'wayleaves' and 'easements' (or 'servitudes' in Scotland). These are legal 'rights of way' granted to permit major pieces of infrastructure to run through or above land, and to allow a right of access to maintain equipment, such as electricity cables running across a site. Landowners receive payments for granting such rights.

Any such rights that others enjoy over your site should be flagged up by your solicitor's search of the title documents. Taking time to chat with neighbours can also elicit useful information, and there may be clues on site, such as gates leading to adjoining land.

If any such legal rights are discovered, it needn't be as scary as it sounds. Not all easements are a problem. Some merely grant occasional access and many others work in your favour and 'run to the benefit of the land', giving you the right to connect to drains and other services.

However, as noted earlier, pipes and cables may be subject to 'sterile zones' either side that can't be built on, such as where construction is prevented within 3m of buried cables. It's always worth enquiring about the possibility of diverting them, because some can be renegotiated, or bought off. Or if there's no one around to enforce them, they can normally be insured against.

Even the most idyllic rural sites may conceal legal restrictions.

Photo: Peter Kent Architect

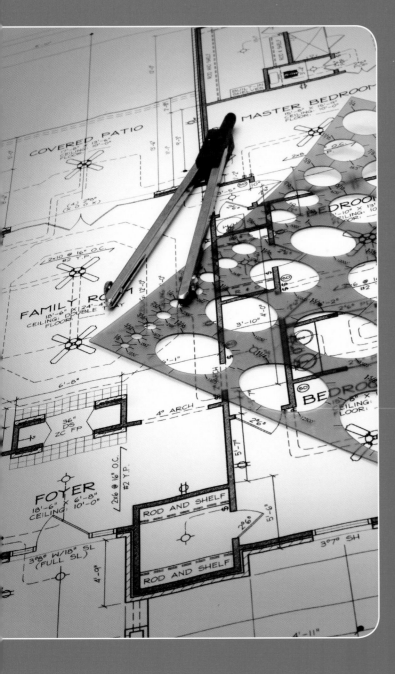

5 THE DESIGN

Photo: Velfac

Photo: oakwrights.co.uk

With the purchase of your plot now well and truly under way, the design of your new home can really begin to take shape. By this stage you'll know how much money you'll have left after buying the site, and where the rest of the funding is coming from to pay for the build.

Of course, until you've exchanged contracts it's still possible that things could fall through with your land purchase.

Photo: oakwrights.co.uk

But there's plenty of useful design work that can be done in the meantime before appointing an architect. You probably already know the broad style and size of house that you want but because plots impose limitations, as well as opportunities, it's best to start afresh with a blank canvas.

At the end of the design process, you should have a finished set of drawings showing the outside appearance, the layout of each floor, and how the building will occupy the site. But before taking the next step, there are some key questions you need to ask.

Photo: oakwrights.co.uk

What's the planning status?

The design of your new home will need to be compatible with the planning status of the plot. Most building land is purchased with Outline Planning Permission, which means there should still be considerable scope for negotiation about the detailed design. The extent to which the planning situation is likely to affect your design is discussed in detail in Chapter 7.

Dream home – or the first of many?

Most self-builders want to create the home of their dreams, something to be truly proud of. For others, this project is merely a stepping-stone to future projects. This makes a massive difference to the way you approach the design. Building your dream home means creating a property that's custom-designed to fit your family's lifestyle, a comfortable, stylish abode where you can luxuriate for many years. But as a developer, the objective is to maximise profit, so the market appeal of the finished product is paramount, whilst at the same time minimising costs. The golden rule is 'If you're building for others, don't design it for yourself', which means resisting the temptation to indulge in expensive personal whims.

How long do you plan to live there?

If you plan to put down roots for many years the design should be capable of adapting to your needs as they change over time. For example, if the prospect of raising children is on the horizon, you'd best be prepared for lots of brightly coloured plastic toys strewn liberally around the house. Open-plan designs will find you forever tripping over them, whereas confining children's activities to separate rooms can help preserve your sanity. You might also want to consider potential future requirements such as home offices or even a granny annexe. If you plan to stay for the longer term, the immediate resale value won't be so crucial, so it may be worth spending more on quality fittings to get as close as possible to your concept of an ideal home. On the other hand, if you plan to move during the first few years you need to think more like a developer who puts profit before personal taste.

The number of years that you're likely to spend living in your new property will have major implications when it comes to investing in green energy-saving devices, which typically have high capital costs and long payback periods. You will need to weigh up whether longer-term savings in running costs justify the high upfront installation costs.

Photo: buildstore.co.uk

How realistic is your budget?

There's one thing above all that will ultimately dictate what you can build, and that's how much money you can scrape together. Working out your budget is at the heart of the entire project. A common mistake is to assume that the whole project will run smoothly from start to finish, because in reality not everything does. So a fairly generous contingency sum must be included. If the plot turns out to be more expensive than you'd hoped, having a decent wodge of money in reserve means there should still be enough in the kitty to build the home you really want.

But contingency planning can take other forms. If money's tight, some of the costs can be pushed into the future, to be met later when funds permit. Your home could be designed to be built in stages, evolving over time as finance becomes available. Or you could prioritise space over quality. A reduced specification with cheaper fittings can save money in the short term, allowing you to replace them with better-quality ones in future. Another popular way to eke out your budget and save money is to manage individual trades direct, rather than paying more for the services of a main contractor.

What sort of home do you want?

The way you want your house to look, and the way the rooms are laid out to fit your lifestyle, goes to the heart of the self-build dream. So before involving a professional designer, it's important to be clear about what you want – and what can realistically be achieved on your chosen plot.

Design factors

There are three key design factors:

Appearance

Perhaps the most fundamental decision of all is how you want your home to *look*. The style of architecture and the 'kerb appeal' of your home is a highly emotive subject. One person's dream design might be another's nightmare.

Room layout

One of the first design decisions concerns the internal layout – the size and arrangement of your rooms. Which ones should get priority? And how large should you make the entrance hall, landing, and stairwell, at the expense of adjoining rooms?

Plot

The site is often the major limiting factor. Your design needs to make the most of positive attributes, such as views, whilst overcoming negatives like road noise, being overlooked, and planning restrictions. A good design should work with the landscape rather than against it. So where a site has a particular constraint, such as a steep slope, a creative design can sometimes turn it to your advantage, moulding it to make an interesting and stylish home.

Design factor 1: Appearance

When it comes to deciding what your new home is going to look like, it depends on whether you prefer traditional or contemporary architectural styles. The majority of self-built homes are modern versions of traditional designs, inspired by history. Just about every decade in the last 200 years had its own distinct style, and we all have our favourites. Some may aspire to a Gothic lifestyle complete with gargoyles and stained glass. Others may prefer a sleek white modernist box.

Photo: Canada Wood UK

For most of us, designing a fantasy home means incorporating ideas borrowed from interesting buildings we've spotted – perhaps a stylish entrance porch from one property blended with roof tiles or windows from another. A good way to approach this is to take lots of photos and collect images of designs that tickle your fancy. Once you start thinking along these lines, you suddenly notice an abundance of design inspiration all around.

But no matter how far your imagination runs riot, at the end of the day the external appearance of your design is likely to be strongly influenced by the planners. This in turn is likely to depend on the local architecture, which will often

dictate the style of design deemed appropriate for the plot. Of course, if you happen to be located in a Conservation Area or within the grounds of a Listed Building, the planning controls will be far stricter. This may explain why most self-builders end up sticking broadly to convention, producing houses that, on the outside, may not appear that different to those of mainstream developers, albeit of superior quality.

That said, one of the big attractions of self-build is the freedom to create a stylish home of real quality, even where the local vernacular does not accord with what you want. A building doesn't have to match the existing architecture to enhance a street. A well-designed modern building that's different from those around it can often complement the character of its surroundings. But you have to be prepared for local residents to be a little wary, given the sheer volume of ugly, overbearing new developments that have blighted our towns in recent years and the more quirky or overbearing a design, the more resistance you're likely to encounter from neighbours and planners.

The value of your home will be strongly influenced by its 'kerb appeal', *ie* how attractive it looks from the road. This may not matter if you plan to live there for many years. But if you choose to defy convention, and win the planning battles, bear in mind that it may ultimately prove harder to sell to a conservative homebuying public.

Primary design

One of the first design considerations is the basic shape of the building, sometimes referred to as its 'massing'. The simplest shape is obviously a basic box, popular with developers. But an inherently boring shape can be transformed simply by applying a few smart design tweaks:

- Roofs: several smaller roof areas, rather than one big roof, can add interest.
- Levels: try varying storey heights with a mix of single- and multi-storey.
- Outdoors: courtyards or internal gardens can add style and create privacy.

Once the basic shape has been decided, the architectural design ideas start to flow. Different roof styles can be experimented with, mixing and matching hips, gables, and alternative positions for dormer windows. Perhaps a subsidiary flat roof or balcony would look good, plus a couple of chimneys.

Getting the 'primary design' features right also means considering how the shape is likely to affect the occupants of the building. It's important to achieve a balance so that a distinctive exterior appearance doesn't cramp your interior style by dictating the layout of the accommodation – or vice versa. This is when it can be well worth trying out architectural computer programs that let you explore different design styles and materials. It's fun to create a 2D plan and then watch as it magically reconstructed into a 3D model on your screen.

But computer simulations have their limits. Architects employing modern non-traditional materials are sometimes accused of ignoring the effects of ageing and weathering

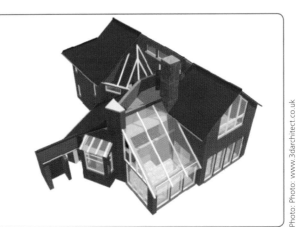

over time. Artificial materials generally wear less well than natural materials. Over the years, concrete can become streaky, stained, and blotchy, and plastic often becomes warped and faded. The planners may therefore dictate that you dress a modern building in traditional materials to help it 'blend in' to the streetscape. But in practice such materials may limit what you can build. For example, traditional plain clay tiles normally can't be laid to a roof pitch much shallower than 35°.

One of the major determinants of the look of your house is the position of its windows. You need to consider precisely where they should go so that the building feels just right. Get this wrong and the house could end up looking rather odd. Key decisions need to be made about their size, as well as the appropriate shape and style – there's no law that says windows have to be rectangular. Of course, the windows also have a major effect on the quality of the interior – see below.

Secondary design features

NATURAL LIGHT
There's more to picking the right windows than making the house look good. Optimising the view out of them is clearly a major factor, and encouraging natural light into your home

is key to all good designs. It's claimed that people feel more relaxed in a room that has windows on more than one side and, given the choice, will gravitate to rooms that have light on two sides. Daylight entering from more than one direction certainly makes a room feel brighter and more interesting.

Photo: oakwrights.co.uk

In terms of maximising the amount of natural light, it's claimed that roof windows (ie skylights flush with the roof slope) can provide up to three times more light than the same-sized window fitted in a wall.

Other window-related design factors to consider include security, privacy from nosy neighbours, and ventilation.

THE MAIN ENTRANCE

The way you approach and enter the building dictates the character of the whole property. The trick is to combine a visible and welcoming entrance for visitors with the need for security and privacy. If possible, place the main entrance where it can be seen immediately from the approach, as a prominent visual feature. Good designs often incorporate a simple porch with raised landscaping that combine to create a pleasing entrance that flows naturally as you approach. Architects don't always get this right. Some modern designs even position front doors so that they're deliberately hidden!

Photo: Caststone

WALLS AND ROOFS

One of the key decisions that will determine the appeal of your house is your choice of materials for walls and roofs. The main construction options are discussed in Chapter 6. But whether your walls are of conventional masonry, timber frame, or something more radical, it's important to determine early on what visual style you want to create. Consider the colour and texture of the facing brick or stonework, render, or cladding. Similarly, picking the appropriate roofing materials can make all the difference to your design.

GARAGES AND OUTBUILDINGS

A garage isn't essential, although space for parking and turning normally is. Attached garages are cheaper to build and consume less space than the detached variety, but the size needs to be proportionate to the house – for example, a slim single garage may look a bit mean sitting next to a grand six-bedroom house. Building a double garage from

Photo: potton.co.uk

new presents a golden opportunity to make use of the roof space at minimal extra cost, perhaps as a den or office, or even a spare bedroom. This would basically require a couple of skylights, beefed up insulation, stronger rafters and floor joists, plus a power supply.

Where plots are large enough, you may want to construct outbuildings or annexes, perhaps linking them to the main house via a glazed lobby. Building garages or outbuildings in advance of the main project means they can be usefully employed as storage space for deliveries of materials and equipment. Alternatively, where budgets are tight construction could be postponed until funds are available.

LANDSCAPING

Good landscaping has real power to enhance even the most uninspiring properties. The topography of your plot, the local climate, and the architecture of the house can all help determine what sort of surroundings will best enhance your home.

Photo: oakwrights.co.uk

Design factor 2: Room layout

Although the planners may try to dictate the way your home looks on the outside, when it comes to designing the interior the self-builder is king of the castle. There's nothing to stop

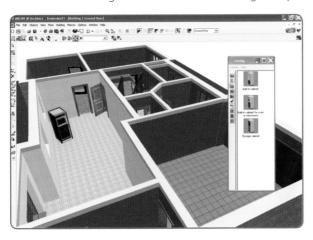

you putting the rooms wherever you want them. You may even decide to reverse conventional wisdom and place your living rooms upstairs to take advantage of views, and the sleeping areas downstairs. Of course, being different means that when you eventually come to sell your property, for every prospective buyer who loves the design's individuality there will be several who are put off by a departure from the norm.

To introduce a real sense of style, some larger rooms can utilise split-level flooring, where different zones are separated by a step, and different levels defined with contrasting floor coverings. Incorporating design features such as split levels, mezzanines, and galleries can turn an ordinary dwelling into something more playful and intriguing with a variety of viewing points. But multiple levels may not be appropriate where the occupants are physically impaired.

The best approach is to start with a blank piece of paper and imagine what you'd really like to have. Perhaps an en-suite for every bedroom, and a studio or games room in the loft? Or an outbuilding with a proper gym and sauna? Then take this list and decide which ones you really want, and which are a non-essential luxury.

Rooms and lifestyle

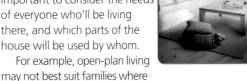

To be certain that your new home will tick all the right boxes when finished, it's important to consider the needs of everyone who'll be living there, and which parts of the house will be used by whom.

For example, open-plan living may not best suit families where different members have conflicting lifestyles. Keeping any noisy activities well separated – perhaps screened off behind sound-proofed walls – means that TVs, PlayStations and the like needn't disturb others trying to sleep or do homework. Even family pets need their space, and a generously proportioned utility room can be a godsend. Where rooms are predominantly used in the evenings, such as some of the reception rooms, maximising daylight and views may not be so important.

But as well as considering how each room can best accommodate the activities that will take place within it, you also need to determine the optimum way in which the rooms in the house will connect and relate with each other.

Interconnections

Deciding how all your rooms are going to link to each other is key to designing the overall layout of your house.

Start by sketching out some rough plans. The way the connections between the rooms develop will depend on your lifestyle. One of the main flashpoints for arguments at this stage is whether to have a spacious, open-plan kitchen/dining area, or to opt for a traditional layout with separate rooms. A good compromise may be to have your living room, kitchen, and conservatory all linked, but with dividing doors in between. This way, it can be opened out into one big space or divided up as the mood takes you.

The next question is which rooms should be located at the back or front of the house. This will depend to some extent on your plot – there may be views you want to take advantage of or busy roads to shelter from. At the back, rooms can enjoy a view of the garden and benefit from privacy, enabling parents to keep a watchful eye on young children playing. Rooms facing the front allow you to observe the world passing by, and to benefit from early warning of approaching visitors.

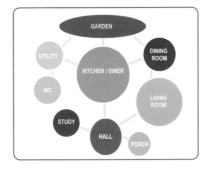

Contents and storage

A major influence on the required size and shape of your rooms is the furniture that goes in them. One way to visualise this is by drawing experimental room sizes on graph paper to see what kind of room shapes work best.

If you're the proud owner of antique wardrobes and monster-sized chests and dressers, you may need to accommodate them by designing generously proportioned door openings and loftier ceiling heights, particularly to the stairs. If your collection extends to snooker tables, the logistics of delivering one up to a loft room will need careful planning, not least to ensure that the floor will take the weight. To play snooker or table tennis comfortably, you need at least 2m space all round.

One consistent complaint about new developer-built homes is that storage space is minimised to make the rooms larger. In some properties you're lucky even to get an airing cupboard. Provision of larders and storage space is often something of an afterthought, and yet is something that makes day-to-day living far more convenient.

Future-proofing

It doesn't cost much to build flexibility into your design. So if you plan to live in your new home for many years, the design should ideally take account of future needs as family circumstances change over time. The option of ground-floor bedrooms could be a godsend should you ever need greater care or an elderly relative move in. Wider doors and spacious bathrooms that can accommodate wheelchairs make good design features in their own right. It makes a lot of sense for your design to allow for a future extension or a room in the roof that can be easily added when funds permit, or if you anticipate that your home will shortly be overrun with children.

There's not a lot you can do to guard against changes in taste over coming years. After all, chocolate-coloured bathroom suites were once considered the height of style. But when choosing fittings and interior design it's worth remembering that anything that's very fashionable tends to become very unfashionable very quickly!

Feature lighting

Some systems can alter the mood of a room at the press of a button – from 'party' to 'relax' mode, and everything in between. Consider fitting pre-programmed lighting with a choice of different settings such as bright 'task lighting' for reading, mellow background lighting, and wall-mounted uplighters for entertaining. But the final decision of precisely where individual lights go can be left until first fix stage.

Focal points

Anyone designing their own home will want it to be stylish and attractive. Kitchens and living rooms in particular lend themselves to the inclusion of 'focal points' that can transform the plainest design into something really special.

A fine farmhouse kitchen can define a building's character, with a traditional range and copper canopy along with the tasteful application of natural materials such as exposed brick or stonework. But focal points need to be related to a central theme, so you don't end up with an bizarre mishmash of conflicting styles.

For many self-builders, one of the key design features in the house will be a stunning fireplace, although this needn't necessarily have a traditional theme. Modern hole-in-the-wall gas-fuelled fires with 'living flames' flicking over decorative shapes can work well. But with anything quirky, bear in mind that fashions can change very rapidly. The staircase is another major design opportunity that can be worth lavishing money on. But don't overlook the importance of paying careful attention to smaller design details – things like stone window sills and traditional brick arches can also help create the desired effect.

Fire and safety

Statistically, more fatal accidents occur in our homes than anywhere else. So it makes sense to design-in basic safety precautions, such as fitting more than the minimum number of smoke alarms, and having more than one escape route out of the house in case of fire. Domestic sprinkler fire systems are increasingly popular, as they minimise the chance of being fatally trapped in a house fire, whilst reducing the amount of physical damage to property and possessions.

Passive solar gain

Using the sun to heat your house is the simplest way to save energy. It's also free. This can be harnessed quite simply by having more south-facing windows – although to prevent

over-heating in summer you may need to install shutters or blinds. Ideally, the rooms that are used most should be on the south side to take advantage of the solar gain. Rooms used mainly in the morning, such as kitchens and breakfast rooms, will get maximum benefit from the sun by facing south-east. Less frequently occupied rooms such as utilities and bathrooms can be located to the north, along with the circulation areas – halls, landings, stairs etc.

This all makes perfect sense except for the fact that there will inevitably be conflicting requirements dictated by the local environment. For example, you may want a particular room to take advantage of a stunning view. And even the best-laid plans can come unstuck thanks to the dictates of the planners, where a design needs to fit into the existing street scene. Getting this balance right is a big part of the designer's job.

Designing each room

ENTRANCE HALL AND CIRCULATION AREAS
No matter how big your house, you never seem to have quite as much space as you'd wish. So deciding how much elbow room to allocate to 'circulation areas' can be a tough call. By committing a generous amount of space to the entrance hall and landing it's possible to create a luxurious light and airy feeling. Psychologically, if it appears that you can afford to 'waste space' on communal areas, it somehow gives the impression of an expensive 'non-estate' property.

But if you want your entrance hall to

really make a grand gesture, designing an open double-height space can create a bright open void with lots of natural light. But devoting the full two storeys to a skyward expanse will, of course, mean having to sacrifice some of the upstairs floor space. For this to feel right, you need a minimum hall width of 1.5m, preferably with part of the hall being wider still. Otherwise a double-height hall can feel uncomfortably cramped and narrow, as if peering up from the bottom of a crevasse. Instead, you may prefer to maximise the size of your rooms and to sacrifice most of the 'circulation area'. But the risk is that the resulting cramped hallway and stairs may look mean and pokey. Either way, the design should encourage natural light to flood in, brightly bathing the stairwell, perhaps via roof windows or even a full height 1930s-style ladder window.

One increasingly popular option is to do away with the 'halls versus rooms' distinction altogether. The original Middle Ages design solution comprised a large hallway that could double as a 'dining hall'. Alternatively, you could opt for a modern open-plan layout that dispenses with a separate entrance hall altogether, giving the illusion of space and conveying a contemporary feel (unless, of course, you choose to roast an ox in the inglenook). But as with all open-plan designs, it helps if you're fastidiously tidy and not too bothered about privacy.

Stairs
The way you link one floor to another can provide an opportunity for some exhilarating design work. Imagine your new home with a grand staircase sweeping elegantly down to reveal finely crafted curved handrails and hand-turned balusters. Or perhaps you prefer the clean modern lines of a stunning polished steel and glass creation that perfectly enhances the entrance hall. Whatever your taste, you'll need to allow plenty of space to accommodate such a focal point, as well as a sizeable budget.

There are some key issues to consider at this stage, such as whether you want the staircase to be freestanding or, as is more common, adjacent to a wall. Would you prefer a simple straight flight, or stairs that turn, with landings or winders?

Photo: oakwrights.co.uk

Photo: Kermi

Photo: Peter Kent Architects

Bedrooms

How many bedrooms should you have? Whatever the answer, the temptation might be to squeeze in an extra one, as this can take you into a higher house-price bracket whilst the build costs remain substantially the same. Perhaps this explains why a consistent complaint with new developer-built houses is that the bedrooms are too small. However, the planners may restrict the number of bedrooms, so for the best of both worlds you could always design larger bedrooms that can be subdivided when you eventually come to sell.

Most self-builders want generously proportioned bedrooms with en-suite bathrooms and ample storage space. Well thought-out design can prevent noise issues arising. For example, it makes sense to have the bedrooms separated from each other by bathrooms, cupboards, or a strategically placed landing or a thick masonry wall between you and the source of noise can work wonders. And if you have younger children of similar ages, for the sake of family harmony it's a good idea to give them bedrooms of the same size, so that they can't claim 'it's not fair'.

Bathrooms and WCs

Bathrooms are a place where self-builders can really unleash their design fantasies. After all, why settle for a bog-standard bathroom when you could have a replica Roman spa, complete with sauna, steam-room, and jacuzzi? You may even want different themes for each individual bathroom or wet room.

But how many bathrooms should you have? If you're building a four- or five-bedroom home you would normally expect to have en-suites to all the double bedrooms. Even in less grandiose designs, squeezing in an extra bathroom can allow you to have a 'guest suite'. Bathrooms can be planned so they're fairly compact, which frees up space for larger

bedrooms. If space is at a premium, one possible compromise is to have a single en-suite bathroom with access from two bedrooms via separate lockable doors.

With bathrooms and WCs, good sound insulation is clearly advisable, so that dinner guests seated below an upstairs bathroom aren't treated to the sound of periodic toilet flushing.

The Building Regulations require at least one WC on each floor, and also stipulate that rooms should be spacious enough for WC's to be used by someone in a wheelchair. But there's some debate about where best to locate the downstairs cloakroom/WC. Siting it immediately adjacent to the front door probably isn't ideal, because then it's one of the first things that visitors see or hear as they approach the house. A loo under the stairs can be a good way of utilising space, although it can feel a bit pokey.

All in all it's generally better to have downstairs loos tucked away to the rear or side of the house, off the utility area, perhaps via a small lobby, making them easily accessible from the garden.

In terms of ventilation, there's no requirement for windows to WCs or bathrooms (although extractor fans are usually necessary), which makes juggling the positions of all the rooms easier. Cloakrooms can even be sited so that they open into a kitchen – something that used to be prohibited.

Reception rooms

A spacious living room with rear doors that open out to the garden is an essential part of a new home for many self-builders. But in most designs the focal point of this room is the fireplace, perhaps a cavernous inglenook complete with increased seating arrangements. But to feel right it's important that the style of the fireplace chimes with the overall feel of the house, and is proportionate in size to the room it occupies. It's very easy to get carried away at the design stage with a baronial monster that, when built, looks slightly comical, dwarfing the rest of the room.

Traditionally, the second main reception was the dining room. But in most new houses today these have been substituted by large open-plan kitchen/diners. But depending on your lifestyle there may still be good reasons for including a second reception. Most homes need a quiet backwater area for homework or just for reading the papers. And if you've got young children you need such a space to deposit those mountains of toys, otherwise they'll end up strewn throughout the hall and kitchen.

A popular alternative solution is to include a 'family room'. This is an all-purpose space that means different things to different people – perhaps a playroom for the children or a relaxing space where you can eat, watch TV, or put your feet up and read. Either way, the second reception room works best opening off the kitchen, ideally with large folding doors that open into the garden.

If you do opt for a traditional dining room, check that it can comfortably accommodate a full-sized table for six or more guests, along with any substantial pieces of furniture. Otherwise, when someone gets up from the table to leave the room the remaining guests will need to shuffle their chairs and breathe in sharply to let to let them past.

Study

Even if you don't work from home, the study has become an indispensible space – somewhere you can retreat amidst the maelstrom to manage the household accounts, or a compact den in which to pursue hobbies. If your study is for serious office use, consider making it larger with a separate entrance from the side, to avoid any clash between business visitors and family life.

Kitchen

Kitchens are no longer just somewhere to rustle up a meal. They've become the place to socialise with family and friends, and accordingly need to be generously dimensioned. The location of the kitchen in relation to the rest of the house needs to be carefully considered. Direct access from the entrance hall is normally required, but you also want glazed doors that lead out to the garden, and windows positioned to take advantage of any views. Adjoining the kitchen you may want a dining room, family room, or a utility area. The ideal arrangement might be to design a kitchen/diner that incorporates a highly glazed extension opening to the garden (or a pukka conservatory).

The downside to combining cooking and dining in a single open-plan room is that all your dirty dishes and clutter are permanently on display. So some discreet screening can be a worthwhile addition.

Photo: potton.co.uk

Photo: oakwrights.co.uk

But whatever your chosen layout, there's always a huge amount of detail to get right with kitchens. So before finalising your plan, it's worth sketching out some more detailed designs to coordinate the electrics, lighting, plumbing, and waste, as well as the various appliances. This also helps focus the mind on the positioning of windows, so that they don't clash with sinks, cooker hoods, and wall units. While you're at it, don't forget to include plenty of storage space – traditional pantries and larders are desirable features.

Utility room

The more you think about it, the longer the list becomes of the things that a utility room needs to accommodate. It's a place to dump wet dogs, muddy boots, paintbrushes, litter trays, and feeding bowls. Then there's all the assorted Dysons, rubbish bags, and overflowing collections of spare light bulbs, batteries, and shoe-cleaning kits. It's a space for drying washing and hanging coats – not forgetting a home for appliances such as boilers and washing machines. It can be tucked out of the way, with access to the garden – perhaps incorporated into a rear lobby.

Conservatories

What's freezing in winter, stifling in summer, and painfully expensive to heat? Some conservatories are considerably better than others, but a cheap plastic one can be a turn-off for some people. When building from new, it's better to build a highly glazed extension, designed to look like part of the house rather than an afterthought. The walls may comprise large folding glazed doors, but the roof can be conventional and fully insulated, perhaps incorporating a couple of skylights. This helps block searing heat from a high summer sun, whilst allowing in weaker heat from low-level winter sun.

Incorporating the garden into your home

A good design can make the garden feel like a natural extension of the living space, rather than a separate entity. But you need to decide which rooms benefit most from having direct access to the garden. This means picturing the gritty reality of children and pets with muddy paws traipsing in and

out in wet weather. If garden space is limited, or overlooked, an internal courtyard design may provide a better solution. Courtyards are ideal on tight sites since they allow light and ventilation into rooms that would otherwise be starved.

A garden that slopes down as you move away from the house can be largely lost from view, something that elevated decking or a raised patio can help remedy. Conversely, a garden that rises up from the house lends itself to being terraced and planted, so that natural beauty is visible from all levels of the home.

Loft rooms

When building from new, the additional work and expense of adding extra accommodation in the loft is minimal

Photo: Velux

compared to the benefits. Making the most of loft space that would otherwise be wasted can be a subtle way to increase the interior size of the house where the planners have restricted the height of the roof.

However, you can't just start sawing bits off ordinary roof trusses to create sufficient space. The best option is to specify special 'room-in-roof' (RiR) trusses that provide a ready-made framework for loft rooms. The alternative is to run steel beams to beef up a traditional timber 'rafter and purlin' roof design.

Loft rooms are perfect for bedrooms with en-suites, and for studies, but not for noisy activities that could drive people mad

in the rooms down below. From a design perspective it's important to provide sufficient headroom, especially where the stairs and landing enter the loft. Where space is tight, a strategically placed dormer window can sometimes provide a neat solution.

The simplest arrangement with loft stairs is to locate them within the same stairwell as the main flight below. With a new house it's a mistake to try to save money with pull-down ladders or compact loft stairs that are steep and narrow. But as soon as your design has three or more storeys the fire regulations require that there's a suitable 'escape route'. This normally comprises the upstairs landing and main stairs leading down to the ground floor entrance hall and out to the front door. So most cases, the only change you're likely to require is to fit fire doors to all the rooms en route (except bathrooms).

Basements

In much of Europe and North America new houses are often built with basements as a matter of course. But somehow basement living has never quite caught on in Britain. Perhaps it's because of the extra cost, or the fact that underground accommodation isn't ideally suited for use as bedrooms or living rooms. However, where plot space is tight for the size of accommodation you want, constructing a basement can make a lot of sense when building from scratch. It's also worth considering on sloping sites and where extra-deep

Photo: buildstore.co.uk

foundations are required, because a lot of the excavation work will already have been done.

The design of basements is affected by a number of factors, not least soil conditions and the water table. Effective waterproofing is probably the biggest design challenge. One well-known approach is 'tanking', where a waterproof layer is

Basement excavation – poured reinforced concrete construction.

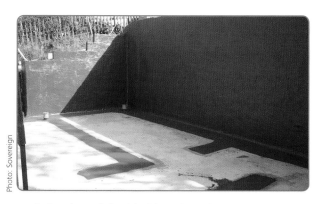

Photo: Sovereign

applied to the walls both inside and out. Some systems are designed to capture any moisture that accumulates, and harmlessly channel it away behind waterproof internal panelling to a below-floor sump, which then pumps it out to a drain. Basement walls need to be reinforced and integrated with the flooring, and to prevent condensation all wall and floor surfaces need to be fully insulated.

Fitting high-level windows and light tubes can solve problems of sourcing sufficient natural light and ventilation, so that the basement space doesn't feel like a dark and dingy *oubliette*. With habitable subterranean living space, you will need to provide a secondary means of escape, such as an external staircase.

Because of the deep excavation required, narrow sites with adjoining properties close by can pose problems. This may necessitate the use of more expensive excavation methods that minimise noise and vibration. Be sure to take detailed photos of the adjoining properties before work commences in case of subsequent arguments over any damage.

There are several different ways to construct basements:

- ■ Solid masonry using massive blockwork.
- ■ Hollow blockwork with reinforced concrete infill.
- ■ Hollow ICF interlocking polystyrene blocks with reinforced concrete infill (see next chapter).
- ■ Poured concrete (the walls are shuttered with plywood formwork and cast with poured concrete and reinforcement rods).
- ■ Pre-cast concrete panels (*eg* 'ThermoneX' concrete panels on raft foundations).

ICF polystyrene blocks – see chapter 6.

The quickest solution is normally to use pre-cast concrete panels that can be installed in a few days, considerably quicker than for masonry construction. However, there is a simpler alternative method of building basements that's been around for donkey's years. The traditional Victorian 'semi-basement' overcomes many awkward design problems. This is because the surrounding earth is excavated away, reducing external ground levels. In effect you have a deep trench, at least a metre wide, surrounding the walls, which means you don't need to worry about tanking. But the outer sides of the trench need to be built as retaining walls, as they effectively hold back the garden. This design has the added advantage of allowing light and ventilation into basement rooms via conventional windows.

Pre-cast concrete panel basement

Photos: ThermoneX

Photo: Peter Kent Architects

Design factor 3: The plot

There's no such thing as the perfect building plot. All sites impose constraints on the design, although some are far more demanding than others. Many sites also offer real benefits, such as stunning views or a south-facing garden. So before getting too engrossed in design details, you need to take a long, cool look at the plot, to figure out it's full potential, and 'design into the site'.

Local properties

One of the first points to notice is the style of neighbouring properties. This is likely to play a major role in determining the size and appearance of your new home so that in planning terms it's 'in keeping' with the locality. See Chapter 7.

Size and shape

A small site will obviously impose limits on what can realistically be achieved. But an awkward shape can also pose dilemmas. For example, a long, narrow plot with a small frontage will impose severe restrictions on the direction that your main rooms can face.

It's also essential to consider how vehicles will get to the dwelling. There may be an opportunity to create a sweeping driveway with parking space for several vehicles.

Site orientation

Making the most of any attractive views will be a major influence on the direction your house ultimately faces. But for your design to take full advantage of natural daylight you need to identify the sun's path throughout the day, noting east and west points for sunrise and sunset.

Ideally the morning sun should beam down to greet the occupants of bedrooms, kitchens, and breakfast rooms, whereas reception rooms benefit most from afternoon and evening sun. Imagine enjoying the last few hours of evening sunshine dining in your garden, under a magnificent orange and red skyscape as the sun sets. You may even be lucky enough to have a south-facing garden.

Natural features

Natural features such as old stone walls and mature trees can often be absorbed into your design, adding instant character. But where it's not possible for them to be utilised or worked around, you'll need to explore how best to remove them. As we saw earlier, cutting down large mature trees can sometimes destabilise ground conditions, potentially endangering the foundations of adjoining houses, and should therefore be avoided where possible. They may also be protected by Tree Preservation Orders (see pages 51 and 174).

Site exposure

Remote sites with wonderful views may seem perfect on a tranquil summer's day, but when building 'out of season' such exposed sites are likely to experience harsher conditions. Designs must take account of extreme wind and weather – for example, roof tiles need to be nailed at every course to protect them against violent gusts, and cavity walls should retain an air gap and not be fully filled with insulation to resist the spread of penetrating damp. However, it may be possible to moderate the worst effects of the wind with sheltered landscaping and by planting banks of fast-growing trees, which can also be useful for screening any nearby eyesores from view.

Noise

Noise from nearby roads can be reduced by siting garages and outbuildings so that they form a barrier, while the house itself should be oriented so that it faces away from the noise source. If possible, you want to position a main wall without doors and windows so its facing the noise. By locating the bedrooms furthest away from the problem side, the mass of the building can provide sound insulation, with non-habitable rooms such as cloakrooms, utilities and bathrooms, nearer the noise source.

The shape of the building also has a role to play. A courtyard or U-shaped design facing the noise will trap and amplify the sound. But turned round to face away, it can act as a shield.

Triple glazing can substantially reduce airborne sound levels, as can well-designed secondary glazing. By fitting a heat-recovery ventilation system, windows on the noisy side needn't open. Walls and ceilings can be insulated with special acoustic plasterboard.

Landscaping can also make an important contribution. Most construction sites have to dispose of large amounts of excavated soil, rubble, and spoil, at considerable expense. So it makes sense to recycle it as an earth bank or landscaped 'bund' wall forming a barrier between the house and a busy road. Planted with dense rows of evergreen trees this can absorb sound, although trees tend to reduce light, so the neighbours may not be too keen.

Security

Any security risks, such as potential access from nearby public footpaths, will need to be addressed in the design. Positioning the more frequently used rooms to the front of the house allows you to keep an eye on who's coming up the garden path to your front door.

Privacy

Stand at the centre of your plot and check how many neighbouring windows you can see. Most sites are to some extent overlooked by the neighbours. This may be more of an issue in dense urban areas, where glass roofs and skylights may not be acceptable, although screening with large garden shrubs can sometimes offer a solution. Unless you have an exhibitionistic streak, the position of some of your windows may need to be staggered so as not to be directly visible from those next door. The planners may, however, be more relaxed when it comes to applying rules on how close windows between houses can be, because of pressure to build at greater densities.

Flat spots

Is there a naturally flat area within a largely sloping plot where the house can be sited? Or will a lot of expensive excavation be needed to create a flat area for the house to sit on? Sometimes a large amount of earth needs to be

removed, not just for the house but for garages and gardens as well. Intelligent planning now could save hours of hard graft with wheelbarrows later.

The challenge of sloping sites

Most plots have a certain amount of slope, but steeper sites aren't popular with developers because they make the design process more complicated, requiring extra expenditure for landscaping, drainage, or even stepped foundations. Which is why they often fall into the hands of self-builders.

Such challenges can actually present a unique opportunity. With a touch of creative flair you might just come up with a design that boasts real 'kerb appeal' and an interesting split-level interior layout.

The normal approach to building on a slope involves carving out a level base from the hillside, thereby forming a platform to build on – a process known as 'cut and fill'. The resulting spoil can then be recycled to build up levels where the ground slopes away on the lower side of the plot.

Fortunately, with modern suspended beam and block floors a deep void under the floor needn't be a problem. The only significant costs with the superstructure relate to the increased amount of wall construction below floor level. An alternative method of building on steeply sloping sites is to construct the house on top of a series of columns or stilts. This cuts out the need for a lot of expensive foundation excavation, because it leaves the ground relatively untouched.

SITES THAT SLOPE UP FROM THE ROAD

A house built up a hill will typically be designed so that the front half facing the road has an additional 'lower ground floor' to take advantage of the lower slope. This lower storey at the front may only extend back about half the depth of the house into the hill (unless you go to the expense of building it all the way into the hill as a rear basement). But this space can easily become a kind of poorly lit 'no man's land', comprising utility rooms and garage

Photo: Canada Wood UK

Photo: protein.co.uk

space, so you end up with a huge garage door dominating the front of the house. Where this is the case, their visual impact can be reduced by setting them back a little. Being recessed into the building they are overhung by the upper floors, perhaps making a feature of a protruding 'Juliet balcony' or a prominent jettied gable. If you need to construct a detached garage on a very steep site it can sometimes be done by burrowing into the hillside front garden.

SITES THAT SLOPE AWAY FROM THE ROAD

A house built down in a dip poses the opposite problem to that described above. Here you have an extra storey to the rear, creating a lower ground floor. But where such a house comprises only two storeys to the rear, the front elevation facing the road would effectively appear to be a small bungalow. This would hardly do justice to what might actually be quite a large detached house, and the stunted kerb appeal would be reflected in the property's value.

To resolve this shortcoming, you could position the house so it's possible to see from the front that it extends downwards, perhaps building it at an angle to the road (although this may invite a certain amount of overlooking). Or you could draw attention to the larger accommodation within by fitting small dormers to the front roof, even if there are no loft rooms.

SITES THAT SLOPE SIDEWAYS TO THE ROAD

These are probably the easiest type of sloping sites to design into. With the house facing the road, one side will be buried in the slope of the hill. Again, the ground floor could be

extended into the upper part of the hill in the form of a basement, or you could make a feature of it by with stepped mezzanine floors.

RETAINING WALLS

On sloping ground, a considerable amount of excavation is normally required in order to level parts of the plot. This means the remaining adjoining areas of high ground need to be shored up by building special retaining walls. These have to withstand substantial ground forces, and require an engineer to design them, making them very expensive to construct. But there may be a cheaper and more pleasing alternative. Instead of constructing one single massive wall, you could build a series of shorter walls, stepped up the slope like a hillside vineyard. In recent years the use of wire cages filled with decorative stones has become popular; alternatively hollow interlocking blocks can be filled with soil and planted.

DRIVEWAYS

The maximum gradient for driveways is normally about 1:10. On a fairly steep slope this may necessitate a curved driveway, winding round like a miniature mountain pass. Although this will consume a large amount of garden space, a winding driveway presents an interesting design opportunity, adding an element of mystery, with the house nestling behind neat walls and dense shrubs like a villa in the Hollywood hills. On steeply sloping sites this arrangement can also solve the problem of pedestrian access, by avoiding long flights of steep steps that deter all but the hardiest visitors, especially in icy winter conditions.

The designer

The majority of self-builders will need to employ someone who can turn their ideas into a real building. But before formally appointing anyone it's essential to wait until you have exchanged contracts on your plot, because there's still a risk that the purchase could fall through.

This, however, is a good time to consider exactly what you want from your designer. Not all self-builders require the 'full service'. Much will depend on how much you can contribute personally from your own knowledge and skills. But even if you're the world's most confident designer, your mortgage lender may have imposed a loan condition that requires you to employ an architect throughout.

Services offered by designers

SITE APPRAISAL
Your designer should visit the site to get the feel of it before putting pen to paper. This is important to assess the features and constraints of the site in relation to the client's budget.

SKETCH DESIGN
Preparation of draft sketches will help you to jointly develop the final design, showing options for different site layouts, room layouts, and external appearances.

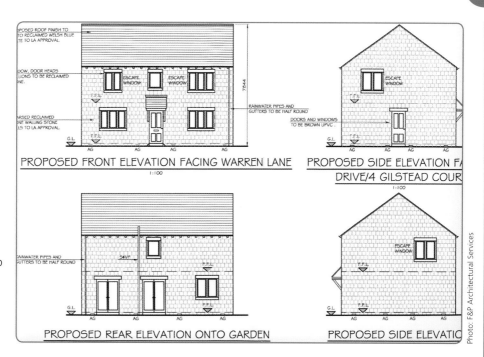

PROPOSED FRONT ELEVATION FACING WARREN LANE

PROPOSED SIDE ELEVATION FA DRIVE/4 GILSTEAD COUR

PROPOSED REAR ELEVATION ONTO GARDEN

PROPOSED SIDE ELEVATIO

PLANNING APPLICATION
Before submitting drawings, the architect will liaise with the Planning Department to be clear about what's likely to be acceptable. Before the application is submitted, key decisions need to be made about the appearance of your house. See Chapter 7.

BUILDING REGULATIONS APPLICATION
Your architect should submit the application and deal with any queries. This application comprises technically detailed drawings, engineer's calculations, and specifications. See Chapter 8.

TENDER PACKAGE
Inviting tenders is the best way to obtain accurate prices from building contractors. This requires clear drawings and detailed specifications based on those used for the Building Regs application. Your architect should produce a list of suitable firms and provide contract documents once the contractor has been selected. See Chapter 9.

PROJECT MANAGEMENT
Managing the project on site means regular visits are made to check that the works are being carried out in accordance with the approved drawings and the contract. See Chapter 10.

CERTIFICATION
Mortgage lenders usually release money when work has been completed at agreed stages. Some require receipt of a Professional Consultants' Certificate (aka 'CML Certificate'). Formerly known as Architects' Certificates, these state that the work has been completed to a satisfactory standard. It's not just architects who can sign these – chartered surveyors and other qualified professionals with Professional Indemnity Insurance cover (PII) can equally do so.

Stages in the design process

■ **Inspiration** – The ideas stage, based on photos and sketches of other buildings you like.

■ **Consultation** – an initial face to face meeting with a professional designer to discuss your project – and for you to interview them.

■ **Working up the design** – taking it from the ideas stage up to basic plans.

■ **Design details** – discussion of materials etc.

■ **Submission** of plans for Planning and Building Control.

■ **Supervision** of construction – certifying each key stage for release of funds from lender (optional).

■ **Completion** – ensuring that the Building Control completion certificate is obtained when the construction is finished.

Types of designer

There are several types of designer:

ARCHITECTS

You can't call yourself an architect unless you're a qualified member of RIBA (Royal Institution of British Architects) and registered with the Architects Registration Board (ARB).

Architects are known primarily for their ability to design. So although they're trained in the practicalities of construction on site and managing contractors, their creative skills are often their real strength.

Some specialise in green building, such as sustainable design or straw-bale construction.

ARCHITECTURAL DESIGNERS

Anyone can call themselves an architectural designer or architectural consultant, but most will have relevant qualifications, such as surveyors or technologists (see below). An experienced local designer who may not be an architect could be the right person for your job.

TECHNOLOGISTS

Members of CIAT (Chartered Institute of Architectural Technologists) are qualified in design but with a more practical, and less creative, edge. This may be a better option for clients who are more concerned about 'nuts and bolts' construction input rather than styling.

SURVEYORS

Anyone can call themselves a surveyor, but qualified *chartered* surveyors are always MRICS or FRICS, members or fellows of the RICS (Royal Institution of Chartered Surveyors).

Chartered surveyors come in many shapes and sizes, most of whom don't design houses. So look for those with experience in 'architectural services'. The strength of surveyors also tends to be more practical and hands-on, with less creative ability than architects.

RIBA
Chartered Practice

RICS

aSba

Chartered Institute of Architectural Technologists
the qualifying body in Architectural Technology since 1965

'Turnkey' projects

A 'turnkey project' is one where the entire process is managed for you, including the design. Some firms even help provide plots. Your involvement as client is then limited to providing the money and design guidelines. This may be suitable for those with minimal time to devote to their project, and for whom creative design isn't important. Popular options include:

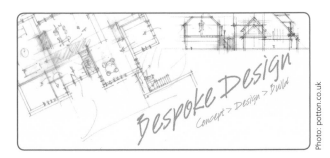

Bespoke Design
Concept > Design > Build

Design & Build contractors

Some building contractors can provide a 'do it all' service, offering a full design and build package once you've secured your plot. Whilst a lot of self-builders might feel that having everything done by someone else would rob them of the glory of running their own project, others see this as a price worth paying to get the home they want with the minimum hassle.

But there are some other possible disadvantages. Being tied in to using a single designer-contractor means there's no competitive edge, and is likely to cost more. You may also find yourself steered in the direction of a standard design. But by making the process relatively easy it may be the best option for first-timers or busy clients who aren't bothered about too much personal input.

Package deal companies

Most package deal companies provide off-the-shelf timber frame homes in kit form that are delivered to site to assemble, rather like giant flat-pack furniture. Some even

claim to 'build in a day'. However, there are other firms who can offer a customised service, taking your ideas and converting them into plans, then organising and managing the entire project on your behalf.

The attraction of package deal firms is that they offer to 'hold your hand' all the way through, starting with an assessment of your site and preparation of initial drawings. Based on this, they then provide a quote for their services, which usually includes supplying their kit or materials. Only once the quote is accepted will they proceed with planning and Building Regs applications.

If you buy a standard house kit some suppliers can provide an architectural service to customise the design to your requirements. For many self-builders this 'free' design service is the carrot that induces them to sign up to the rest of the deal. But you can't cherry-pick the free drawings and

then not buy the kit! Most firms offer a pretty good service and are familiar with the requirements of self-builders. But there is a price premium for the convenience and relative certainty on outturn costs and the final product.

How to find a designer

The ideal way to track down an architect/designer is through personal recommendation from a satisfied-looking self-builder. However, your choice may be restricted if your mortgage lender insists on an architect issuing interim certificates at key stages of construction. If so, check that your chosen designer's signature will be acceptable. Check out the following for recommendations:

Other self-builders

If you come across a building site with a mobile home on it, there's a pretty good chance it's someone building their own house. Self-builders are usually more than happy to share their experiences and can be a mine of useful information, so pop in for a chat. At the very least you should come away well informed about the designers and the trades they've dealt with.

The ASA

The name of the Association of Self-Build Architects is self-explanatory – you can find their website at www.asba-architects.org.

Local authorities

Though Planning Officers and Building Control Surveyors are not at liberty to make direct recommendations, some may be happy to give out a list of names of designers in your area, and it's unlikely such a list would include anyone with whom they've had serious problems.

Take a look also at local planning registers, which can be viewed online, for recent applications for single-house developments. Look under 'agents' and check out the quality of the designer's drawings.

Magazines and exhibitions

There are several well-known monthly self-build magazines, and a number of regular exhibitions, such as the National Self-Build and Renovation Centre.

Picking the right designer

Appointing a designer is a bit like getting engaged. They may seem perfect now, but can you be sure the relationship will survive when the going gets tough?

After mulling over the pros and cons of various designers, write a shortlist of three or four names. You may have a 'gut feeling' that a person is the right one (or totally the wrong one), but the best choice will normally be someone experienced in self-build, with plenty of enthusiasm and good ideas, who seems to understand your requirements.

Most designers should be willing to devote at least an hour or so of their time, free of charge, to discuss your project. Some will even make a preliminary visit to the site with you.

Nine key questions to ask your designer

- Do they offer a free initial consultation?
- What qualifications do they have?
- What is their experience of self-build, and how many one-off houses have they designed and managed?
- Who will personally be working on your project and what is their experience?
- What's the cost and how are fees calculated? Does it include all fees for Planning and Building Regs applications?
- Can they supervise the project to completion if required?
- Are they happy to show examples of completed jobs and let you talk to former clients?
- How soon can they start?
- What Professional Indemnity Insurance do they carry?

What to look for in a good designer

Design flair

When it comes to creative flair it's often said that you've either got it or you haven't. Architects can be very talented creatively but may in some cases be less interested in the practicalities on site. The worry with creatively gifted people is that they can sometimes be so smitten by their own genius that it can come across as arrogance. You don't want someone who regards their client simply as an irritating obstacle to realising some long cherished master-plan. Avoid those who seem determined to hijack your dream, demanding total creative control funded by you. Fortunately, most experienced architects are reasonable people, not megalomaniacs.

Easy to work with

No matter how brilliant a designer, if they're unwilling to listen to you or to incorporate some of your better ideas they're the wrong person. A friendly, approachable manner counts for a lot when you're working closely with someone over many months. A helpful attitude and the ability to listen to clients is essential for a successful working relationship. True, some of your suggestions may turn out to be unrealistic, but your designer should be prepared to explain why. If they make you feel intimidated or uncomfortable, forget it.

Small and local

Large practices typically don't deal with one-off house projects, so the chances are you'll be dealing with a small firm or a one-man band 'sole practitioner'. Such one-to-one arrangements are fine as long as they have the necessary skills, experience, and Professional Indemnity Insurance.

It's often better to pick a designer based close to where the building work is taking place, so that frequent site visits aren't an issue. Being local also means they should be familiar with the whims of local planners and Building Control.

Qualifications and insurance

Check that your chosen designer has suitable qualifications and membership of a relevant profession as discussed above. Most professional bodies have dispute resolution procedures, which can be useful if things go wrong. All professional designers must have full Professional Indemnity Insurance cover (PII) as a mandatory requirement for membership, so if the designer happened to make a serious mistake which cost their client money the insurers could provide compensation. Also, banks may not release stage payments unless the person issuing certificates has the necessary PII cover.

Photo: Mark Kingsley Architects / JM Collingwood

Experience and track record

It's important that your designer has specific experience of one-off housing, because this type of project is a very different animal from larger developments. Many architects specialising in self-build are members of ASBA – the Association of Self-Build Architects. But if the only private house an architect has ever built is their own or their parents' it means they won't have the necessary experience of working in partnership with clients.

If possible, track down previous clients and ask them if the designer's cost predictions at the start of the project turned out to be reasonably accurate. As *Grand Designs* viewers will know only too well, it's not unknown for budgeted figures to be wildly out. If that happens to you it could spell disaster, and even the loss of your home.

Clear and accurate drawings

Ask to see drawings from previous jobs. Make sure they're clear and easy to understand. Cluttered or incomplete plans will jeopardise Planning and Building Regs applications and will drive you mad on site.

Photo: Buildstore

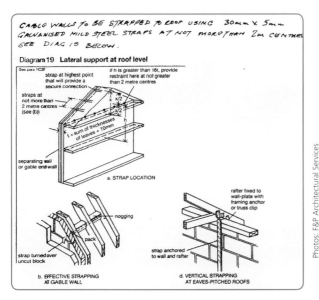

GABLE WALLS TO BE STRAPPED TO ROOF USING 30mm × 5mm GALVANISED MILD STEEL STRAPS AT NOT MORE THAN 2m CENTRES SEE DIAG 19 BELOW.

Diagram 19 Lateral support at roof level

a. STRAP LOCATION

b. EFFECTIVE STRAPPING AT GABLE WALL

d. VERTICAL STRAPPING AT EAVES-PITCHED ROOFS

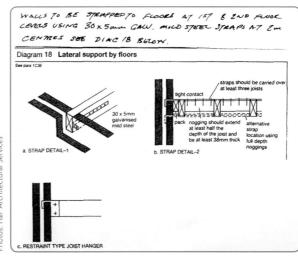

WALLS TO BE STRAPPED TO FLOORS AT 1ST & 2ND FLOOR LEVELS USING 30 × 5mm GALV. MILD STEEL STRAPS AT 2m CENTRES SEE DIAG 18 BELOW.

Diagram 18 Lateral support by floors

a. STRAP DETAIL–1

b. STRAP DETAIL–2

c. RESTRAINT TYPE JOIST HANGER

Photos: F&P Architectural Services

Additional builders' drawings are often needed to explain the more complex construction details, such as where a roof abuts a wall, or how floor level changes are to be carried out. This means the builders won't have to improvise tricky details on site. You shouldn't be charged extra for this, so agree that any necessary explanatory detail will be included.

However too much detail on drawings can be as bad as not enough. Builders have to deal with a variety of drawings, and crowded drawings can be confusing and can obscure important points. It's often better to have a number of separate drawings of the same part of the building rather than cram everything on to one piece of paper. For example, separate ground floor plans can highlight a key part of the design – floor joists, insulation, electrics etc. Also, a separate written specification helps keep the plans uncluttered and easier to read. That way nothing gets missed and any special requirements should be easily recognised.

Flexible fees

You need a proper explanation of the fee structure, so that you know exactly what you're getting for your money. Avoid designers who are unwilling to discuss fees. Anyone who dodges the issue should be treated with suspicion.

Many designers quote a fixed fee for the whole job. As long as you know exactly what's included, that's probably the best arrangement. Some architects charge by the hour, which obviously leaves the client wide open and is best avoided. Others quote a percentage of the total build cost, typically between 6% and 10% of the contract value for a full service. Percentage fees are fine as long as you're clear what exactly the fees are a percentage of. The snag is, you rarely have a firm idea of the build costs at the start. And what would happen if at some later stage you raised the quality of materials, perhaps switching to dearer hand-

made clay roof tiles? There wouldn't be much extra work involved for the architect (or the builder), so it's worth agreeing that you shouldn't pay more in fees.

To be perfectly clear, you might want to agree a separate price for each part of the project – for the initial drawings, the Planning drawings, Building Regs drawings, working drawings, the specification, tender documents, and project management on site (if required). Architects' fees for doing all the drawings can be as much as £6,000. This includes engineer's fees and SAP calculations (see page 93), but excludes site supervision.

Awareness of your budget limitations

Some architects regard budgets as a trivial inconvenience, expecting unquestioning clients to be able to lay their hands on large reserves of cash and accept sky-high costs. So it's best to avoid any designer who goes ahead with the drawings without first discussing your budget. Some overly-trusting self-builders have slipped up badly by allowing the drawing and planning stages to proceed unchecked, only to discover later that builders' quotations then come in way above budget. So it's best to inject a dose of realism by talking prices from the start.

Not all architects are particularly conscious of build costs, some regarding any discussion as an insult to the free expression of their self-evident creative genius. Fortunately, those specialising in self-build tend be more down to earth when it comes to awareness of prices for materials and labour. Build costs aren't a state secret – several self-build magazines publish regular updates.

Project management ability

Many self-builders only want help finalising the design, rather than the full project management package through to completion. This is an important decision because it can

What *clients* need to contribute

At least half the battle of any successful project is down to the client clearly communicating what they want.

A CLEAR IDEA OF THE BUDGET

It's important to make your budget clearly known from the word go and stress that it cannot be exceeded. Monitor progress at each stage of the drawing process by asking 'Are we still on budget?'

A CLEAR LIST OF REQUIREMENTS

You've probably got a collection of photos of designs that appeal, or some rough sketches showing broadly what you want. If they scoff, as if to say 'It's not your job to interfere,' head straight for the exit.

THE ABILITY TO COMPROMISE

There always have to be compromises. You're never going to get 100% of what you want. In order to make the overall project affordable, it may be that one of your cherished ideals will need to be sacrificed. Don't waste time fighting with your designer about things that are ultimately up to the planners. And if you're told that something simply can't be done, you at least deserve a proper explanation.

STAY IN CONTROL

Always remember that it's you who's paying for the project. Being new to the world of designers, planners, and Building Regs can sometimes make you feel that things are outside your control. Builders are sometimes reluctant to do things in a certain way, perhaps because they know a better way from experience. So it's important to take your time and not let yourself get rushed or flummoxed. There may be another solution, and as you're the person paying you have the final decision.

Photo: Piers Taylor Architects

cause problems if you later change your mind and decide you need a project manager. A different architect may not trust someone else's design.

Confirming the appointment

Once you've completed the purchase of the site, you're free to formally appoint your designer. This should be confirmed to them in writing, together with a list of the key tasks you want carried out for the agreed price. Prior to the first meeting your designer should have made at least one site visit.

The first design meeting

The object of the first design meeting is to brief your designer. This is basically about 'placing your order' and highlighting your most important requirements. The best way to do this is to produce a written design brief that summarises the overall size and style of house that you want, identifying the points you consider essential, although at this early stage it can only be an outline. The key points you need to make clear are:

- Your total budget
- The timetable for the design process
- The design brief:
  - any design features that you especially like, or strongly dislike
  - any preferred construction method (*eg* oak frame)
  - accommodation that's essential

Photo: oakwrights.co.uk

Most of these points will have already been briefly covered at the initial consultation, but your designer may be working on several concurrent projects, so it's a good opportunity to refresh the basics now that they are officially on your payroll.

Before exploring design solutions in detail, the architect should embark on an information-gathering process and ask what you want. So start by discussing photos and sketches of other houses you like the look of. You may simply have one overriding concept, such as 'We want our home to be in the style of a country cottage'. Either way, it's important to clearly project your ideas well before any drawings are started. At the very least, make known any particular likes and dislikes, such as a passionate hatred of flat roofs or plastic windows.

Good designers don't blindly follow their client's brief – they question it. So expect some constructive criticism. Try to be receptive to their suggestions as to how your ideas could be modified to improve the design – as long as it's within budget.

Whilst it's a mistake to insist that all your ideas are slavishly followed, don't allow yourself to be steamrollered into accepting some weird design concept they've dreamed up. Architects are highly creative people and some may see this as an opportunity to let their imagination run riot. Whilst such an approach may suit clients who are primarily paying for design inspiration, most self-builders have their own personal vision, hence the importance of a design brief to keep things on track.

A fundamental part of the design process is to generate ideas, and sketch out modified designs. This should be an enjoyable process, trying out different concepts on paper, or on screen, before they appear in glorious 3D brickwork. Some ideas may turn out to be impractical once drawn to scale. The budget and site restrictions will always impose limits on what can be achieved, so inevitably some compromise will be necessary. This is where a designer's skill and experience can help make the most of these restrictions – perhaps turning some into advantages. At the end of the day, a successful design is one that meets your requirements within these limitations.

As the design process develops, it's tempting to keep making the property bigger, adding more and more desirable features and expensive fittings until you've finally created your ideal house. With some draft floor plans sketched out you'll be able to get an idea of the total floor area. This is normally when you discover to your horror that the projected build costs have skyrocketed far beyond the available budget. At this stage, just about everyone finds that they have to compromise, deciding which rooms are essential and which can be reduced in size or taken out altogether to bring the project back in line with your original budget.

So before getting any further this is a good time to update your budget calculations, checking the estimated construction costs per square metre.

Design brief

Preferred materials and styles.

OVERALL THEME

Options: contemporary, farmhouse, period cottage, Victorian Gothic, neo-Georgian, Art Deco, Modernist, etc.

STYLE

Options: two-storey detached house, chalet bungalow, three-storey town house, etc.

ACCOMMODATION

For example, five bedrooms, three en-suites, three receptions, large kitchen/diner/conservatory, utility room, study, loft room, basement, etc.

EXTERNAL WALL FINISH

Options: brick, rendered blockwork, exposed oak post and beam, timber-clad, tile-hung, etc.

ROOF CLADDING

Options: plain clay tiles, interlocking concrete tiles, glazed pantiles, natural slate, artificial slate, etc.

STRUCTURE

Wall options: modern timber frame with masonry outer leaf, traditional brick and block, oak post and beam, SIPs, ICFs, etc.
Roof options: traditional cut-timber, room-in-roof trusses, etc.

EXTERNAL WINDOWS AND DOORS

Material options: hardwood, softwood, UPVC, steel, etc.
Style options: sash, cottage-style casement, etc.

INTERNAL DOORS

Material options: softwood or hardwood, solid, hollow, etc.
Style options: four-panel, six-panel, flush, etc.

OUTBUILDINGS

For example, detached double garage, timber frame conservatory, etc.

OTHER IMPORTANT FEATURES

For example, green design using natural materials, low maintenance, easy-to-build, etc.

Photo: oakwrights.co.uk

6 CONSTRUCTION METHODS AND GREEN TARGETS

Photo: oakwrights.co.uk

Photo: claysLLP.co.uk

Photo: kingspan.com

Self-build homes are generally more eco-friendly than those produced by housing developers. This is because self-builders want a better quality lifestyle are prepared to consider new innovations. Which is just as well, because today being green is no longer a matter of choice – all new homes are supposed to be 'carbon neutral' by 2016, meaning that a house must use less energy than it generates, adding no carbon to the atmosphere over the average of a year. In fact, emissions from all energy use in the property must be zero carbon in total. This applies to fixed equipment (e.g. boilers for space heating & hot water, plus lighting and ventilation) - but does not count retrospectively installed appliances (e.g. for cooking). So to comply with the Building Regs there will be some pretty stiff targets that your new house will need to meet.

The good news is that you don't have to live in a wacky-looking box with a tree stuck on the roof to benefit from green design. When you're building from scratch, there are some basic common-sense things you can do to cut running costs and slash fuel bills.

Having minimised the building's consumption of energy and water and shrunk its 'waste footprint', you can then complete the eco-design by generating some of your own power. This means selecting the best combination of solar, wind power and air/ground source heat to produce your own renewable energy. But first you need to make what is probably the most fundamental decision of all – the method of construction.

Construction methods

Traditional brick and blockwork is still the most common type of wall construction for new homes in Britain. But many self-builders find timber frame to be a better alternative to conventional masonry walls. Timber frame construction now accounts for more than a quarter of new UK housing. But the fact is, both methods have a lot in common. Both cost pretty much the same and provide similar levels of thermal efficiency, plus there's no difference in the value of the finished properties (although traditional oak framing can command a premium). However, these aren't the only options. As we shall see, there are plenty of tried and tested alternatives, popular with self-builders. All have their pros and cons, and the one that's right for you will depend on the circumstances of your development.

Nonetheless, it's important to bear in mind that picking relatively non-conventional construction methods or materials means you'll need to provide Building Control with detailed technical information to demonstrate compliance. Plus the more extreme your design, the fewer contractors you're likely to encounter who are familiar with the necessary construction techniques. Whatever the method of construction, the fact is most new homes take the same length of time from start to moving in – typically 6 to 12 months depending on the procurement method and project management.

Photo: Solar Century

Photo: oakwrights.co.uk

Conventional construction

Traditional masonry

Masonry cavity walling is one of the most widely understood and least expensive options. Cavity walls comprise an outer wall of brick, stone or blockwork tied to a separate inner leaf

normally built from concrete blocks, with a gap in between. External blockwork can be finished with render or clad with timber, tiles, slates or even UPVC. The outer leaf is not load bearing, but plays an essential role in stabilising the wall by being tied to the inner leaf with steel wall ties.

The inner leaf does most of the hard work supporting loadings from roofs and floors, so it's important to select the correct type of blocks. Heavy, dense concrete blocks have a high load-bearing capacity but a low thermal conductivity rating for insulation. Lightweight 'thermalite' blocks, on the other hand, have a much better thermal efficiency but are not as strong.

Your designer and engineer will specify the type of block with the necessary strength to support the loadings and simultaneously resist the cold with a decent thermal rating. Their compressive strength is expressed in Newtons per square millimetre ('N'), and a typical minimum strength for inner leaf blockwork walls would be 3.5N. The precise figure will, of course, depend on the loadings being supported, but as an example, load-bearing blockwork to the lower storey of a three-storey house is likely to be at least 7N.

To improve the property's thermal efficiency, walls are usually dry-lined with plasterboard internally and the cavities are insulated. However, there is some debate about whether cavities should be fully filled with insulation in case it allows damp to 'bridge' across. Clearly this is more of a risk in locations exposed to force ten gales and driving rain. Insulation is discussed later in this chapter as well as in Chapter 11.

A masonry structure gives a house a feeling of solidity. The density of the blocks provides an element of 'thermal mass' acting as a background heat store. It also offers a higher level of acoustic mass helping to deaden noise from outside. And

unlike many timber frame kits that have a waiting period of around 12 weeks, the materials are readily available.

From an eco viewpoint, natural clay bricks are preferable to anything made from concrete, and reclaimed bricks set in lime mortar are even better.

Timber frame

Shakespeare would have recognised timber frame construction, at least the traditional 'green oak frame' variety that's currently enjoying something of a renaissance.

Photo: Piers Taylor Architects

But the type of timber frame buildings that are especially favoured by self-builders are modern pre-manufactured timber frame kits. These also happen to be the most popular method of house construction in Scotland and Scandinavia – which may have something to do with the weather. In a cold climate, any system that can offer high performance thermal insulation qualities, combined with minimal time spent standing in the freezing rain on site, has to be a winner. Some modern 'twin wall' systems can achieve amazingly efficient U-values of less than $0.1W/m^2k$ (the heat loss in Watts for every square metre of a material's surface).

It's normally possible for the basic timber frame structure to be erected in a day or two, with a fully weather-resistant building ready in a couple of weeks. However, greater foundation accuracy is required than with conventional masonry walls where discrepancies can usually be made up for by the brickies. Foundations must be level to within 20mm, and square to within a 12mm tolerance.

PACKAGE DEALS

There is one overriding reason why so many self-builders opt for timber frame, and that's the tempting prospect of the package deal. As we saw earlier, some firms can offer a full 'turnkey' service, where they even help locate suitable plots and find the right builders. House designs are based on a range of standard models. Having agreed the design, they can handle the Planning and Building Regulations applications, and assist you throughout the project with support and guidance during the build. Of course, all this comes with a price tag. The client pays in the mark-up on the timber frame kit and the labour on site bolting it all together. But overall, such packages usually provide reasonably good value for

money. For novice self-builders, having dedicated support at hand can make all the difference.

But you don't have to choose the full 'holding hands' service. Some timber frame manufacturers offer less expensive reduced-packages, leaving the initial design and planning to you. Based on your own approved plans drawn by your architect, they'll produce a manufactured kit that's ready-to-build.

DIFFERENT TYPES OF TIMBER FRAME

There are two basic types of timber frame construction: modern 'stud frame', where large panels are joined together to form the load-bearing inner leaf of the wall; and 'primary frame', where the framework supports the loadings, such as in traditional oak 'post and beam' buildings.

STUD FRAME PANELS

Once built, modern timber frame houses look just like any other house. The outer leaf of the cavity wall is made from the same brick or blockwork as traditional properties. It's the wooden 'stud frame' inner leaf that takes the main structural loadings.

There are several timber frame packages to choose from, so it's important to be sure that the quoted prices include the full cost of the completed building with no extras. But there's another potential trap. Some are described in terms of their gross floor area, including the wall thickness, which can make them seem around 10% larger than other products described in terms of the net usable floor area of the rooms. But the major difference is whether they're standard 'open panel' design or the thicker 'closed panel' system.

Photo: potton.co.uk

OPEN PANEL

With a standard 'open' system, wall panels are manufactured from a framework of treated 50 x 100mm softwood timber studs lined with large sheets of plywood sheathing (or OSB) to provide stability. Thicker 125mm or 150mm studs can accommodate deeper insulation between the studs for better energy efficiency.

With your foundations and ground-floor structure already in place, construction of the inner leaf walls begins by erecting the large ready-made timber wall panels. Special timber baseplates ('sole plates') are first fixed in position over a mortar bed and a DPC, using special anchors direct to the ground-floor structure (rather than to a wall base as you might expect). It is essential that the sole plates are accurately laid, as any faults at this stage will be magnified as each storey is erected.

The wall panels are positioned with their plywood sheathing side facing outwards into the cavity and the panels are then joined together. A waterproof breather membrane is fixed to the outside face of the sheathing as a vapour barrier. This provides a protective covering against any moisture that may get through the brick outer wall and cavity. The panels are left open on their inner face (the room side), hence the name 'open panel'.

Once the house is weathertight and the electrical and plumbing carcassing has been completed, the insulation can be installed between the open studs (ie the side facing the rooms). For obvious reasons, with this type of construction there is a strong emphasis on keeping out any moisture,

such as from condensation. So a polythene sheet is laid across the insulated wall surface as a vapour-proof barrier prior to being plasterboarded. Timber frame wall panels also incorporate small barriers in the cavity to prevent fire spread. The upper wall panels are normally erected once the upper floor deck is installed.

Timber frame suppliers usually include the complete prefabricated roof structure within their package. Instead of conventional roof trusses, thickly-insulated, factory-produced structural panels (SIPS) can be delivered ready to assemble on site. Once the framework for the walls and roof is fully in place, the outer leaf of brick, block or stone can be built up as normal, with window and door frames fitted on site. See Chapter 12.

This masonry skin is tied to the timber panel inner leaf using special wall ties, leaving a cavity of at least 50mm, which is always kept clear.

CLOSED PANEL

'Closed panel' systems are similar to open panel, but with considerably more pre-fabrication carried out in the factory. This is the method favoured by Scandinavians, where entire wall panels are delivered in virtually finished form. Having the insulation already in place, the windows and doors pre-fixed, and the ducting for services installed saves time and money on site because the whole thing can be put together in a few hours. With timber frame 'cassette' systems, whole rooms can be craned into position and stacked on top of each other. The advantage is an airtight structure, with site work reduced to a minimum. The disadvantage is that you have to make detailed decisions about fittings and services at a very early stage of the design.

Post and beam

This is a traditional form of construction that the Neolithic builders of Stonehenge famously employed to lasting effect. One of the simplest, yet most durable forms of construction, this load-bearing framework comprises a pair of upright posts supporting a beam laid horizontally across their tops. Whilst timber has long been the material of choice, steel or reinforced concrete components can perform the same task.

Post and beam can be used where a design requires large

Photo: oakwrights.co.uk

openings in the walls, because unlike conventional masonry or timber panel systems the sizeable spaces between the load-bearing posts can be left free to take an infill material of your choice – perhaps a large expanse of high-performance glazing.

OAK FRAMING

Welcome to the Middle Ages. The choice of traditional oak-frame construction can make a new house look like it's been standing for centuries. You may wish to perfect the deception with some strategically placed traditional open trusses and curved windbraces. Or how about a double-height room overlooked by open galleries? Oak is hard to beat for quality and character, but comes with a price tag to match.

The frames themselves are fabricated in the workshop, where huge green oak timbers are cut, planed, and shaped to form a massive skeleton. The joints are finished off by hand so it can all be held together with traditional mortise-and-tenon joints. With the frame erected on site, the joints can be fully secured using tightly cleft oak pegs or wooden dowels and hammered wedges. The post and beam framework is usually built one storey at a time, known as 'platform framing', as opposed to 'balloon framing', where the wall frames are first constructed to the full height of the building.

One of the great strengths of oak is its natural resistance to weathering and insect attack, so it doesn't need treating with preservatives. Green oak is simply oak that has not been seasoned, with a typical moisture content of 20–25%. However, as it dries out it shrinks across the grain, causing the joints in the structure to tighten up. This makes the frame even stronger, but in the process creates minor splits along the

Photo: oakwrights.co.uk

frame, known as 'shakes'. Although this may appear slightly scary, it is part of the natural drying process. You can expect the frame to shrink by around 10%, so the build will need to allow for this movement, e.g. with expanding seals that keep the home airtight and weathertight.

Traditionally the gaps in the framework would be in-filled with wattle and daub or clay brick and lime. Being relatively flexible, such materials could accept a certain amount of shrinkage movement without cracking. Today high performance rigid polyurethane panels are more likely to be used to meet demanding thermal standards. These are normally waterproofed on the outer face before being rendered. But modern infill materials still need to be able to accommodate a certain amount of movement and prevent gaps around the infill panels, for example by incorporating sliding joints. Externally untreated oak will naturally weather to a silver grey, which some people like; otherwise to keep it looking fresh a preservative (e.g. special wax oils) can be applied, but will need periodic re-application.

When used for traditional single-skin wall construction, the huge oak timbers can be left exposed inside as well as out, to get the full visual effect. Alternatively, you could clad the outside with Structural Insulated Panels (SIPs – see below), leaving the oak timbers exposed only on the inside face.

The downside to building in oak is the relatively high cost – at least a third more than a similar softwood post and beam system.

Hybrid designs can provide a sensible solution that keeps the budget within the earth's orbit whilst retaining the charm

Photo: oakwrights.co.uk

Photo: potton.co.uk

of oak. Costs are reduced by restricting the oak framing to the more visible parts of the house, using less expensive softwood for concealed stud walls and the roof structure, with sympathetic period-style brick or stone construction elsewhere. Some frame kits may only feature 20% oak.

Alternatively, kits made from Douglas fir, a highly durable softwood, can make a lot of sense and are available from major manufacturers such as Potton. These capture the traditional charm of exposed internal timber beams throughout the house but without the daunting price tag of oak. To save on skilled labour costs, joints are made using bolted metal plates, which are largely hidden.

MODERN POST AND BEAM

One drawback with traditional 'aisle frame' construction for some self-builders is that supporting posts sometimes need to be located within the main rooms. Depending on the

Photo: potton.co.uk

layout of the house, these have a tendency to 'get in the way'. To overcome such perceived drawbacks, modern post and beam designs have been developed using stronger materials capable of longer unsupported spans. For example, some kits use engineered 'Glulam' beams to form the structural skeleton of the building. However, unlike traditional oak buildings, these single-skin frames cannot simply be in-filled and left exposed externally, and are normally given a rendered outer finish, or clad with conventional brick or decorative timbering. Internally the walls are finished with standard insulated timber panels.

There are a number of contemporary timber post and beam designs on the market that couldn't be more different from the classic English 'Olde Worlde' house. One of the best known is the German *Huf Haus* prefabricated construction system, dubbed 'leader of the flatpacks', a term

which probably doesn't do justice to these visually stunning, highly glazed kit houses. Freed from the constraints of conventional load-bearing and dividing walls, these designs use large expanses of highly insulated glazing and have spacious layouts with distinctive dual-pitched roofs.

The kits are made to order in Huf's factory before being shipped in component parts, like a giant jigsaw puzzle, to be rapidly assembled on site. You can select kitchens, bathrooms, and most design details from a menu. The open-plan layout provides excellent wheelchair accessibility. The downside is the fairly substantial price tag and the potential for planning battles (issues with overlooking etc). Ordering a prefabricated house also means having to make a lot of very detailed decisions prior to manufacture, such as the position of all the plug sockets.

THE 'SEGAL METHOD'

Photo: Piers Taylor Architects

Segal post and beam construction was devised in the 1960s specifically with self-builders in mind. It is economical and straightforward to construct. Structural softwood timber posts are supported on simple individual concrete pad foundations 600mm square and 900mm deep, which can be dug by hand. The gaps between the beams can be in-filled with panels made from a variety of materials – anything from brick to straw or hemp.

Being supported on beams such a house effectively floats above the

Photo: Piers Taylor Architects

landscape, which saves on levelling the site. The post and beam frames are bolted together flat on the ground using dry fixing methods such as bolts and screws. Once erected, galvanised steel plates secure the beams to the posts before the floor and roof joists are installed.

Non-conventional construction

Self-builders are far more open to innovative methods of construction than large housing developers. Indeed, many innovations that are now standard practice – such as underfloor heating – were originally pioneered by self-builders. There are a number of alternative wall-construction methods that have all been successfully proven and are

Photo: claysLLP.co.uk

worth considering. However, be aware that innovative designs and cutting edge products can have a downside – you may find that few trades in your local area have relevant experience, which could negate some of the advantages.

SIPs

Many architects regard SIPs systems (Structurally Insulated Panels) favourably, despite being a fairly recent arrival on the scene. These exceptionally strong, lightweight panels are formed by bonding a thick core of expanded polystyrene foam insulation between two rigid sheets of plywood or OSB, like a large sandwich. The main load-bearing walls of a house can be swiftly constructed simply by slotting together a

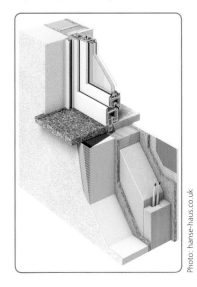

Photo: hanse-haus.co.uk

Photo: claysLLP.co.uk

series of these panels. Alternatively, SIPs can simply be applied as cladding to a conventional structure, such as post and beam.

SIPs panels can also be used for roof construction, and don't require any support from roof trusses. Being lightweight they can be swung into place relatively easily and rapidly secured, aiding the rigidity of the structure. With a couple of roof lights installed it's a fairly simple task to form extra rooms in the loft. The boards can be cut with a saw, so it's easy to carve out window and door openings.

As a structure SIPs are very efficient, and are increasingly being used as the main load-bearing wall element of package-deal kit houses. Ready-designed SIPs houses can be purchased with the entire structure factory-made, delivered to site for assembly. A wide variety of external wall claddings can then be fixed to the structural panels.

SIPs score very highly on thermal insulation values, which means that Building Control may allow you a 'trade-off' with lower insulation elsewhere, for example permitting heat loss from larger glazed areas. They also have the advantage of swift speed of construction and not having to be kept dry whilst awaiting erection. The panels can usually be delivered quite quickly once ordered. The main downside is the fairly high cost and their relative unfamiliarity to the trades on site.

Photos: hanse-haus.co.uk

Photo: Beco

Thin-joint masonry

Thin-joint blocks are specially engineered lightweight blocks that are glued together with special adhesive, rather than being conventionally bedded in cement mortar. Laying thin joint blocks requires less skill and they can be laid very fast, at a pace rivalling timber frame construction.

This system reduces drying times because, without

mortar, there's less retained moisture. It also provides a smoother interior finish that's claimed not to need a thick backing coat of plaster. Walls can be finished by simply spraying a fine coat of plaster or render on either side. Thin-joint blocks can be used in a traditional cavity wall, or in a thicker version as single-skin construction.

As with many alternative building systems, this has been proved successfully in the demanding conditions of Scandinavia. But in the UK special training is required even for experienced bricklayers, and as yet the system is nowhere near as popular as conventional masonry.

ICFs and PIFs

Insulated Concrete Formwork (ICF) is a modern method of solid wall construction that's now fairly well established. Sometimes referred to as Permanently Insulated Formwork (PIF), this is a building system that uses hollow polystyrene blocks of white polystyrene (EPS) that slot together like a giant Lego set.

It's a relatively simple task to assemble a wall from these lightweight interlocking blocks, which lock easily together without glue or mortar. Once stacked up into the shape you

require, the hollow blocks are filled with concrete pumped in from above, sometimes reinforced with steel rods placed vertically through the structure. Lintels for windows and doors are delivered as part of the package, pre-manufactured to suit the opening sizes for your building and ready to interlink with the other components.

When set, the polystyrene is left in place, providing superior insulation qualities that help meet 'zero energy' targets. The polystyrene can then be rendered on the outside, though many self-builders prefer to add a brick skin. The insides can be conventionally plastered or dry lined. And being naturally watertight, ICF is particularly useful for basements and foundation walls.

Many self-builders are attracted to ICFs because the system is relatively quick to construct, only requiring semi-skilled labour. However, in reality it can be more complicated than it looks to build correctly – even pouring the concrete isn't entirely straightforward – and it can work out relatively expensive.

Clay blocks

In much of Europe terracotta-coloured clay blocks are widely used for house construction. Clay blocks are a natural alternative to brick and block cavity walls, and are built as a simple single skin. Internal walls can also be built with them. Blocks are made from fired clay and have a honeycomb cross-section with lots of hollow air channels, making them thermally-efficient.

The large block size, coupled with the fact that cavity wall construction isn't required to satisfy the Building Regulations, means that the main walls can be erected at almost twice the speed of conventional masonry. The blocks are laid with a horizontal mortar joint 8–12mm thick using a special lightweight mortar, but the vertical 'perp' joints are tongued and grooved and slot together without mortar. Special renders and plasters can be applied directly to the completed blocks.

Concrete wall construction

Walls assembled from hollow concrete blocks are also common in Europe but relatively unknown in British housing. Steel reinforcing

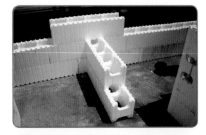

Photos: Beco

rods are placed down the hollow cores of the block walls, which are then filled with poured concrete to make a strong, solid wall. One attraction of this system is the claim that self-builders can build them without employing brickies, but in reality you would need good experience of masonry construction to do a good job.

Steel frame and concrete beam

Although widely used for offices and shops, steel and reinforced concrete frame buildings are more difficult to source and are comparatively expensive for complete house structures. However steel framing can be very useful for constructing key parts of the design such as complex roof structures. Long spans allow creation of large open spaces free of columns and load-bearing walls. Pre-manufactured components can be assembled quickly on site. On more challenging sites, steel can facilitate designs with reduced points of contact to the ground, requiring less excavation.

Eco-building

For those seeking maximum eco-credibility, there are a number of alternative 'back to nature' construction methods. Most can be combined with a conventional post and beam framework.

Straw

Most of us would be a little wary of building a house from straw, but we probably wouldn't bat an eyelid at thatched

roofs. Whilst it's possible for straw bales to support their own weight up to a limited height, it's a lot easier to use them as an infill material, thanks to their excellent insulation properties. Contrary to popular belief, compressed straw bales are too dense for rodents to show much interest, and their density also makes them hard to ignite.

Bales can be laid on top of one another like giant bricks in a stretcher bond pattern (ie all laid lengthways), creating thick walls around 600mm. Alternatively, they can be laid on their edges about 360mm thick as infill within a primary post and beam frame. For strength, hazel pins are driven down through the bales.

As you might expect, the foundations need to be raised to a sufficient level to prevent the straw becoming damp. Bales can be coated with 'breathable' lime render, but under load straw can settle over time. If this causes render to crack

Wall structure

1st	Post and beam timber frame
	Local stone
	Recycled bricks
	Earth, straw, or hemp
2nd	Modern timber frame with outer
	masonry leaf
	Sand-lime bricks
	New clay bricks
3rd	Hollow concrete blocks
	SIPs
	Concrete blocks with recycled
	aggregate
	New clay bricks
Not recommended	Dense concrete blocks

Comments

From a green perspective, framed buildings are generally preferable to masonry. The exception is reinforced concrete frames, which are not good environmentally. Lime mortar is preferable to cement but can take several weeks to dry.

Timber should be untreated if possible.

it can let damp in, allowing rot. So bales are sometimes pre-compressed using steel rods and planks of wood. However, an external rainscreen cladding such as timber weatherboarding is normally advisable.

Hemp

Industrial hemp, although famously related to cannabis, has for centuries been used to make rope, paper, and clothing. The centre of the stalk of the hemp plant contains woody material known as hurds, which can be processed with hydraulic lime to make lightweight 'hempcrete'. This can be packed into plywood formwork for use as infill between a post and beam timber frame structure. When compressed, it sets to form a strong rigid wall that can be finished in lime render.

Hemp is an excellent insulator, and although water-resistant must be able to 'breathe' so that any trapped moisture can evaporate. Its thermal performance is better than conventional masonry.

Hemp walls are built on top of a brick plinth above the splash zone. Costs are equivalent to conventional construction. And if there's a house fire at least you'll die with a smile on your face.

Modern earth building

Probably the lowest energy method of them all, as well as the cheapest. Earth has high thermal capacity: although such buildings takes a long time to heat up, they store heat within their structure, creating comfortable conditions in winter. See website for details of cob, rammed earth, and 'modular contained earth' methods.

Green targets

Green design is no longer an optional extra. All new homes now have to meet increasingly stringent energy-efficiency targets – something that has become a major part of the Building Regulations compliance process, with Part L setting minimum thermal performance targets for each element of the building, as well as dictating efficiency standards for boilers, heating controls, and insulation. This means your building has to be designed with energy efficiency in mind from the outset.

Photo: British Eco

Photo: potton.co.uk

However, investing in eco-friendly design can be more of a dilemma when funds are tight. Where there's a straight choice between affording an extra bedroom or solar-heating, it's the green option that tends to lose out. Fortunately for most self-builders, meeting these targets shouldn't be a problem because part of the ethos of building your own house is to achieve the highest possible standards within budget limitations. But it's probably worth bearing in mind that the two main factors that affect domestic energy consumption in the real world are the number of people working from home and how many teenagers live in your house! In other words, it's the occupants as much as the building that determine a building's energy performance.

Photo: Solar Century

Eco design

Eco design means first reducing a building's demand for energy as far as possible and then replacing some of the energy from fossil fuels with 'clean' power generated from renewable sources. Most Planning Authorities insist that new houses meet at least 10% of their energy needs from on-site renewables, where feasible.

But as with all things green, it's a question of how far you want to go beyond simply complying with the basic regulations. The more radical options tend to be expensive, with longer payback times. So the first step is always to make the most of passive design factors, optimising the building's siting and layout. A lot can be done at the design stage for minimal cost, such as putting more windows on the south-facing side of the house to benefit from heat-gain from the sun, thereby reducing the amount of energy needed to heat the house in winter. To prevent heat leaking away, the design needs to include high-performance glazing and insulation as well as sealing air gaps in the structure. Your eco-credentials can be further enhanced by specifying green materials with a minuscule carbon footprint.

Zero carbon targets

Between now and 2016, when all new homes must be zero carbon, the green target that new houses have to meet increases in steps. These are described in *The Code for Sustainable Homes*, which sets minimum standards for energy and water use. Your overall design will be given a star rating for sustainability performance using a one to six points allocation system. By demonstrating reductions in CO_2 emissions you earn points towards your Code Level. A maximum Level 6 rating is equivalent to zero carbon, required from 2016. Until then builders need to achieve Level 5. This can normally be achieved by a combination of passive design measures (reducing energy demand) and renewable energy generation. To get a Level 6 rating, your design needs to score 90 points (with a minimum 17.6 from energy and 7.5 from water). For full compliance details see website.

SAP ratings

To meet energy performance targets, the main elements of a building's fabric – its walls, roof, windows etc – have to conform to minimum 'U' values (the measurement of heat loss in Watts for every square metre of a material's surface). The lower the U-value figures the better.

However, compliance with the Building Regs is no longer simply about meeting individual U-value targets for each element of the building. What matters is the overall energy and carbon

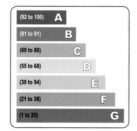

(92 to 100)	A
(81 to 91)	B
(69 to 80)	C
(55 to 68)	D
(39 to 54)	E
(21 to 38)	F
(1 to 20)	G

Conserving water

Photo: Roca Baths

Part G of the Building Regulations sets targets for the efficient use of water, and outlines where rainwater harvesting or grey water recycling may be used to reduce the demand on the mains water system. New homes must now be designed so that 'potable' drinking water use will not exceed 125 litres per person per day. But before you can start to save water, it helps to know where most of this precious commodity is consumed.

Baths and showers

The average four-person home will typically consume about 185m³ of water a year, the equivalent of over 2,000 baths. In new houses, it's showers and baths that account for almost half of all water use. As you might expect, showers generally use only about a third of the water used to run a bath, the exception being 'power showers', which can actually use more. One way to economise on water use is to install 'flow regulators'. These are devices fitted to pipes to provide a fairly constant flow to the taps, irrespective of mains pressure. Shower flow regulators can be fitted between the mixer and the shower hose in order to limit the flow rate. However, electric showers should not be restricted. It is now a requirement to limit the temperature of hot water from bath taps, as well as showers, in new homes to 48°C, to prevent the risk of scalding.

WCs

It is estimated that around a third of total water use in homes goes straight down the toilet. Hence the maximum flush volume for new WCs has now been reduced to six litres.

APPLIANCES

Washing machines and dishwashers use the most water in the house after WCs and baths – as much as 20%. The machines labelled the most energy efficient tend to also be the most water-efficient.

BASINS AND SINKS

Nearly 10% of domestic water use is from washbasins and sinks, but most can be saved by fitting spray taps or 'tapmagic' inserts (or simply not leaving taps running when brushing your teeth).

Who can do a SAP report?

Only authorised Domestic Energy Assessors (DEAs) can carry out SAP calculations. Probably the simplest course of action is to approach Local Authority Building Control, some of whom, for an additional fee, can provide a full range of services including:

- SAP assessments (Part L)
- Code for Sustainable Homes assessments (Part L)
- Air pressure testing (Part L)
- Acoustic testing (Part E)
- New homes structural warranties

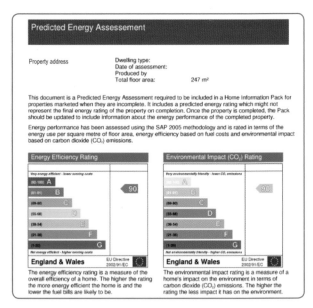

performance of the whole house. This is measured by carrying out a SAP assessment (Standard Assessment Procedure), which calculates the property's likely energy costs (for space heating and hot water) per square metre of floor area. The score is expressed as a rating of 1 to 100 – the higher the number the lower your energy consumption, with 100 representing the holy grail of zero energy cost.

One of the attractions of SAP ratings is that they provide an easy way to understand and compare the running costs and environmental impact (carbon footprint) of buildings. They also allow you a fair amount of freedom with elements of your design (eg to have large expanses of glazing) as long as the overall house meets the necessary energy performance standard.

There is, however, scope for conflict to arise with SAP calculations – for example, by specifying the optimum insulation materials to meet sound proofing targets you may find the insulation fails in thermal efficiency terms.

DESIGN SAP ASSESSMENT

All new properties require a full SAP Assessment at the design stage. The purpose of this 'Design SAP' calculation is to demonstrate to Building Control that your house will achieve compliance by meeting energy efficiency targets. The goal is to obtain a document known as a 'compliance report' upon completion of the build. This is rather like an MoT certificate that officially confirms that your house meets all the various energy targets. Most architects now instruct a report from an energy consultant to verify their design's compliance with Part L.

Building Control Departments won't usually approve plans until they've received a 'Design SAP' with a satisfactory Predicted Energy Assessment (PEA). Other Councils make it a condition that compliance is confirmed prior to commencement of work on site. Where Building Control don't insist on an SAP assessment up front, there's no point leaving it and carrying on regardless, because they may demand costly changes at the end of the build, even requiring parts of the house to be rebuilt before they can issue the completion certificate.

Currently, it's not uncommon for designs to fail at the first attempt to meet the required standards, and the SAP assessor will advise how the design can be modified to pass, perhaps by providing thicker insulation, a more efficient heating system, or superior glazing. Extra credit may be

given for fitting low energy lighting throughout. Or you could boost your score by adding a secondary heating source, such as a log-burning stove.

AS-BUILT SAP

Having a design that's squeaky clean in energy performance terms is all very well, but of course a lot can change 'twixt drawings and finished product. Suppose the builders can't be arsed to put in all that tedious insulation? Or what if they decide to substitute some cheaper concrete blocks when no one's looking? So to check that the completed property actually performs as well as was promised at the design stage, a final 'as-built' SAP calculation will need to be carried out on the finished house, normally by the same energy assessor who did your 'Design SAP' calculations. Armed with this final SAP calculation, your Energy Performance Certificate (EPC) can then be produced in the familiar form of a colourful bar-chart that all new buildings now require (also known as an On Completion Energy Performance Certificate, or OCEPC). The compliance report can also be filled out at completion once your air test results and commissioning certificates are available.

DERs versus TERs

SAP reports also include a 'Carbon Index Rating' (or 'Environmental Impact Rating'). This tells you the maximum amount of CO_2 that your new house is allowed to emit, a figure known as the 'Target Emissions Rate' or TER. This is based on a notional building of the same size and type of property (measured in kilograms of CO_2 per square metre of floor area per year).

In itself, this is pretty meaningless, so the SAP also calculates the *actual* CO_2 emission rate of your proposed building, known as the 'Dwelling Emission Rate' or DER. This figure is largely determined by the U-values of all the building's 'thermal elements', such as the floors, walls, roofs, and windows, as well as taking into account the efficiency of the heating system, insulation, ventilation, and lighting – see the boxout below. To achieve compliance, your design must be equal to, or better than, the target figure. In other words, your DER must be no worse than your TER.

At the design stage before work begins your DERs & TER calculations will need to be submitted to Building Control. On completion of the build, they can normally accept a certificate of compliance produced by an accredited energy assessor, with all necessary 'as-built' info provided, to confirm that the key specifications have been followed on site.

Photo: excelfibre.com

TARGET FABRIC ENERGY EFFICIENCY

Target Fabric Energy Efficiency (TFEE) is the minimum energy performance requirement for new dwellings. It measures the building's annual 'energy demand' (in units of kilowatt-hours per m2).

New homes in England are assessed on this as well as the Target Emission Rate (TER) .

How well you score on TFEE depends on 3 main factors
- U- values
- Air pressure test result
- Thermal bridging

The Building Regs include a 'notional' building specification with suggested U-values, plus an air test result and psi values. If these are followed then the build should meet the TFEE. But there's some flexibility - you're free to pick and choose which areas to exceed the standards so you can get a bit of leeway elsewhere.

	'Notional' U-values W/m2K (to meet TFEE)
ROOFS (all types)	0.13
WALLS	0.18
FLOORS	0.13
GLAZING	1.40
Doors (solid)	1.00
Doors (with glazing)	1.20

AIR PRESSURE TEST   'Notional' target (to meet TFEE)

5m³ per hour /m²

Remember TFEE is only the *target* rate – i.e. based on your design on paper. What really matters is the actual performance of the completed new dwelling, referred to as the *Dwelling Fabric Energy Efficiency* rate (DFEE).

Before work starts, the designer must predict the DFEE rate of the dwelling as designed, to demonstrate it will not be greater than the TFEE rate, and give this calculation to Building Control.

When construction is complete, the designer or builder must notify Building Control and advise them of any changes to the approved specification that was used to calculate energy performance at the outset.

Specifying green materials

To be truly green means looking at the materials you plan to use, and their broader impact on the environment. Natural local products that are renewable are, of course, preferred. However, these are not always available at realistic prices and may not meet required performance standards. So to help select eco-friendly materials, in the following chapters we've included 'Green Choice' boxes that list the top three options for each part of the build. In the meantime, let's take a quick look at which materials can justifiably claim to be green.

The five steps to SAP compliance

To demonstrate compliance with energy efficiency targets in the Building Regulations, the SAP Assessor will check five key criteria:

1. That the predicted rate of carbon emissions, the Dwelling Emission Rate (DER), is no greater than the Target Emissions Rate (TER).

2. That the performance of the building fabric (*ie* the actual construction when built) will be be no worse than predicted at the design stage in meeting target U-values (see Target Fabric Energy Efficiency).

This step also involves checking the heating and hot-water systems as well as the internal and external lighting. The boiler must meet a SEDBUK efficiency target (Seasonal Efficiency of Domestic Boilers), and pipes and hot-water cylinders have to be insulated. Renewable-energy sources such as solar water heating will also be taken into account. In addition, your lighting must comply with energy efficiency rules (see Chapter 13).

3. That there will be no excessive overheating in summer, and that the design includes appropriate 'passive control measures' to prevent excessive solar overheating. This can be done by a combination of window size, orientation, glazing, and solar protection through shading.

4. That the building's fabric and fixed services perform no worse than set design limits. This means that, upon completion, the building's fabric is checked to verify that:

There are no gaps or significant thermal bridges in the insulation or at key junctions (such as wall/roof, wall/floor, and around window and door openings).

The permeability target in the DER has been achieved, confirmed by air-testing.

The heating and hot-water system have been fully commissioned.

5. That instructions have been provided to ensure efficient operation and maintenance. If the people living in the finished house aren't provided with proper operating and maintenance instructions it could make a mockery of all that expensive green design, so you need to show that 'provisions have been put in place to ensure efficient operation'.

Photo: oakwrights.co.uk

Timber

Wood is the most important renewable raw material used in construction. Most timber used for construction is softwood. Home-grown spruce or pine can be used for non-structural joinery, such as floorboards and doors. The best softwoods for structural purposes are Douglas fir and European larch (classified as 'moderately durable'). For hardwood, locally-sourced, UK-grown oak and sweet chestnut are good for construction, or alternatively oak from Europe or wood such as red cedar from Canada. When it comes to tropical hardwood, the only reliable guarantee of sustainability is wood certified by the FSC (Forestry Stewardship Council).

Wood preservatives are one of the major sources of toxicity in new homes. NHBC no longer require interior timber to be treated. Where timber does need to be treated the best advice is to specify factory treatment rather than site application of preservative, or to use water-based treatments, rather than solvent-based ones.

Engineered timber beams

'Composites' are lighter and stronger than solid timber. They're also generally eco-friendly because they recycle poor

GREEN CHOICE

TEN TIPS FOR GREENING YOUR DESIGN

- Fully insulate your home and build thicker walls.
- Put more windows on south-facing side than north-facing, to get free heat from the sun.
- Elsewhere, don't use too much glass, putting it mainly where it will reduce the need for electric light.
- Use materials made from recycled waste, such as eco-concrete, eco-blocks, and softwood 'I-beams'.
- Use renewable materials, such as softwood rather than hardwood.
- Specify an energy-efficient boiler.
- Generate your own energy with solar panels, PV cells, heat pump systems, or micro wind turbines.
- Fit low-energy lighting throughout.
- Fit intelligent controls for heating, lighting, and appliances.
- Consider rainwater harvesting and grey water recycling.

quality timber and wood waste. There are several composite products:

- **Laminated timber** was one of the first composites to be developed. Here, thin strips of timber are glued together horizontally in layers.
- **LVL** (laminated veneer lumber) is manufactured from thin veneers glued together vertically to make beams or sheets that are similar in appearance to plywood. Due to its composite nature it is much less likely than conventional timber to warp, twist, bow, or shrink.
- **'Parallam'** or **PSL** (parallel strand lumber) is made from strands of timber rather than veneers. Not as strong as LVL, but has a pleasant texture and can be used to achieve a fine finish.
- **I-beams** have been developed for use as joists or rafters. Looking similar in profile to steel beams, they have a thin vertical centre web sandwiched between a flat top and bottom flanges. This lightweight centre web can be made from OSB or masonite (a structural grade hardboard made by compressing sawdust under high pressure and temperature, relying on the natural cellulose to bind it together). The flanges can be softwood or LVL for greater strength.
- **Eco-joists** use light steel struts to connect the top and bottom flanges. This gives them the advantage of being able to accommodate pipe and cable runs, and even large diameter ducting for ventilation systems.

Timber panel products

Chipboard and MDF (medium density fibreboard) are made from waste wood. Similar OSB (oriented strand board) uses higher quality wood chips.

Glues

Formaldehyde is irritating to skin, eyes, and lungs, and there is some evidence of an associated cancer risk. It is used in glues and some wood composites. Specify non-formaldehyde glues, chipboard, and MDF.

Reclaimed materials

Reclaimed timber tends to be high quality and well-seasoned with few knots. Second-hand bricks and tiles are often handmade, with a depth of character and texture.

Paint

Solvent-based paint, even when dry, can continue to give off toxic VOCs (volatile organic compounds). Use organic or water-based paints instead. See Chapter 14.

Insulation

Rigid foam boards made from polyurethane are widely used in cavity walls, floors, and roofs. Alternatively, thinner 'multi-foils' may be worth considering. These comprise thin blankets of mineral wool sandwiched between shiny reflective sheets of foil that act as 'radiant heat barriers'. Being only around 30mm thick these multi-foils are well suited for insulating new roofs, but are currently not acceptable in all LABC areas, so check first.

However, from a green perspective natural materials such

Photo: Actis Insulation Ltd

Photo: Actis Insulation Ltd

Photo: Penycoed-Warmcel

Photo: Penycoed-Warmcel

as sheep's wool and cellulose fibre (*eg* 'Warmcel', made from recycled newspaper) are preferred, as they're both renewable and sustainable. Cellulose fibre is delivered loose or in compressed bales. It can be blown into a timber frame wall or between sloping and flat roof joists by trained BBA-registered operatives, or can be used DIY as loose loft insulation. It's also available in the form of batts, which are more expensive but don't require special equipment.

Wool is pleasant to handle, unlike fibreglass or mineral wool, which require protection for skin, eyes, nose, and mouth. Of the fossil-fuel-based materials, polystyrene (EPS) is the least environmentally damaging. For information on insulation materials see www.home-insulating.com.

PVC

Plastic is one of the worst materials environmentally. Although hard to avoid for some uses, in most cases there are greener alternatives:

Plastic materials	Non-PVC alternatives
Guttering and downpipes	Stainless/galvanised steel or aluminium
Underground drainage	Traditional clay pipes
Above-ground pipework	Polyethylene
Vinyl floor coverings	Lino or wood flooring
UPVC windows	Treated softwood
Electric cables	Low-smoke cables

Windows, glazing, and external doors

The most environmentally damaging option is UPVC, followed by aluminium. Research shows that UPVC windows, often thought to have low maintenance costs, are

Photo: velfac.co.uk

easily damaged and difficult to repair, and actually have a relatively short life of 30 years or less.

Well-designed windows made from durable softwood are one of the best options. The maintenance of timber is less critical if good quality wood is used. A factory-applied microporous paint finish can be expected to last 15 years before it needs to be redecorated.

Many of the best timber window units are imported from Scandinavia. Custom-manufactured in automated factories, they come with high-quality, factory-finish paint sprayed on before glazing so that it goes into all the rebates. Doors can be ordered ready-hung in their frames, and factory-fitted with secure triple-locking systems. The downside of pre-ordering custom-manufactured units is not being able to change your mind and submit later revisions.

To meet U-value targets, windows need to be high performance. This means sealed double- or triple-glazed units with low-emissivity coating on at least one face of the cavity. See Chapter 12.

Green roofs

Having a meadow on your roof isn't everyone's idea of a dream home, but green roofs planted with sedum are now well established and popular with eco-builders. Although they're most widely found on flat roofs, pitches as steep as 30° can accommodate turfed or seeded coverings, but may require netting or battens to prevent the soil from sliding.

Contrary to popular opinion, you don't need to mow a grass roof – properly designed, they don't need maintaining.

There are several benefits to green roofs. As well as absorbing noise, dust, and pollution, they soften a building's visual impact – which may just swing a controversial planning application in your favour. And because the earth acts as a 'thermal buffer' it helps keep the building warmer in winter and cooler in summer, reducing heating costs. Rainwater is also temporarily absorbed, which slows water run-off, unlike hard surfaces that produce large volumes of storm water that in some areas can flood drainage systems. See website for further details.

Green energy

Making your home eco-friendly means you can dramatically slash future power bills whilst securing some energy independence.

The challenge is to design homes that intelligently exploit free green energy. This means maximising heat gain from the sun, minimising wasted heat leakage, and generating your own renewable energy.

Passive solar gain

It is estimated that solar energy (both passive and active) can provide up to a third of the space heating, and more than half of the hot-water requirements, for a family house. It can also provide up to 70% of the electrical power.

To make the most of the free warmth from sunshine or 'solar gain', the starting point is always the site layout. At its simplest, this means facing the building south to capture passive 'solar gain', the easiest way to boost space heating. Heat absorbed during the day is stored in the mass of the building (*eg* concrete walls and floors) and later released naturally when the air temperature drops below that of the building fabric. At night, shutters or heavy curtains can help retain the heat.

Buildings with high 'thermal mass' are best at storing natural warmth. For this you need thick masonry walls. Conventional cavity external walls are around 300mm thick, but this can be extended to around 450mm by building an inner skin of

Photo: Living roofs

Photo: velfac.co.uk

Photo: velfac.co.uk

concrete blocks laid flat on their sides. Some even advocate filling cavities with concrete to form castle-like thick solid walls. If space permits, internal walls can similarly be built to twice the usual 100mm thickness, and perhaps also a deeper than normal concrete floor screed.

To reduce thermal gain and brightness in summer when it's least wanted, strategic positioning of deciduous trees on the south side can provide shade from the high angle of the summer sun. This helps prevent overheating of south-facing rooms and conservatories. And because these trees lose their leaves in autumn, weaker winter sunlight can still provide warmth.

However, you need to bear in mind that a property's orientation can also affect the performance of some renewables. For example, a shadow falling on an array of PV cells (see below) can significantly reduce their effectiveness, whereas shading on a solar hot-water panel is much less critical.

Minimising heat loss

As well as passively capturing warmth from the sun, the aim of an energy-efficient building is to minimise heat *loss* through the shell or 'envelope' – the walls, roof, floors, and windows.

A sealed environment with no gaps for warm air to seep out means houses are cosier, cheaper to run, and with lower carbon emissions are better for the environment – hence the requirement for air testing (see Chapter 14).

Intelligent design can reduce heat-loss by minimising the area exposed to cold winds , for example by sheltering the building behind earth mounds. Also, porches and 'solar space' conservatories function as attached heat stores, providing a useful buffer space and acting as airlocks when people go in or out of the house, minimising heat loss and reducing the amount of cold air penetrating your home.

Renewable energy

A key aspiration for most self-builders is to create a warm, well-insulated property with extremely low running costs, even generating a profit by selling power back to the grid. This, of course, has nothing whatsoever to do with Government policy. It's all about the desire for independence, self-sufficiency, and a better quality of life – combined with a glow of satisfaction at beating the system. Hence the interest amongst self-builders in renewables – also known as Low and Zero Carbon (LZC) technologies. This generally means solar panels and PV cells ('active solar power'), wind turbines, and heat pumps.

One advantage of such 'micro-generation' systems is that the energy is generated at the point it's consumed. But as everyone knows, the downside is the long payback period. Which makes the availability of grants a critical factor. In the UK, the amount of Government grants available has proved pitifully inadequate. Grant levels are constantly changing and so far seem to be aimed more at generating eco-headlines than electricity. To qualify for a grant, the system you've selected has to be within a recognised scheme and must be fitted by an approved installer who is registered with that scheme.

The good news is that generating your own energy can earn you 'clean energy cashback'. See boxout on page 103.

Solar electric roof tiles

Photo: solarcentury.co.uk

Solar water-heating (SWH)

Solar water-heating systems collect heat from the sun to generate hot water that's piped to your taps, as well as being used for space heating. Sometimes referred to as 'Solar Thermal' (ST) or 'Solar Hot Water' (SHW), these systems basically comprise a solar collector and a hot-water store such as a conventional hot-water cylinder with an additional solar coil.

There are two main types of collector – flat plates or evacuated tubes. Tubes are generally more efficient, but are more expensive. The best-performing flat plates have double-glazed plate collectors with a microscopic film applied to the collector plate to increase the amount of solar radiation absorbed. The rest of the system comprises a transfer unit (a high recovery coil), a hot-water cylinder to store the heated water, and a controller that manages the flow through the panel and accommodates heat expansion.

Even in overcast Britain, solar water heaters work remarkably well. On cloudy days they can still collect sufficient radiation to generate a useful amount of heat. On warm days enough energy is collected to heat water to normal bath temperatures, while on dull days they can still heat the water to lukewarm temperature (heating water from cold to lukewarm uses considerably more energy than from lukewarm to hot). They are therefore used in conjunction with a boiler or other heat source as a backup.

Solar hot-water systems also need frost protection. This

Above: Mixed solar electric and thermal tiles

Right: Solar thermal

can be in the form of a direct system with a frost-resistant collector and plastic pipework, or, more usually, an indirect system with a separate water circuit from the collector to the hot-water cylinder, which has anti-freeze in it.

The collectors don't have to be positioned on a roof – they can work well at low level against a wall or in the garden. When mounted high up (or incorporated into the roof finish) an additional pump and controller are necessary.

To work effectively, the collectors don't have to face perfectly due south. They can face anywhere between south-east and south-west, at a tilt of between 10° and 60°. This should still achieve 90% of optimum performance (the optimum tilt is 32° facing slightly west of due south). In terms of size, for a typical family house a tube system will need to be at least $3m^2$ or $4m^2$, and a matt black plate collector around $5m^2$.

Solar hot water systems can typically supply up to half your hot-water requirements, but are not so good at providing hot water for conventional radiator central heating. They are better suited to linking up with underfloor heating systems, since these run at lower temperatures. However, a secondary energy source will still be required. Installation costs vary from £3,000 up to £5,000 and payback periods for solar water heating are currently estimated at 7 to 15 years (depending on Renewable Heat Incentive payments).

PV solar electricity

Photovoltaic (PV) panels make electricity from light, as opposed to solar thermal panels that absorb heat from the

Solar tubes

Photo: British Eco

Solar PV

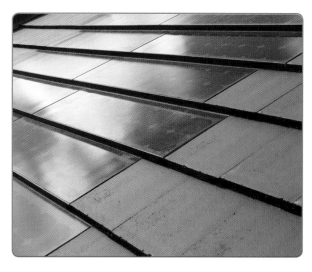

sun. PV panels incorporating silicon cells are the same technology used to power millions of pocket calculators and remote telecoms and traffic signals. Each panel is made up of several modules of tiny cells, and several panels grouped together are known as an array.

PV panels silently convert sunlight to DC electricity, which then needs to be converted to 240V AC. This requires a small 'inverter' box, which is the only item that takes up any internal space. A cable connects the panels to the inverter and the system is connected to the mains. PV panels don't require direct sunlight and are still effective even on cloudy days, but as noted above, overshadowing from trees etc can significantly reduce power output.

PV is the easiest way to generate renewable electricity in the home, but it is still relatively expensive. A full array of PV panels could cost between £7,000 and £10,000. Because there are no working parts, these are low-maintenance installations that should last 20 to 30 years. Panels are normally mounted on south-facing roofs, although any roof facing south-east to south-west should perform relatively well, ideally at a slope between 30° and 40°. They can also be mounted on flat roofs but need to be angled on brackets. Silicon mono-crystalline cells are the most efficient (and expensive) type of mass-produced cells.

PV cells are also available in the form of roof tiles, and although more expensive can look far more pleasing, although the variety of solar tiles currently available is fairly limited. There's also the enticing prospect of PV glass currently being developed that could replace large glazed areas in future.

To meet an average annual household demand of around 4,800kWh, a 6kWp (kilowatts produced at peak) system would be required. To supply 100% of electricity it would normally need to be used in tandem with other alternatives such as wind turbines. One obvious drawback with PV panels is that they generate most electricity during the day, when demand is low, the opposite of what's required.

So unless you install a bank of batteries (see wind turbines below) PV systems need to be connected to the grid so that power can be supplied when the panels aren't working, and surplus electricity can be sold back to the supplier. This requires two-way metering, which adds to the cost. Payback periods are currently estimated at around 10 years (depending on Feed In Tariff payments).

Wind turbines

It is claimed that a domestic 'microwind' turbine can produce enough electricity to power the lights and run the electrical appliances in a typical home. But individual wind

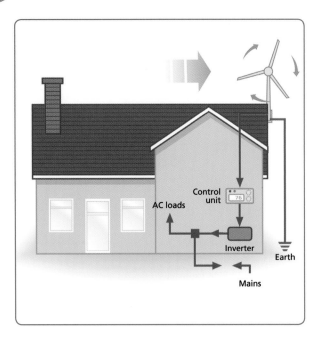

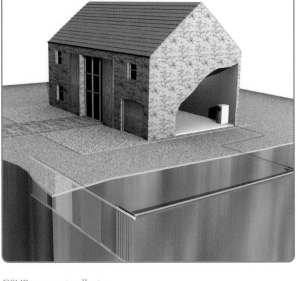

GSHP compact collector

Photo: Worcester Bosch

turbines vary considerably in size and power. Roof-mounted domestic wind turbines are typically around 1kW to 2kW in size. Larger mast-mounted wind turbines, free-standing in gardens, are around 2.5kW to 6kW.

Assuming reasonably breezy local conditions, a 6kW machine around of 1.5–2m diameter could potentially slash a few hundred pounds a year off electricity bills of a single dwelling when some of the power is sold back to the grid. Nonetheless, wind generators usually need to be backed-up by solar PV or a diesel generator for use when wind speeds are low.

Having a sufficient amount of wind is obviously critical, particularly when you realise that doubling the speed of the wind increases the power output an incredible eight times. In fact, no electricity at all is produced when wind speeds are below 3m/s (metres per second). A typical urban area will experience wind speeds of between 3m/s and 5m/s and a turbine of between 1.5 and 2m diameter will provide around 10% of the average annual power use of around 4,700kWh. But in a rural area blessed with good wind speeds it's claimed you can generate up to 70% of your electricity. To find the wind speed in your area see www.bwea.com.

The main enemy of wind turbines is turbulence, which reduces efficiency and causes noise, vibration, and wear. To minimise turbulence, height is critical. Turbines should be a minimum of 10m above the roof and clear of obstructions from trees and buildings within 100m. But applications for tall masts may be opposed by planners and neighbours, especially in Conservation Areas. This makes wind power better suited to remote locations where there is no mains supply from the national grid.

As with solar PV, most wind systems generate DC electricity, so if it's not connected to the grid you would need battery storage and an inverter to convert your DC electricity

to the AC used for mains electricity. Because most systems are connected to the grid, no batteries are required. But if you do want to store electricity in batteries they'll need housing in a ventilated area; plus a controller is necessary to ensure that they're not over- or under-charged.

Turbines typically have a life of around 15 to 20 years but require service checks every few years. A domestic 2.5 kW wind turbine costs around £5,000 for the turbine itself, rising to around £15,000 when you include an 11m high tower, cabling, electronics and inverters. A turbine of this type can produce around 4,000 kWh per year. The payback period is likely to be around 16 years (depending on Feed In Tariff payments).

Ground source heat pumps (GSHPs)

Heat pump sales in Britain have started to rise in recent years, yet still remain small by comparison with Sweden, where up to 90% of new homes are fitted with them. Ground source heat pumps work by extracting warmth from the ground under our feet, which is heated by the sun to a typical average temperature of about 12°. So it's really just another way of using of solar energy, albeit stored in the earth.

Ground source heat pumps work by using a series of coolant-filled coils or 'slinkies' buried about 1.5m to 2m deep in a trench. They require about 10m run of trench per 1kW of heat energy produced. The coils contain a refrigerant liquid or gas which absorbs the heat, which is then compressed to make it a lot hotter, working like a fridge in reverse. This heat is then 'exchanged' to your water supply, delivering hot water for space and water heating, before the cooled refrigerant is then re-circulated to collect more heat.

Because the water produced is at a relatively low temperature, these systems work best combined with underfloor heating, which runs at lower water temperatures than conventional radiator systems. A secondary heat source

Photo: Worcester Bosch

Air-to-water heat pump

is normally needed to raise the temperature of water so that it's hot enough for domestic use.

Although heat pumps consume electricity, they typically supply three or four units of heat for every one unit of energy used to run them (ie a 'Coefficient of Performance' or CoP of 3 or 4).

The costs of installing a typical system are fairly high – ranging from about £10,000 to £15,000 for a ground source or £6,000 to £10,000 for an air source system, not including the cost of the heat distribution system. But with minimal maintenance, the running costs are around half those of oil-based heating systems. Incidentally GSHPs are not the same as 'Geo-thermal', which involves drilling very deep wells and pumping in water under pressure to benefit from deep geological heat.

Photo: Charnwood stoves

Using similar technology, air source heat pumps absorb heat from the outside air, even when the temperature outdoors is as low as –15°C. There are air-to-water systems that use the captured heat to warm water, and air-to-air systems that produce warm air that's circulated by fans to heat your home. Although cheaper to install than GSHPs, their efficiency is lower with a typical CoP of around 2.5.

Biomass

Biomass means any material grown to produce fuel. The burning of renewable fuels might be caveman technology, but the reason it has now been officially pronounced green is down to the fact that plants consume tonnes of carbon whilst growing, which cancels out the carbon released during combustion. The usual biomass fuel is wood pellets, but some boilers can consume just about anything you can set alight. One major attraction is that you can harness 'free energy' whenever you feel like it by nipping down to the woods and shoving some of nature's offcuts into a sack. Most biomass stoves feed themselves semi-automatically with an augur and hopper system, cutting out most of the hard labour involved in stoking old-fashioned boilers. An individual pellet stove will cost around £4,300 including installation. For boilers, an automatically fed pellet boiler costs between £14,000 and £19,000 including installation, flue, fuel store, and log boilers between £11,000 and £23,000.

The bottom line

For most self-builders solar hot-water heating is likely to be the most cost-effective green energy option, compared to generating electricity using PV panels or wind power. For remote properties without access to the national grid, the maths is dramatically more favourable. However, the introduction of cashback payments means installing renewables is now becoming far more attractive from a financial perspective. And in future, as fuel prices rise and equipment becomes cheaper and more efficient, renewables will make increasing economic sense.

For more information see the Haynes *Eco House Manual.*

Clean Energy Cashback

Homeowners who generate their own electricity can earn a guaranteed tax free income on top of savings on energy bills. Feed-In Tariffs (FITs) pay homeowners for each KWh of energy generated (even if they use it themselves). This applies to low-carbon technologies such as wind turbines and solar panels.

The Renewable Heat Incentive (RHI) similarly pays towards the costs of generating renewable energy to heat your home, e.g. for solar thermal installations or air/ground source heat pumps.

These payments mean that payback periods are dramatically reduced. And when it comes to selling a house with an income, there should be a positive impact on market value. But payment rates are constantly changing - see Decc.gov.uk

7 THE PLANNING MAZE

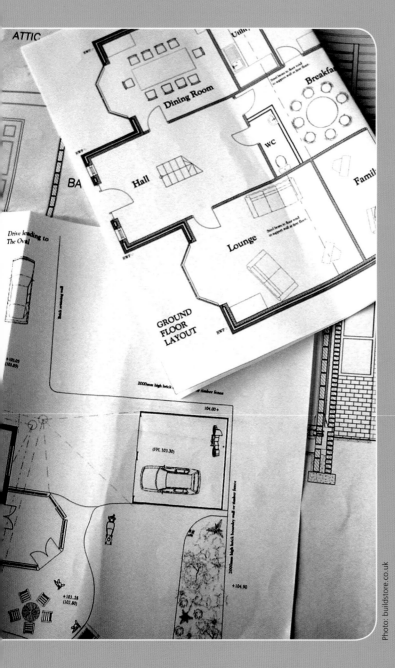

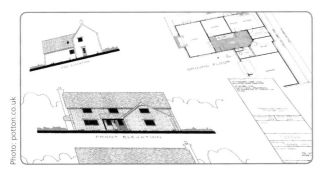

Photo: potton.co.uk

for detailed planning approval for your specific design. But of course, most vendors wouldn't be prepared to hang around for the necessary two or three months while your application was processed. Consequently, most self-build plots are acquired just with OPP. This means that consent has been granted for a new dwelling to be built on this plot *in principle*. But you can't start work on site because most of the design details have yet to be approved. This includes essential matters such as the property's precise dimensions and layout, the method of construction, the choice of

Photo: claysLLP.co.uk

Back at the plot-hunting stage we looked at some site-specific planning issues that can determine what you're allowed to build – such as whether the plot is located in a Conservation Area or in the Green Belt or has got a protected tree growing smack in the middle of it. Assuming that any such concerns have since been ironed out and the plot is now legally yours, the next step is to focus on how to successfully negotiate with the planners and get consent for the design that you really want.

Outline Planning Permission

Outline Planning Permission (OPP) is concerned with the use of the site, not the detailed design. So in an ideal world, before finalising the purchase of your plot you would apply

Photo: potton.co.uk

building materials, and positions of windows etc. Together these key aspects are known as 'Reserved Matters' and a further application will need to be made to get all the details approved – see below.

Outline consent is always granted subject to conditions. It is therefore essential to check whether there's anything in these conditions that could restrict what the planners might later accept for the detailed design, and could prevent you from building your dream home. If the original application included a 'Design & Access Statement' this may also contain valuable clues as to what design details are likely to be permitted.

Approval of Reserved Matters

If you've already got outline consent you will need to 'finish the job' by seeking Approval of Reserved Matters with an application for 'detailed consent'. You normally have a time limit of three years from the date the outline consent was granted to make your Reserved Matters application and get the detailed design approved, otherwise you'll lose it. If you leave it to the last minute before submitting your application, the time taken to process it could take you beyond the three-year deadline. If it then fails, your original outline consent could expire.

Just because you've got OPP, getting approval of Reserved Matters can't be taken for granted. However, if your application is refused it doesn't negate the original outline consent (as long as it hasn't expired). The worst-case scenario, if your detailed design should be rejected outright, would be having to put the plot back on the market – and in the process swallow all the acquisition and sale costs, not to mention the bill for the design work.

Once approval has been granted for the outstanding details, you then have a further two years to start work, or five years from the date of the original OPP approval, whichever is the longer. After this date, unless you've started work on site, the permission will expire and you're back to square one. A reapplication would then be needed, and there's no guarantee that consent would be granted a second time – a potentially expensive mistake!

Approval of Reserved Matters is restricted to the following:

- **Layout** – The relationship between the main buildings, access routes, and open spaces, including those beyond the plot's boundaries.
- **Scale** – The height, width, and length of each building.
- **Appearance** – The visual impression the building makes.
- **Access** – Access routes into the site for vehicles and pedestrians and circulation space within it.
- **Landscaping** – How you propose to enhance open spaces with the planting of trees, shrubs, hedges, lawns etc, and screening by fences and walls.

Full Planning Permission

This combines Outline Planning and Approval of Reserved Matters in a single application. Sometimes sites come on to the market with full planning approval already in place for a detailed design, so that no further planning applications are needed. This may sound like an ideal situation, but someone else's design is unlikely to match your idea of a dream home, so you may well want to change it.

To modify an existing detailed consent means having to make a fresh application for the design you actually want to build. Fortunately, in most cases – as long as it's not radically different from what's already been approved – you should be OK.

Where a site is marketed with full planning permission it may be because the planners wouldn't accept an outline application. This sometimes happens where they want a high degree of control, for example in Conservation Areas. In such cases, getting a revised design approved can be more difficult, as any significant changes may be frowned upon.

Once full permission has been granted, it's normally valid for three years (or in Scotland five years), during which time construction must begin on site. In other words, use it or lose it.

The most likely situation where a self-builder might need to start from scratch and make a full planning application is where a piece of land doesn't have any planning permission.

As we saw earlier, such a purchase should always be conditional on consent being granted. In such cases it often makes sense to make one full application, rather than muck about doing it in two separate stages – unless, of course, you're up against the clock with a looming deadline to secure the plot, in which case a simple outline application should normally be quicker.

Getting planning on someone else's land

As we saw in Chapter 4, you don't actually need to own a piece of land to make a planning application – you don't even need to get the owner's permission. But you do first have to serve the proper notice on the landowner. Planning permission always benefits the land, rather than the person who applies for it, so if you went around making planning applications on land that you don't yet own, the owners might well rub their hands in delight, and then sell the plot to someone else once consent was granted. So the only way this could make sense would be to tie things up legally in advance, obliging the owners to sell to you in the event of a successful planning application. As described earlier, that can be achieved by exchanging contracts on the land subject to receipt of satisfactory consent, or with a legal 'option' that gives you the right to purchase within an agreed timeframe at an agreed price, or at a price to be determined by a later valuation.

Demolition and replacement

Demolition and replacement of older buildings on attractive sites has become a major source of land for self-builders – a process quaintly known as 'bungalow eating'. So the first question that comes to mind, when you spot a suitably decrepit hovel perched on a glorious plot, is normally 'What size replacement can I build?'

The extent to which the size of a replacement dwelling can be increased will depend on local Council policies, but an increase of 50% by volume above the size of the existing house is widely accepted. So if you happened to find a

Negotiating a size increase

Your discussions with the planners may not look too promising if they say they'll impose a strict condition limiting the size of the replacement development, or will only accept a single-storey design. However, it may be possible to negotiate a size uplift and exceed the Council's size policy. One way you can do this is by demonstrating your knowledge of Permitted Development Rights.

Although, strictly speaking, PDRs only apply to a dwelling once it's occupied, there are two main arguments that can be used to justify building a significantly larger 'one-for-one' replacement dwelling.

You could point out that once the existing property on the site has been demolished and your newly constructed house occupied, you would then be in a position to extend it

bungalow with a 100m² footprint (gross floor area) you may only be able to replace it with a maximum 150m² property, even if the plot is enormous.

Permitted Development Rights

Permitted Development Rights (PDRs) allow homeowners to extend their properties by a fairly generous amount without the need for planning permission – what you might call your 'free allowance'. When it comes to replacing an existing building with a larger one, a knowledge of PDRs can make all the difference.

You're normally allowed to extend a property to the rear by at least 3m over two storeys, and to the side you can build a single-storey extension up to half the width of the original house. You're also permitted to occupy the roof space and to extend it by a generous 50m³, for example by adding large dormer windows.

In most cases you should be able to construct additional outbuildings – sheds, summerhouses, garden offices, swimming pools, ponds, and tennis courts – all without planning permission. Needless to say, the devil is in the detail, and sometimes PDRs may have already been used up or withdrawn. See website for full details.

further using your PDRs. So it would make sense for the planners to approve a bigger property now, rather than later permit a smaller one with a mishmash of outbuildings and extensions added over the years.

Alternatively, you could argue that the existing property could have been bigger had the owner chosen to enlarge it and use up their PDRs (unless, of course, they've already extended it). In fact, there's nothing to stop you doing just that, and actually building a simple extension to the existing old bungalow to use up your full allowance, perhaps adding garages and outbuildings while you're at it. Then the replacement dwelling could be 50% bigger than the enlarged building. The only snag is, you'd then have to demolish what you've just built!

A smarter variation of this theme might be to start by getting consent for the maximum size replacement that the planners will accept. But before demolishing the old bungalow, you could use your PDRs to build a carefully designed extension and a 'games room' outbuilding located a few metres away. Then when you later come to implement the new planning consent you could knock down the original bungalow, leaving the new extension and outbuilding, so they can be incorporated into the new dwelling.

Future expansion

If you want to utilise the loft area as living space or have rooms in a basement, but the planners are keen to restrict the building's overall habitable floor area, it may be possible to turn this to your advantage. By excluding these areas from the total floor area at this stage, you may in future be able to enlarge your living space with a loft or basement conversion (subject to Building Regs). This means your planning drawings should describe loft and basement areas as 'storage' or 'void' space, rather than defining them as 'occupied space'.

It is normally acceptable for homeowners to occupy converted lofts and basements without the need for planning consent under Permitted Development rules, as long as there's no effect on the external size and height of the dwelling. Although some Planning Authorities may take the strict view that creation of extra internal living space in future is contrary to the originally approved plans, it's unlikely that you'd later be prevented from making such improvements, as long as they're built to a habitable standard.

Demolition

Demolition is classed as Permitted Development only for very small residential buildings of up to 50m³ (measured externally, including the walls). As you might expect, properties in Conservation Areas are subject to severe restrictions, and Listed Buildings are sacrosanct. However, demolition will normally be included in your application for redevelopment of the site.

Even where you're free to unleash the wrecking ball, it's essential to plan your next move very carefully. Once the old building's gone, there's technically no obligation on the planners to grant consent since there's now nothing to replace. Although in most cases you would get permission, you can't always be certain. For example, if the site is in the green belt, or for some reason the planners would prefer

Photo: buildstore.co.uk

there to be no house there, they could refuse to grant consent for a new dwelling. So the golden rule is: never knock the old building down until consent for its replacement is in place.

Local development policies

The 'Local Development Framework' outlines the planning strategy for the local area, and directly affects planning applications in the short term. It comprises a number of key documents that outline the Council's key development goals:

LOCAL DEVELOPMENT SCHEME

This is the starting point where you can find out about the Council's planning policies for a particular place or type of development.

DEVELOPMENT PLAN DOCUMENTS

There are three main documents that define areas where new housing can be built, as well as locations where it's prohibited:

- **Core Strategy** – This sets out the Council's general 'spatial vision' and objectives, and can also include broad 'strategic site allocations'.
- **Adopted Proposals Map** – This map illustrates all site-specific policies. It also identifies areas to be protected, such as green belt land, Conservation Areas, and land considered of high landscape value or archaeological merit.
- **Area Action Plans** – These focus on specific locations and areas subject to conservation or significant change. This could include a major regeneration project or a growth area. It outlines protection for areas sensitive to change and aims to resolve any conflicting objectives for areas subject to development pressures.

Local Development Framework

The starting point for judging any application is to see how it measures up against the Council's stated planning policies. These were traditionally contained in 'Local Plans' and 'Structure Plans', which are now being superseded by new 'Local Development Frameworks' (LDFs) that define the Council's planning intentions. In addition, there are new Regional Spatial Strategies (RSSs) that set out a broad, long-

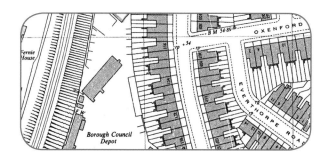

term planning strategy for how a region should look in about 20 years time.

Planning policy

If you want to find out whether a piece of land is zoned for any particular use, or what the Council's specific policy is on new development, the LDF is the place to look. You can view it at the Council offices or online at their website.

Before any planning application is granted or refused, it will first be checked against the LDF for the area, and its policies will be vigorously defended by the planners. Consequently if your application conflicts with these policies, you'll be at an immediate disadvantage, and it's likely to be refused. So before making any application, first find out how your plot has been zoned, to see what the Council's reaction is likely to be.

To maximise your chances of success, it pays to study the Council's planning guidelines. Policies can be interpreted in different ways, so try to couch your proposals in these terms in order to support your application. Refusal notices normally state reasons explaining how your application would have contravened specific planning policies. The LDF should also be checked for possible future developments such as new roads, estates, or industrial parks next door to your site.

Grey areas

Suppose you have a site that's got consent for a block of flats – or perhaps for a pair of semi-detached houses. If you decided to make a fresh application for a detached house with a smaller overall size and a less intensive use, you'd imagine it would receive a warm welcome and sail smoothly through the system. However, where Councils consider that there are already too many big houses and not enough affordable homes in a particular area, they sometimes have policies that restrict such development. As a result, they may refuse your application. In extreme cases there may even be a presumption against any new housing, perhaps where the local population has declined and there's a surplus of existing homes in need of renovation. But Local Planning Authorities also have to take account of national, regional, and County planning policies, and Government pressure to meet housing development targets has been intense in recent years.

Just to make things more complex, planning policies sometimes contradict each other, or can be subject to different interpretations. So nothing is ever 100% certain. Planning policy always seems to be in the process of being reviewed or amended. As old plans approach the end of their term, new draft plans gradually become more influential. This can sometimes work in your favour, for example where an area is no longer zoned as green belt. On the other hand, if your killer argument is that the council approved a similar house over the road to the one you want to build, the Planning Officer may simply turn round and say that it was done several years ago and would now be 'contrary to current policy'.

Land use

It's not just the style and size of the building that planners are interested in. One of the first questions concerns the bigger picture, whether a proposal for a new dwelling would conflict with the any 'designated use' for which the site has already been earmarked.

In certain strategically important districts, land might have been zoned for industrial use or reserved for shops or offices. One of the Council's key objectives may be to provide employment for local people, so this could torpedo any application for a house, no matter how neighbourly it looks. Of course, this can be enormously frustrating, for example where a derelict factory is left to decay because the planners are refusing to grant a change of use.

Land use is a politically sensitive subject because it affects the wider community – changing employment and traffic patterns. The worry may be that a change to residential use could hamper a Council's ability to attract new businesses to provide local employment. Which is a reasonable concern. However, when taken to extremes an obsession with zoning has sometimes led to bizarre planning decisions.

A true story...

Shakespeare's Globe Theatre, rebuilt in the 1990s on London's south bank as a faithful replica, a stone's throw from the original site, is now one of Britain's most loved heritage attractions.

But it had a tortuous planning history. Southwark Council originally agreed to sell the vacant riverside site to the Shakespeare Globe Trust, but the Council Planners subsequently decided that they could not, after all, give planning permission for the new Globe. The reason? Because the site was currently in use as a Council yard where street sweepers stored their barrows and brooms. The North Southwark Community Development Group took the view that a new Globe would be contrary to the needs of the 'traditional working class community of North Southwark'. The Globe Trust finally took the Council to Court for reneging on a previous agreement. Eventually, in June 1986, the High Court found in favour of the Globe Trust. Southwark Council had to pay millions of pounds in costs and appease various claimants. The Globe was opened in 1997 with the Council assuring everybody that their 'commitment to heritage was ongoing'.

[For a more detailed account read Gillian Tindall's, *The House By The Thames*, 2006.]

Planning considerations

The nuts and bolts of submitting a planning application are discussed later in this chapter, but first we need to look at how best to prepare the ground. You're off to a flying start if your proposed development is in tune with the 'big picture' of Council policies. But you're not in the clear yet. There are number of site-specific issues the planners will want to consider.

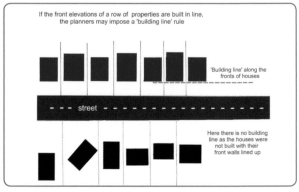

If the front elevations of a row of properties are built in line, the planners may impose a 'building line' rule

'Building line' along the fronts of houses

- - - street -

Here there is no building line as the houses were not built with their front walls lined up

Planning history

The first thing the Planning Officer will do is to dust down any bulging old files and check the planning history for the plot and its immediate area. It's important to research any past planning applications before buying, because there's a potential risk that you could be wandering into sensitive territory, unwittingly reviving memories of bitter battles fought with developers. It's possible that a planning decision may have been overturned at appeal, angering the local community and leaving the planners nursing sore feelings. For all they know, you could now be secretly acting for a developer, the thin end of a wedge bent on flooding the area with ugly boxes. So to help defuse any such suspicions it's worth declaring your position as a self-builder who wants nothing more than to live in peace in the family home. Once a degree of trust has been established, Planning Officers should be able to offer valuable advice about how your plans could be modified to overcome past objections.

Building lines

Planning policy often aims to prevent new houses from being built further forward than the existing ones in a street, insisting that you respect the established 'building line'. But this is only likely to be an issue where you have a plot set in a neatly laid out housing estate or in a suburban street where all the neighbouring houses form an immaculate line. In such cases, it's unlikely that your design will be allowed to step forward. But in many streets the houses aren't particularly regimented. So where they're spaced reasonably far apart, adding a new house that's set forward or back

from the neighbours should be acceptable, as long as it doesn't block highway visibility.

Local character

The character of the local area usually has a major influence on what the planners will accept. A new design usually need to show 'sympathy' to adjoining buildings in terms of their style, shape, height, size, and mass. This requirement to respect adjoining buildings and minimise impact on the environment will clearly be greater near Listed Buildings and in Conservation Areas.

But designs don't always have to be slavishly 'in keeping' with the local area. A deliberate contrast can sometimes work well. Obviously, a radically different design will be more likely to encounter opposition from the local community as well as the planners, and the degree of resistance

Photo: weheart.co.uk

New gabled designs harmonise with existing architecture.

encountered will usually depend on the 'tone of the area'. Unusual architecture is always going to be less controversial on a bland urban site than in a quaint village.

If your heart is set on building an overbearing, alien-looking design, you'll need to start gathering evidence to support your case. Look for clues as to what's been held acceptable to the planners in the past, including recent local extensions. Of course, you don't have to look far in the average British town to find glaring examples of breathtakingly brash, unsympathetic buildings. But just occasionally a contrasting modern steel and glass design somehow manages to fit harmoniously within an historic townscape.

If size is the main issue, clever detailing can help disguise a bulky design. The use of matching materials can likewise

Photos: weheart.co.uk

help gain acceptance. In terms of how much of the site you can cover, some planning authorities are more relaxed than others. This sometimes boils down to how much 'amenity space' must be left after building, and permitted distances between the windows of adjoining houses.

Neighbours

Most plots adjoin other buildings, so what you build will affect existing residents. People are naturally suspicious of change, and when faced with the possibility of a new development next door the first reaction is often to resort to naked NIMBYism ('Not in my back yard') accompanied by the blowing of loud raspberries.

In planning language, the impact of a new development on its surroundings is described in terms of 'local amenity', a slightly vague expression that refers to the quality of the immediate environment. In practice, this means that the planners will be particularly concerned about the appearance of a new building and whether it will detract from the privacy of neighbours and their right to enjoy their homes. Although there is no automatic right to sunlight, if your proposed development significantly reduces such amenities for surrounding properties the planners could have grounds for refusal because of 'overlooking and overshadowing' (see below). Similarly, there is no legal right to a view, and strictly speaking if you obstruct next door's view there's nothing in planning terms that they can do. Except hate you forever.

New houses are regularly built within large gardens of existing homes – a process sometimes referred to as 'intensification of use' or 'backland development'. Some Councils may have policies that restrict this, refusing such applications because of the 'impact on neighbouring properties'. Also in a bid to restrict 'garden grabbing' government

planning guidance now defines gardens as 'greenfield' rather than as 'previously developed land'. But increasing pressure for more housing means that in some areas higher densities are permitted, with spacing requirements between dwellings relaxed.

When it comes to building close to boundaries, there are no hard and fast planning rules, although in some areas there may be minimum requirements. However, most new house designs include a pathway down at least one side for access (handy for wheelie bins) and drain runs. Some Planning Authorities don't allow guttering to overhang a neighbour's land. But since walls must be built centrally on a conventional strip foundation, in practice this means the closest you may actually be able to build the new wall to a boundary would be about 150mm.

Overlooking and overshadowing

Most new homes have to be slotted into an existing street or townscape, so it's hard to avoid some degree of overlooking or overshadowing neighbours. The question is, what

will the planners interpret as acceptable? As a rule, planners and neighbours are likely to consider features such as balconies and first-floor conservatories highly provocative where there's any risk of overlooking.

But when it comes to windows and doors, there's a curious anomaly in the planning laws. A new design may not be accepted with windows or doors that overlook the neighbours. But Permitted Development Rights mean there's no obligation to seek planning consent for inserting new windows and doors into an existing building, even where this that would be unacceptable in a fresh application (unless your PDRs have already been used up or withdrawn).

Highways and access

Most new houses will need a driveway and parking space. This is something of a specialist area, so the planners

normally consult the county Highways Authority, and their opinion can make or break an application.

How busy a road is and the speed of the traffic will clearly be major factors, and the planners will usually want you to adopt the safest solution. If there's already a dropped kerb, this will normally be the required position for the new access, unless you can come up with a safer alternative.

Where a plot is on a busy road, it's unlikely that you'll be allowed to reverse out, straight into the path of oncoming juggernauts, so in order that vehicles can leave the site facing forwards, sufficient turning space must be provided on site.

Drivers must have good visibility in both directions as their

Before and after sightline improvement

vehicles cross the pavement to enter or exit your site. This can be judged by sketching a triangular 'visibility splay' on your site plan either side of the drive where it meets the road. The two triangles represent the areas of land that must be kept clear of obstruction. These also show the degree of visibility or 'sight lines' for pedestrians and passing traffic. In some cases, the sight lines may be restricted by trees, boundary walls, or due to a curve in the road.

As we saw in Chapter 4, a neighbour's garden may occupy one of these triangles, and they may be intensely fond of their hedge! But the fact is, without adequate visibility it may be impossible to get consent to develop your plot. So a generous incentive may be required.

If your site is in a remote location, it may only be reachable via long unmade tracks or up a steep slope. Where lorries are unable to get access to deliver materials, all your supplies may have to be moved manually – a back-breaking task that can significantly add to your costs. Concerns about unmade roads are sometimes quoted as additional reasons justifying refusal of planning, although for a single new dwelling this would be unlikely. However, it may be possible to negotiate a compromise and improve the track with a simple hardcore sub-base that's suitable for low-volume traffic, rather than going the whole hog with an expensive tarmac job.

Garages, parking and driveways

Most Councils like to see a minimum of two parking spaces for a new family home, but policies vary from area to area. In some urban districts street parking may be acceptable. But normally, the busier the road, the greater the requirements for off-street parking and turning space will be. Although there is usually no planning requirement for a garage, for detached houses a double garage adds a certain degree of status. Of course, you could always compromise with a simple carport, but unless well designed these can look cheap and nasty.

Whether the garage is attached or detached, it should complement the house, perhaps designed in traditional brick or tasteful barn-style timber cladding. Integral garages rob a house of space, but can be converted to accommodation at a later date. It's probably true that planners are more likely to accept detached garages in rural areas if they reflect the local character, whereas attached or integral garages are the favoured solution in urban areas.

Negotiating with the planners

For a successful application it helps to speak the same language as the planners. Planning Officers are trained to take an impartial approach so that they're seen to be fair. So if your brilliant design receives a less than enthusiastic response, don't be downhearted. Misunderstandings sometimes arise because planners are governed by a set of rules that to the rest of us may not seem entirely logical. So forget about hidden agendas and conspiracy theories. To boost your chances of a successful result the best policy is to *understand* these rules.

When it comes to applying planning policies to specific developments, there is often scope for interpretation. There may even be varying shades of opinion between different Planning Officers at the same Council. This is because the issues raised by your application have to be judged not just in relation to Council policies (that may be evolving over time) but also in the light of objections from neighbours and Parish Councils. Add to the mix the fuming pit of local politics, the risk of setting undesirable precedents, and in some cases a distinctly murky site history, and you can understand how perceptions can often vary even between planning professionals.

The good news, however, is that the complexity of the system can sometimes be made to work in your favour, if handled intelligently. To get consent for your new home, you need to be willing to compromise, even if it means ditching those cherished roof gargoyles you had your heart set on. You need to think strategically and be prepared to lose the occasional small battle in order to win the war.

The initial meeting

Whether you like it or not, the planners are absolutely key to the success of your project. So before submitting a planning application, it's best to arrange a meeting to run through your proposals. Ideally the planners should be consulted before you buy the site, and then again at the design stage, so that you're fully aware of any concerns before final drawings are prepared. Having presented your ideas and talked through the drawings, be sure to listen carefully to what they have to say. They may, for example, require the preservation of certain trees or hedgerows on the site.

Your application may be life and death to you, but to the planners it's just one of many. There's an art to approaching these key discussions in the right way – see below. It's important to make it clear at the outset that you're going to deal fairly and honestly with them to win their trust. Once you've spoken with a Planning Officer you should come away with a pretty clear indication about the likely outcome of making a formal application, although there's no guarantee.

Some Councils are more customer-friendly and helpful than others. Strictly speaking, their only duty is to properly consider an application and to make a decision – there's no obligation to discuss proposals in advance with applicants to help them get approval. But it's unlikely that you'll be met with a blank response, because the planners know it saves time later by weeding out daft projects before they hit the drawing board. It's best to avoid making 'blind' applications, with little or no indication of the chances of success. Where cases are likely to prove challenging, it's essential to get the advice of a professional designer who's familiar with the local planners' policies and concerns.

Negotiating tactics

It can sometimes pay to ask for more than you want. By making an initial application for a larger house that's turned down by the planners, you may then be able to substitute one of slightly reduced size that's acceptable but still bigger than they would otherwise have accepted. Cynics might say this helps placate the planners by letting them bite a few chunks out of your initial plans, slimming it down to the size you actually wanted.

Another tactic popular with developers is to submit two simultaneous applications for the same site, one that's over-

How about a low profile, green roof, 'Teletubbies' design?

Photo: velfac.co.uk

sized and contentious, alongside a separate application that in comparison looks relatively benign. This can save time, because if the first one's rejected the second one is already in the system.

Even if both are refused and it goes to appeal, the focus tends to be on the difference between the two applications, rather then whether it should be built at all. But to do this requires expert advice, otherwise you could simply end up alienating the planners. Housing developers sometimes negotiate valuable trade-offs by demonstrating that their proposals will improve local amenities. A similar approach could help swing things in your favour.

Offer to accept conditions

Getting planning approval with a few restrictions is still a massively better outcome than a refusal, so if you're willing to accept a development subject to some conditions it can be a deal-maker.

For example, the planners might be reluctant to grant permission if they think you would later use it as a foot in the door, leading to further development. To put their minds at rest, you could volunteer to accept consent with conditions limiting your freedom to extend in future without planning consent – ie taking away your Permitted Development Rights. Similarly, on a large rural plot you could accept a condition that defines most of the garden as 'paddock', or sign a legal agreement binding on future owners that the land will not be built upon.

Consider the threat of appeal

A negotiated solution is normally far better than having to go to appeal. Appeals take up a lot of time and only about one in three are successful, so they should really only be a last resort. But they also take up a lot of the planners' time and deprive them of control. So where a Planning Officer is wavering, a subtle reminder that you would reluctantly be prepared to appeal against a refusal can be a useful negotiating tool. However, be careful not to adopt a threatening manner. It can be a good idea to mention any similar past refusals that were subsequently overturned on appeal. This could help swing your development as the 'lesser of two evils'.

Lobbying

For major developments, lobbying is a well-established tactic, widely employed at town halls as well as at Westminster. Some controversial development decisions, such as the third runway at Heathrow, have allegedly been influenced by lobbyists acting for commercial interest groups. Even members of the House of Lords have been reprimanded for acquiescing in such antics. In comparison, lobbying your parish council and district councillors is small beer. It is perfectly legal to ask elected representatives for their support as long as no bribery is involved. They should at least be able to tell you the names of the members of the planning committee, so you can then present your arguments in the best possible light.

Smoothing the path

Large numbers of objections from local residents can sway a planning decision against you, so it's usually a good idea to 'charm the neighbours' in advance about your proposals. It also pays to consult the Highways and Environment agencies and try to pre-empt and resolve any concerns they may have. This all helps smooth the path of your application, giving the planners fewer reasons to justify a refusal.

Precedents

Take a good look around the local area and take photos of any similar developments that have recently been approved. Producing such evidence can make it harder for the planners to argue against your proposal. Even if the Planning Officer declares that they 'will not be held hostage to past mistakes', this sort of evidence can be very powerful if the case goes to appeal, as the Planning Inspectorate take note of local precedents.

But precedents can also work against you. As well as looking back in time to see whether something similar to what you want has already been granted consent, the planners also need to gaze into the future. They have to consider whether granting approval to you now could set an undesirable future precedent, unintentionally giving someone else strong evidence for doing something that they're trying to restrict. For example, if they're prepared to compromise by allowing you to have one less car space on site, in future this could be seized upon by aggressive developers keen to minimise parking and squeeze in extra houses. The only way round this is to demonstrate that your case is special, and come up with compelling reasons why it wouldn't be comparable to other sites in the area.

Avoid contentious applications

If an application is clearly at odds with Council policies, no Planning Officer is going to recommend it for approval. The easiest way to get planning consent is to make sure your application doesn't fly in the face of established policies and that it contains all the key criteria for approval. If it looks like the planners will object to the size or scale of your design, then it might make more sense to seek approval for a smaller house, and then extend it later under PDR rules.

It also helps to make the right kind of application. Where the broad principle of development is at stake, an Outline Planning application should be made, so as not to waste everyone's time.

Make it clear

An amateurish or confusing set of drawings can harm your chances of approval. Planning drawings don't need to be particularly detailed compared to those for your Building Regulations application.

It can be tempting to save money by producing drawings on your home PC. With time and practice these may be fine. But if they look unattractive it could be a false economy, as they may not persuade the planners in the way that a set of well drawn, professionally produced plans can.

Housing developers often commission artist's impressions of the front elevation and street scene to help convey an attractive image. A photo mock-up or 3D model can be very helpful to positively illustrate what you're proposing. Remember that planning committees are populated by local councillors, who are usually lay people who may be 'plan blind'.

Don't get angry

The process of negotiating a planning application can be extremely frustrating. Sometimes it feels that the Planning Officer's main goal in life is to obstruct your every move. But the fact is they're simply doing their job, and haven't 'got it in for you' personally. They have to interpret the rules and balance a lot of factors beyond the immediately obvious. Also, it's important that they're seen by all parties to be even-handed and objective, which is why they tend to adopt a fairly guarded facade. It's a tough job, because no matter what decision they come to it invariably upsets somebody. There's always someone who wants to take it out on them personally and subject them to a hostile rant. Some interview rooms even have panic buttons.

If you get angry or try to intimidate them it will backfire. Losing your temper is pretty much guaranteed to sabotage your application. It's essential to control your feelings and be

polite. Don't bluff or try to trick them into agreeing something. They've seen it all before.

At the end of the day, a junior Planning Officer trying to be nice can be more trouble than one who tells you straight that they'll reject your design. There's nothing worse than being led to believe your application will be accepted, only to discover a couple of months later that it's about to be refused.

Consider using a planning consultant

With an ambitious application, it can make sense to appoint a specialist planning consultant to handle your case. Many consultants are former Planning Officers – 'gamekeepers turned poachers' – which should make them well qualified to negotiate the system. However, where they used to work for the same Council it's possible there could be some simmering resentment amongst former colleagues that could actually work against the application.

Professional planning consultants can be tracked down by searching the planning register (online) under the category of 'agents'. These are normally either architects or consultants. If you limit your search to applications that were approved, you should be able to identify those with a consistently successful track record.

Delegated authority

Where an application is straightforward, it may be decided under 'delegated authority'. This is a streamlined process that applies in cases that are 'black and white' – either clearly uncontroversial or obviously unacceptable. This cuts out the need for judgement by a formal planning committee, thereby reducing red tape. Instead, a senior Planning Officer will act as 'judge and jury'. However, this may not suit you if it appears that the Officer in question has set their mind against you. You could try writing to a senior planner or lobbying your local councillor to request that your application is taken to committee. Alternatively, you could request a site meeting, where you are given a further opportunity to argue your case.

Submitting your application

No matter how straightforward your application, you can never take getting planning permission for granted. Even if the Planning Officer is your new best buddy, this is the stage over which you have least control. The planners will need to consult several other organisations and individuals, giving each an opportunity to influence the outcome. A decision is supposed to be reached within eight weeks of the application being registered, but in more complex cases where there are several competing interests to balance it often takes longer.

In some cases the planners may request a time extension because they need a bit longer to consider your application, in which case they may notify you that the time limit has been extended to 13 weeks. There's no point alienating them by being awkward about this. But no matter how long the process drags on, no building work should be started until the appropriate planning consent is in place.

Do your homework

Where an earlier application has been refused, before submitting a new application it's important to study the planners' reasons for refusal. It's unlikely that all the relevant information will be available online, so it may be worth visiting the Council planning offices to have a good old nose through bulging case files.

Objections submitted by neighbours and Parish Councils often raise non-planning issues that shouldn't be taken into account. But make a note of any comments from Highways and Drainage Authorities, as these are often pivotal to the success of an application. Armed with this information you should be able to come up with a new design that overcomes the killer issues that sank the earlier application.

Where a refusal is now several years old, you might get lucky, because changes may have taken place in the meantime that work in your favour. For example, some of the more hostile local objectors may have moved out of the area. Or Council design policies might have changed, and Government pressure to release land for housing may have influenced local policies on green belt zoning. Perhaps, by relieving traffic, a new ring road will have resolved former deal-breaking issues with regard to car access and parking.

What drawings to submit

You need to submit a full set of scale drawings clearly showing the work you propose to carry out in metric measurements. A typical set of plans would comprise two or three sheets of A2 or A3 paper drawn to 1:50 or 1:100

Design and Access Statement (DAS)

Whether your application is for Full, Detailed, or Outline Planning, it will normally need to be accompanied by a 'Design and Access Statement'. This is a short report that supports planning applications and explains the key design principles. Most of this detail will already be shown on the drawings, but it gives you an opportunity to further explain and justify any aspects that might prove contentious. Less information is required for OPP applications.

The following information is required in a DAS:

■ **Amount** – Details of how much development is proposed, *ie* the number of buildings and their floor space.
■ **Layout** – Detailing the way the buildings are to be positioned on site, and how they relate to adjoining open spaces and other nearby buildings.
■ **Scale** – The height, width, and length of all buildings.
■ **Access** – Explain how all users have equal and convenient access arrangements for buildings, open spaces, and local transport. Access for emergency services should also be explained.
■ **Landscaping** – Full details of both hard and soft landscaping, including how the landscape will be maintained in future.
■ **Appearance** – A summary of the architecture, facing materials, colours, and lighting etc.
■ **Context** – An assessment of the impact the development will have on the surrounding area, physically, socially, and economically. This is clearly more onerous for larger developments.

scale. Each drawing needs to be clearly labelled with its title (*eg* 'Front Elevation'), a unique number (*eg* Dream House 001A), the scale, the address of the property, and the contact details for yourself or your architect. You will need to run off at least four sets of drawings, plus a few spares to keep for yourself.

In addition, a 1:1250 scale OS plan must be provided showing the site area outlined in red pen, and any land that you happen to own next to the site edged in blue. OS maps are copyright and can purchased online or from the Planning Department, although your architect should have a licence to supply as many as needed.

Applications sometimes fail because of 'over building' – where it appears you are trying to fit too large a house on too small a plot. A scale plan of the site can help address this issue by showing how the house and garage fit on to the site whilst leaving space for access, drives, and gardens. If the building you've designed is especially prominent, or in a Conservation Area, the planners may request extra information. Your architect may be able to provide 3D computer models, 'artist's impression' photo images, or a scale model.

Although cross-section drawings are not normally required for planning applications, they can be useful for a sloping site to clarify details of retaining walls, car access, and how the design affects neighbouring properties.

The drawings you need to submit therefore comprise the following:

1. Proposed elevations

1:50 or 1:100 scale

The front, rear, and side elevations showing what the finished property will look like.

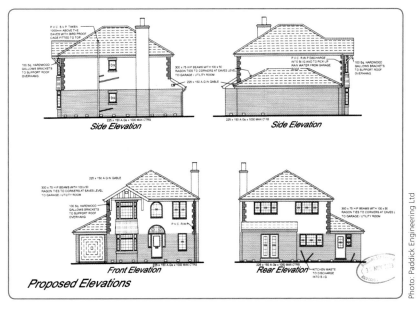

Proposed Elevations

Photo: Paddick Engineering Ltd

2. Proposed layout plan

1:50 scale

The floor layout of each level with the name of each room clearly marked.

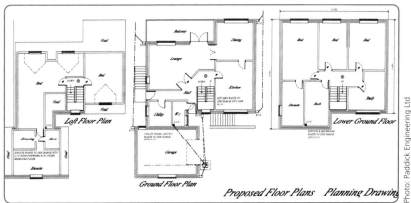

Proposed Floor Plans   Planning Drawing

Photo: Paddick Engineering Ltd

3. OS 'Location' or 'Area' plan

1:1250 scale (not less than)

A copy of the Ordnance Survey plan of the street with your site outlined in red.

SITE LOCATION PLAN
1:500

Photo: F&P Architectural Services

4. Site or 'Block' plan

1:200 or 1:500 scale

It may be acceptable to combine this with the location plan. It is a 'close-up' plan of your plot showing where the new house is going to be built. The position of site boundaries, roads, drains, trees, and adjoining and any other existing buildings should be clear. It shows how the proposed design will fit into the plot and how it relates to your immediate neighbours and the boundaries.

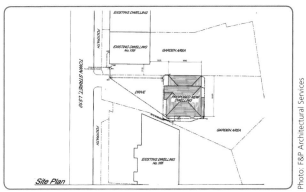

Site Plan

Photo: F&P Architectural Services

What you write on the plans is crucial, because once consent has been granted everything shown on the approved drawings legally forms part of the permission. So if you've written 'natural Welsh slate' on your drawings but later you can only afford cheaper artificial slate, the planners will need to approve the change. This may explain why it's now a common planning condition for materials to be approved before commencement of work on site.

However, the planning drawings don't need to include most of the technical construction detail that will later be necessary for your Building Control application. So it's best to avoid cluttering them up with too much information. Nonetheless, it's important to have the technical design worked out in detail as early as possible, otherwise if you subsequently need to make any major design modifications to comply with the Building Regs a revised planning application could be required!

The application

Along with your drawings and a completed application form and payment, it's a good idea to include a covering letter. This provides a good opportunity to put on record any important correspondence or conversations you've already had with Planning Officers (refer to them by name). This way, if your application ends up being allocated to a different Case Officer, the progress you've already made hopefully won't be overlooked.

You can deliver your application by hand, by post, or online in pdf format (as well as paying the fee online). Once submitted, it will be checked to ensure that it includes all the necessary drawings and documents as well as the right payment. Within a few days you should receive an acknowledgement letter confirming the Case Officer who'll be handling the application and a reference number.

The Council will then post a formal notice near the plot

boundary inviting comments (*ie* objections), and in some cases will write to the neighbours and anyone they think should be aware of the application.

It's best not to start chasing progress too soon. There will be nothing to discuss until comments are received from the parties being consulted, so it's advisable to wait about four weeks. Once you do make contact, if any problems come to light, try to arrange a meeting as soon as possible. Planning departments have turnaround targets to meet and will want to get problems sorted out quickly or, if there's no scope for compromise, to swiftly reject it. If the Case Officer assigned to your application seems inexperienced and unfamiliar with your area there's nothing to stop you requesting a meeting with a Senior Planning Officer.

Objections

A planning application should never be rejected simply because the neighbours have kicked up a fuss, not relishing the prospect of noisy building works, or because there's a lot of huffing and puffing from local interest groups. To be relevant, objections must relate to planning criteria. Anything else is strictly irrelevant. But however invalid in planning terms, local opposition can still delay getting approval. In the messy real world of local politics, things

'Big brother's watching you' – New house overlooks neighbours.

What to include with your Planning Application

The application should include a covering letter and payment along with:

A minimum of four copies of each of the following:
- Application form
- Plan drawings
- Elevation drawings (front, rear, and side)
- OS location plan, with boundary marked in red
- Site plan
- Design And Access statement
- Flood Risk Assessment

– plus a completed certificate of ownership (or, if you don't own the site, a certificate of notification).

NB: Some Councils require more than four copies, and additional copies may be needed for Conservation Area applications.

See website for an example of a completed application form.

aren't always perfectly black and white. If there are a lot of vociferous objections and local councillors get involved, Case Officers can feel pressurised, making them reluctant to adopt a bold position recommending something they know will be highly contentious.

So it's often a good idea to neutralise objections before submitting your application. You can do a lot to help mollify neighbours' concerns, some of which may be unfounded. So nip them in the bud by visiting neighbours well in advance, persuasively talking through your plans, preferably with the aid of an attractive sketch showing the finished house blending in nicely with its surroundings. Otherwise the first the neighbours will learn of your plans is by stumbling across a planning notice stuck to a nearby wall or getting a formal letter from the Council. Without seeing any drawings, the tendency for most of us is to fear the worst and to immediately regard it as a threat. The first reaction may even be one of anger, or feeling insulted that no one had the courtesy to let them know in advance.

So taking an opportunity to meet them should help, and it will put them at ease if they can see for themselves that you're going to be nice neighbours. All the same, it's worth pointing out to them that should your application be refused, the alternative is likely to be a housing developer pushing for a higher density development, with unknown future neighbours. In some ways this is a similar psychological tactic to courteously inviting neighbours you don't especially like to your party (knowing they'll decline the offer), because it will make them less likely to complain about the noise.

Possible objections from Parish Councils and other local interest groups (who will be consulted when an application is made) can similarly be pre-empted. They may be willing to invite you to present your design for discussion before an application is made. Taking the trouble to do this can make all the difference if it helps neutralises people's worst fears.

Strategy

As your application slowly winds its way through the system, it may become evident that there's a distinct lack of enthusiasm and support. By monitoring progress, you can if necessary submit additional evidence to back up your case, and address any concerns that arise before any formal decisions are made. By demonstrating that you're willing to modify the design to accommodate the Planning Officer's suggestions, it should help placate objecting neighbours and councillors.

So at this stage, it's worth considering whether there are aspects of the design that you might be prepared to alter or sacrifice as a compromise. Perhaps you could offer to reduce any overlooking of the neighbours by moving windows, planting trees, or changing the building's position on the plot. You may be able to keep the overall design and layout by limiting the property's impact, for example by reducing roof height, and adding tasteful 'cottage dormer' windows.

It's also worth bearing in mind the power of the magic word 'Eco'. Government and major housing developers have cottoned on to this as a means of helping to swing contentious planning applications. Similarly, by stressing your design's green credentials it may help tip the balance, galvanising the planners into supporting a building that would otherwise be too outlandish.

If after all these tactics, the mood is still unremittingly bleak at Planning HQ, you probably need to be prepared for refusal. But in a final roll of the dice, if you can drum up letters of support from the neighbours it may yet swing it.

Incidentally, in discussions with planners, it's always advisable to get any key comments in writing, because anything they say informally can be disregarded when it comes to making the final decision.

The Committee meeting

A few days before the Planning Committee is due to sit to consider your application, your Case Officer will normally prepare their report recommending approval or rejection. However, as we saw earlier, in straightforward cases decisions can be made instead by the planning staff under 'delegated authority'.

Planning Committee meetings are open to the public, and it's advisable to attend meetings in person. The Case Officer will kick off with a verbal outline of the written report, before the matter is opened up for discussion. Councils often allow applicants and their agents, as well as any objectors, to make brief presentations. Some architects are better at presenting a more persuasive case than others, but if possible ask your designer to attend with you. Committees sometimes defer a decision until receipt of additional reports or pending further enquiries. This is not necessarily bad news – look at it as a second chance to get your scheme approved.

The Planning Committee decision

Approval

If consent is granted, congratulations. But there may yet be a sting in the tail. Planning permission always comes with conditions attached. Construction must not only be carried out in accordance with the approved drawings, but must also adhere to the stated conditions. Otherwise you'll be in breach of the consent.

Most conditions are fairly routine, such as the stipulation for work to begin within a stated period (normally three years). Another standard condition is for the external materials to be approved in advance in writing by the Council, something that frequently gets overlooked and instead has to be approved retrospectively. This is likely to be a more onerous condition where buildings are located in Conservation Areas or are Listed.

Other conditions can be considerably more costly and demanding, such as the need for contaminated land reports or archaeological investigations. If you firmly believe a condition is unreasonable, you can appeal against it – see below. However, on appeal the application is reviewed in full and this can potentially result in new conditions being imposed that may actually be worse. They could even reverse the consent and issue a refusal, but should notify you in advance to give you a chance to withdraw the appeal.

Where you subsequently need to amend the building's external design (or alter the approved accommodation) revised drawings should be sent to the planners for approval. Minor changes can normally be passed by the Council as 'amendments'. An entirely new application is only likely to be required where you want to increase the building's height and size.

Refusal

If your application is refused, you could either submit a revised application, or go to appeal. You are allowed to submit a new application with amended plans within 12 months with no additional fee (assuming it's the same site and applicant).

Refusals usually cite several reasons and these provide the key to formulating an acceptable revised design. The revised design should also take into account relevant comments made by the Planning Committee. However, if it's then refused a second time you don't get another free go. Perhaps the most galling outcome is where the Planning Officer recommends acceptance of your plans, but the Committee ignores their advice and decides to refuse it – for 'political reasons'. All you can do in the case of a shock refusal is to prepare for your subsequent fight-back.

Planning appeals

Going to appeal involves jumping through a lot of hoops.

Photo: Grange fencing

If your application is refused, you have three months in which to appeal. The appeal process typically takes between about six and nine months.

Most appeals are made in writing ('written representation') and decided by a Government-appointed Planning Inspector from outside the area. However, the success rate is low, so it's important to take professional advice from an architect or planning consultant on whether it's worth doing. Normally a reapplication is the better option.

An appeal against refusal starts by making a written submission to the Planning Inspector explaining why you, as the applicant, believe the reasons listed on the Council's refusal certificate are unreasonable. Anything else you write is irrelevant, so you have to restrict yourself to this one area. The Inspector will send a copy to the Council Planning Department, who in turn have a period of four weeks to submit their own views, a copy of which will be sent to you. You then have a further two weeks to submit your reply, a process known as 'exchange of statements'. Councils sometimes don't bother to respond, perhaps because they have nothing to add to their original reasons, in which case the appeal process carries on without them commenting.

Things go quiet for the next two or three months until the Inspector writes to confirm a date for their site visit. This is an important stage, with all parties normally present on site. As the applicant you are allowed to bring along professional advisors, such as your designer, and someone will be present to represent the Council. Neither side is permitted to make any further submissions. Even if you tried, the Inspector would not be at liberty to discuss them.

The point of the Inspector's visit is for them to get a feel for the site in the flesh and how it relates to your refused application. You and your architect are there simply to answer questions that relate to the specific reasons given for refusal, such as concerns about overlooking or visibility splays.

Because of the considerable tension that will have built up over the last few months, site visits can be an emotional powder keg, with real potential for confrontation. But handling it well can present an opportunity to gain the Inspector's support, boosting your chances of success. Above all, it's essential to control any feelings of frustration and anger, and to respect the fact that the Inspector is in charge. So resist the temptation to sound off and launch into a long tirade about the outrageous injustice done to you. And avoid getting into abusive exchanges or punch-ups with the Council representative. If someone representing the other side gets stroppy, it hands you a perfect opportunity to demonstrate your calm, grown-up attitude in contrast to their angry spluttering. This should reflect well on you and may help your case, making the quality of your judgement appear far superior.

The best approach is to treat the Inspector as you might a judge. Respect the fact that they're in charge of the show, and respond to their questions calmly, perhaps subtly getting some additional key points across with your replies.

Finally, about a month after the site visit, you should receive a letter confirming the Inspector's decision. This is final. If you've won, the findings act as your planning permission. If not, then you have to accept that this is the end of the road, and instead marshal your energies towards redesigning your application so that it's more in accordance with the Council's stated criteria.

Unauthorised development

Refusal of your application can obviously be immensely frustrating. But going ahead and starting to build a house without consent is a dangerous game, as well as being illegal. The planners have draconian powers to fine you or demolish unauthorised works (at your expense).

However, there is a potential loophole. Should the authorities fail to challenge any unauthorised development within four years, it can potentially become immune from enforcement. During this period, the building must go unnoticed and unopposed by the planning authority. Thereafter the owner of the property can apply for a Certificate of Lawful Use or Development, which effectively grants retrospective Planning Permission (with conditions). However, it's unlikely that building work would go unnoticed if you've just alerted the whole world to your site with a failed application.

Nor is this loophole quite the 'get out of jail free' card it may seem. Over the years, all manner of wily dodges have been attempted, such as concealing a new house behind a gigantic mountain of straw bails, and even building a new residential property inside an enormous aircraft hanger-sized barn. Neither of these attempts succeeded, and both properties ultimately had to be demolished. The four-year rule also applies to a change of use, for example where commercial buildings are converted to residential. For other changes, such as the breach of a planning condition, the relevant period is a lengthy ten years. And when it comes to Listed Buildings, there is no time limit for unauthorised works.

This 3 bed bungalow was built illegally over 6 years under an open barn camouflaged by straw bales. But walkers alerted the Council who ordered the building to be demolished.

8 THE BUILDING REGULATIONS

Photo: oakwrights.co.uk

Once you've got planning consent (either full or detailed) you are legally allowed to build your new home. But whether it's possible to physically construct it to an acceptable standard is another question. So befoe starting work on site you will first need to submit a Building Regulations application.

The Building Regs are essentially about safety – the kind of stuff that you'd want to get right anyway, like bathroom fittings that don't leak, and wiring that doesn't kill. Their objective is to enforce minimum standards for such things as fire protection, safe access and drainage, as well as making sure buildings are structurally sound and weathertight. Increasingly, however, the emphasis is on meeting energy efficiency standards in order to conserve fuel and power.

It's very much in your own interest to ensure your new home complies with the Building Regs so that the resulting structure can be shown as being safe, and will therefore be mortgageable and saleable in future. Only if the building work has been carried out to the satisfaction of the Building Control Surveyor will you receive official Building Regs approval, in the form of a 'completion certificate' or 'final certificate' upon completion of the build.

But whose job is it to organise this and liaise with the people at Building Control? The fact is, as a self-builder the buck stops with you, unless you're employing a firm of builders as main contractors to whom the task can be delegated. It's important to remind them of this at the outset – and write it into their contract.

The 'approved documents'

Practical guidance showing how to comply with the Building Regulations is contained in the 'approved documents' issued by the Government. As far as your build is concerned, these are the 'bible' and are legally enforceable. The trouble is, the approved documents aren't written in particularly user-friendly language. Some would say they're one of the most complex sets of rules ever devised. They're also revised on a fairly regular basis. As a result, a certain amount of interpretation is sometimes required.

Architects and designers should be aware of the latest requirements, automatically incorporating them into their plans and specifications. But the people who really need to know, the builders doing the job, tend to absorb the practical changes over time as they are introduced on site.

Fortunately there are some helpful summaries of the Building Regs complete with useful diagrams, such as the *Standards Manual* of the NHBC (National House Building Council) and *Builders Guidance Notes*. Both also contain useful span tables specifying the required sizes of timber joists and rafters – see website.

The 'approved documents' comprise:

A Structure
B Fire safety
C Site preparation and resistance to contaminants and moisture
D Toxic substances
E Resistance to the passage of sound
F Ventilation and condensation
G Hygiene, hot water safety and water efficiency
H Drainage and waste disposal

J Combustion appliances and fuel storage systems
K Stairs, ramps and guards (protection from falling, collision and impact)
L Conservation of fuel and power
M Access to and use of buildings
N Glazing – materials and safety
P Electrical safety
(N.B. Part N has now been incorporated into Part K).

If you live in Scotland or Northern Ireland, guidance is provided in the form of technical handbooks. The Scottish Building Standards comprise seven documents ('Section 0' to 'Section 6'), while Northern Ireland has Parts A to V. However, the nitty-gritty areas of concern are largely the same as those for England and Wales. The main difference is that in Scotland you can't start work until a 'building warrant' has been issued, whereas in England and Wales you're free to start building before plans have been approved. Also, in Scotland the fee is based on the construction cost, rather than the flat rate that applies elsewhere, and the completed house cannot be occupied until a 'habitation certificate' has been issued. The rules in Northern Ireland are similar to those in England and Wales. The full documents can be accessed at the website.

Site inspections

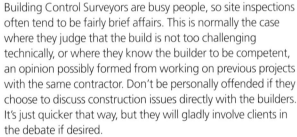

Inspections are carried out by Building Control Surveyors at a number of important stages to enforce minimum standards. It's not the job of Building Control to supervise the works on your behalf or manage your builders. For that you'd need to privately appoint your own surveyor or architect. However, they may be able to help spot any potentially serious problems before they become major disasters.

If the building work doesn't look particularly neat, this also isn't their concern as long as it complies with the Building Regs.

The main contractor is required to notify Building Control in advance of each key stage by making an inspection request. Two days before the work starts on site Building Control must be notified in writing using a commencement notice form, and the inspection fee paid. Subsequent stages can normally be made by contacting your local Building Control office with just one day's notice, usually by phone, other than for the final inspection where five days' notice is now commonly required. Where requests are received before about 10:00am inspections can usually be carried out on the same day, which can help avoid having builders standing around waiting for a visit before they can carry on with the job.

Work must be left exposed for inspection before covering it up and continuing. Should you fail to do this, you may later be required to break open and expose parts of the structure for inspection, which in the case of concrete floors and foundations can make things a little messy. So if the Surveyor doesn't show up within the time limit don't be tempted to cover up and press on regardless – call them!

Building Control stages for inspection	Notice reqd
1. Commencement of the build	2 days
2. Excavation of foundation trenches (before being filled)	1 day
3. Concreted foundations before covering up	1 day
4. DPC level reached	1 day
5. Drains laid but not backfilled	1 day
6. Drainage testing	1 day
7. Completion of construction	5 days

These stages are just the bare minimum, where you must legally provide notice. More than one inspection may be carried out for each stage, and additional spot checks are often required. For example the following works would also normally need to be checked:

■ Structural work – before covering any structural timbers, steelwork or concrete.
■ At wall-plate level (when the walls are finished and the roof about to start).
■ New boilers, flue liners, heat producing appliances, hot-water systems.
■ New waste pipework.
■ Fire and sound insulation, fire doors and stairs.

Building Control Surveyors are busy people, so site inspections often tend to be fairly brief affairs. This is normally the case where they judge that the build is not too challenging technically, or where they know the builder to be competent, an opinion possibly formed from working on previous projects with the same contractor. Don't be personally offended if they choose to discuss construction issues directly with the builders. It's just quicker that way, but they will gladly involve clients in the debate if desired.

Most Building Control Surveyors have a wealth of site experience, often haling from the building trade, as poachers turned gamekeepers. Others have a more academic background, with less site experience. Either way, the overwhelming majority of Building Control Surveyors are informative and helpful.

At the end of the job the completion certificate can be issued, but not before the final inspection has been carried out and all relevant certificates have been received. However, you'll need to specifically request it. This is a valuable certificate, so keep it safe along with the planning consent documents, since it will be required when you come to sell or remortgage.

Policing and penalties

If you build something that does not follow the approved drawings or specification, and the Building Control Surveyor decides that it contravenes the Building Regs, you may have no choice but to demolish and rebuild it. Building Control Surveyors have considerable powers to halt and condemn work. Contravention can lead to being fined or even being sent for a short holiday care of HM Prisons. They have the right to enter sites at all 'reasonable hours' and obstructing them from doing their job is in itself an offence that can result in prosecution.

Action is normally taken against the main building contractor, but the local authority can alternatively serve an enforcement notice against owners, demanding the removal of anything that contravenes a regulation. Or they may force

you to rebuild the offending item so that it fully complies. If you refuse, they can employ other builders to take down non-conforming parts of your house, for which they will send you the bill.

There are a few areas of building work where you don't need to apply for Building Regs consent, for example non-habitable outbuildings, carports, and detached garages (up to 30m^2 floor area), but with an application for a new house it's best to assume that all structures require consent unless notified otherwise.

Local authorities do actually have the legal right to relax parts of the Building Regulations if they believe that the requirement is unreasonable in a particular situation. This is known as 'varying the provisions'. Just don't bank on them being too keen to do this!

Choosing materials

The materials used obviously need to be safe and fit for their purpose, as well as looking right. Where possible, aim to build with better quality materials than the minimum standards required for Building Control, shown in the drawings. Your designer and builders should be familiar with the materials available, and the detailed specification will normally state the quality required (eg 'plywood shall be in accordance with BS 1455').

Products should normally carry a recognised quality branding such as the BSI kitemark (British Standard Institute)

or BBA approval (British Board of Agreement). Products with 'CE' marks are also acceptable, as it means they comply with European standards. If in doubt, ask Building Control what they will accept. The biggest problems with materials usually relate to the way they have been installed. If a dispute later arises with the builders, product manufacturers normally have product advisers available who can help clarify matters.

Building near drains

Some plots in built-up areas will have existing drains buried within them. Having identified any underground drain runs, it may be possible to revise the design so as to avoid building over them, thereby saving a lot of expense and hassle. Maybe you can shift the position of the house over a bit, or make it slightly smaller. If not, then the first thing is to check whether the drains are private or public.

Private drains are the ones nearest a house that don't serve any other buildings. These often join up with other people's drain runs to become 'shared drains' (also known as 'private sewers'), with ownership shared between the various users. Here, all owners are jointly responsible for repair costs (regardless of whose land a blockage occurs in). It will normally then connect to a public sewer, usually located under the road. But sometimes you come across public sewers in gardens – for example, those serving Victorian terraces often run parallel to the rear of the terrace, across all the back gardens.

To build over a shared drain or private sewer you need to demonstrate that it will be suitably protected from any damage, and if possible obtain the prior consent of all the joint owners. But building over a public sewer can be a major undertaking. Even building within 3m of a public sewer requires official consent (the precise distance will depend on the requirements of the utility company) and in some cases may be prohibited. This is because constructing new buildings near sewers increases the pressure on underground pipes, potentially causing the sewer to collapse. It's also likely to restrict future access to the sewer pipe. So leaving a space of 3m either side of the centre line of the underground pipe means you've got a gap 6m wide in total. Should access later become necessary, this should provide sufficient working space to allow a JCB to dig up the drain. If you choose to ignore the rules, the utility company has the legal right to stop your building works. They can even take down any buildings erected without prior agreement.

To prevent such nightmare situations arising, there are three options:

- **Avoid the sewer** – As noted earlier, the easiest and cheapest solution is to modify your plans so the construction will be at least 3m away from the sewer.
- **Divert the sewer** – Where practical you could divert the sewer away from where you want to build. Where pipes are less than 160mm diameter the water company may allow your builder to do this work (subject to written consent).

Storage tanks

As you might expect, anything with the potential to explode in a spectacular fireball is likely to be of considerable interest to Building Control. Hence there are strict rules for properties that require private fuel storage tanks, normally for oil or LPG.

OIL TANKS

Oil tanks with a capacity of up to 3,500 litres must be positioned in the open air on a concrete base at least 50mm thick, and 300mm wider

than the tank. Where sited within 10m of a waterway, tanks should be of twin-walled design, or else surrounded by an impervious masonry 'bund' wall capable of containing a spillage to 110% of the tank's capacity.

Because of the potential risk of fire, the walls of any building within 1.8m (such as a garage) must provide 30 minutes' fire resistance. Masonry walls can easily meet this requirement, but a timber shed or garage wouldn't last long, so a special fire-wall would need to be built. Where a tank is sited closer to a boundary than 760mm, you would also have to build a 30-minute fire-wall, with the tank contained behind it, the wall being at least 300mm higher and wider than the tank. Finally, suitable access for supply tankers needs to be provided.

LPG TANKS

Similar concerns apply to liquid gas tanks as for oil, but because these are pressurised containers with a greater risk of explosion, the rules are a little more demanding. A 1.1-tonne capacity tank must usually be sited at least 3m from the wall of a building or from a boundary. You can reduce this distance to only 2m if you protect the tank by building a 30-minute fire-wall sited between 1m and 1.5m from the tank. The wall must be higher than the tank's pressure relief valve and extend 1m either side. The firms who rent out LPG tanks have specific rules regarding the siting of tanks to allow access for supply tankers. Some tanks are designed to be buried in the garden, which lessens the visual impact. However Part J of the Building Regs discourages locating storage tanks for liquid fuel underground.

- **Go ahead and build over it** (or within 3m) – But before work starts you must enter into a legal 'building-over agreement' with the utility company. There will, of course, be additional charges, and until the sewer agreement is completed Building Control cannot formally approve your plans, which could hold up your entire development.

A 'building-over agreement' usually requires a preliminary CCTV scan inside the sewer by the water company, to check its condition. Any defects should then be repaired prior to building work, and in most cases you should not be charged for this work. Once your development is complete, a follow-up CCTV scan is carried out to tell if your construction works have damaged the pipes. Any repairs at this stage will be charged to you. The agreement will also detail the water company's rights of access for future maintenance and will record the methods you will employ to protect their pipes.

Submitting a Building Regulations application

Before submitting your application in the time-honoured manner to your Local Authority Building Control Department (LABC) you may wish to consider an alternative method. If you prefer, you're allowed instead to submit your application via an independent 'Approved Inspector'. These are private firms approved by the Construction Industry Council to do basically the same job as in-house Building Control staff. For example, NHBC offer such a service. They will process your plans, carry out spot checks on site as the building work progresses, and ultimately arrange for the 'final certificate' once the works are completed to their satisfaction. In other words it's all pretty much the same as the Council would do it, but may suit those who prefer a bit of private-sector involvement.

Alternatively, you can apply to a different Council to provide an 'out of area' plans-approval service. This allows one Council to approve plans for a project being built outside their territory, leaving the site inspections to others (known as the LABC Partner Authority Scheme). This mainly benefits larger developers, but may also suit those who live some distance from the site. But for most self-builders it's generally better to stick with your local Council.

For the sake of simplicity, we will here assume that you're going to apply direct to LABC.

Full Plans or Building Notice?

When it comes to submitting your Building Regs application, there are two options: a 'proper' Full Plans application, or the short-cut method known as a Building Notice. For a major project like building a new house, it's strongly advisable to make a Full Plans application. Where you need to build over, or close to, a public sewer this is the only kind of application allowed. Building Notices should really only be used for relatively simple jobs, like building porches. Compare this to the Scottish system, which, quite sensibly, does not allow work to start before the plans have been fully approved.

With a Building Notice, you're basically making a promise up front that you'll comply with the Building Regs on site, rather than submitting detailed drawings to prove it in advance. The big risk with this method is that a site inspection could later uncover something that doesn't comply with the regulations while the property is being built. And if a problem isn't picked up until late in the day, it could involve considerable extra work for it to be exposed and then rebuilt. All of which would prove highly disruptive, not to say expensive.

In actual practice, skipping the Full Plans process doesn't save as much time and effort as is sometimes supposed. The fee is the same in both cases, regardless of how many site visits are later required, and with a Building Notice application you may still be asked to provide drawings and structural design calculations. In fact, whether you take the Full Plans or Building Notice route, once your application has been checked and accepted there's nothing to stop you from starting work on site after giving the minimum two days' written notice.

About half of all self-build dwellings actually commence work on site before Full Plans approval has been granted. This can make sense if you're confident that your design will comply, especially the initial parts of the job such as the foundations. However, the consequences of being wrong can be very expensive, so a good compromise is to wait a couple of weeks after submitting your application for the initial queries to come back, before starting to build. Once they've checked your engineer's calculations and no major concerns have been raised about the early stages of construction, it should be safe to start digging.

In both cases the building has to be inspected by Building Control Surveyors at key stages and approved as it proceeds. If they approve the work, you may carry on to the next stage. But if they believe the work doesn't comply, even where your plans have been checked and approved in advance, they can still call a halt to the work until it's been suitably amended.

FULL PLANS APPLICATION

With a Full Plans application, you're demonstrating in advance – with the aid of detailed drawings – that the construction will comply with the regulations. It's best to go this route because the process encourages you to carefully think through the details of physically building the house in advance, thereby reducing later problems on site. It also means you'll have an approved set of plans to work to.

To make a Full Plans application you need to submit detailed drawings with a specification describing the work. There are two payments – an initial 'plans fee' paid up front and an 'inspection fee' paid at start-on-site. Once submitted, Building Control should write back to whoever submitted the plans confirming which member of their team will be dealing with your case. Then after a couple of weeks they normally write again to request extra information or amendments.

Your application should normally be passed or failed

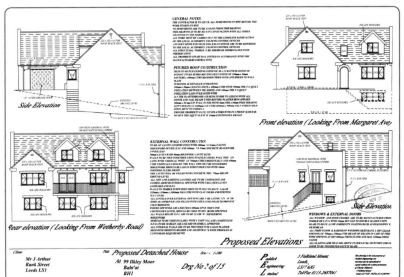

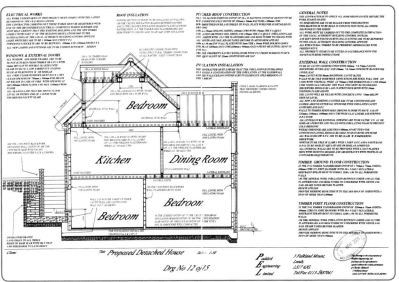

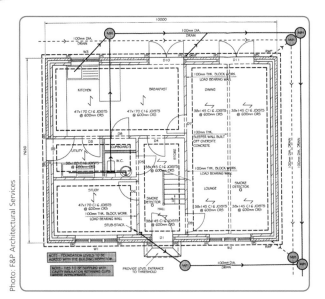

Photo: F&P Architectural Services

Photo: F&P Architectural Services

within five weeks, but applications sometimes cannot be determined within this timeframe, so the period can be extended for up to two months, subject to your consent. Once satisfied, you should receive a 'plans approved' notice, along with a set of plans stamped 'approved' to confirm that the plans comply with Building Regulations. However, approval is often made conditional upon receipt of further information, such as roof truss calculations (which your supplier should be able to provide once your order has been placed).

The Building Control Surveyors will then inspect the work on site at key stages as it progresses. A 'plans approved' notice is valid for three years, and will then expire unless

building work has commenced. It is not the same as a final certificate (aka 'completion certificate'), which should be issued once all the work is satisfactorily completed.

What to submit
Along with a completed application form and payment, you need to submit more detailed versions of your planning drawings plus relevant structural calculations and proof of 'carbon compliance'. The specification describing the construction details (ie the materials and thicknesses, etc) is usually written on the drawings. You normally need to print about four copies of your drawings, plus a few extra for builders to quote from later and some spares. The plans must clearly show the proposed building works and explain the construction details.

You need to submit the following:

1. An OS location plan
1:1250 scale, to clearly identify the site, showing the new building relative to neighbouring streets, houses, and boundaries.

The application form

When completing the application form, most questions are fairly straightforward (see website for a sample form). But some clarification may be useful for the following:

ITEM 5 – Select 'conditional approval'. This allows Building Control to accept your application even though some of the details have not yet been approved (because some information may not be available until later). However, any outstanding matters must be submitted and approved before the relevant work is carried out.

ITEM 6 – Select 'extension of time'. If your application takes longer to process and they run out of time, they'll have to issue a rejection notice unless you've agreed to extend the permitted time.

ITEM 7 – Select 'No'. Most of the fire regulations aren't relevant to new houses.

ITEM 8 – Select 'Yes'. It's essential to formally request a completion certificate. otherwise there's no obligation for one to be issued. This is a vital document when it comes to selling or mortgaging the house in the future.

ITEM 16 – If your house is exceptionally large (more than about 300m^2) the fee will be calculated differently.

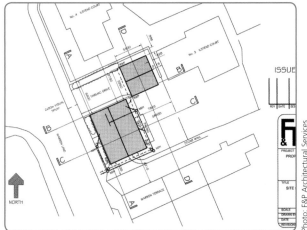

Photo: F&P Architectural Services

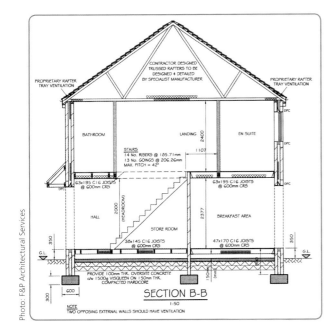

Within the section drawing (labels):

PROPRIETARY RAFTER TRAY VENTILATION

CONTRACTOR DESIGNED TRUSSED RAFTERS TO BE DESIGNED & DETAILED BY SPECIALIST MANUFACTURER

PROPRIETARY RAFTER TRAY VENTILATION

BATHROOM   LANDING   EN SUITE

STAIRS
14 No. RISERS @ 185.71mm
13 No. GOINGS @ 206.26mm
MAX. PITCH = 42°

63x195 C16 JOISTS @ 600mm CRS

63x195 C16 JOISTS @ 600mm CRS

HALL   BREAKFAST AREA

STORE ROOM

38x145 C16 JOISTS @ 600mm CRS   47x170 C16 JOISTS @ 600mm CRS

PROVIDE 100mm THK. OVERSITE CONCRETE c/w 1500g VISQUEEN ON 150mm THK. COMPACTED HARDCORE

SECTION B-B
1:50

NOTE
TWO OPPOSING EXTERNAL WALLS SHOULD HAVE VENTILATION

2. A section plan

A drawing showing a cross-section 'sliced through' the proposed new house (typically drawn to 1:20 or 1:10 scale, but no greater than 1:100). These are the most technically informative of building plans since they expose the details of the new building's construction, showing materials and thicknesses as well as heights and dimensions. This is often the drawing that has most of the specification written on it.

What information to write on the plans

On the section drawing and the elevation plans you need to show:

LEVELS

■ Ground levels – outside ground levels should be drawn true to life.
■ Floor levels – the new floor levels can be shown in relation to the external ground levels.
■ The levels of new drains – show the 'invert levels' where new drainage is to be connected at inspection chambers. You can take your own levels from any existing drains by measuring the depth between the manhole cover and the bottom of the channel, known as the 'invert level'.

THE SPECIFICATION

The 'spec' includes a detailed description of the construction, describing full details of the type and thickness of materials, and should clearly state:

■ The dimensions of all the main components of the building (foundations, floors, walls, roof, windows, doors, roofs etc), explaining their type and showing their position in the building.
■ The thickness and type of insulation to floors, walls, and roofs etc.
■ The drains – their routes, falls, and the position of inspection chambers (and rodding access points) should be shown, together with details of how pipes are protected where they pass under the building.

---

3. Plan and elevation drawings

1:50 or 1:100 scale, together with detailed technical specification notes that fully describe the proposed works.

4. Structural calculations

Calculations will normally need to be produced by a qualified structural engineer for all load-bearing elements. These are to confirm that the key components of the building are tough enough for the job and correctly specified for the loadings imposed on them. These will normally be checked by the Council's in-house engineer, who may then request additional calculations. But to the non-engineers amongst us, structural calculations can be hard to decipher, as they employ all sorts of obscure abbreviations. In Scotland you have to submit a 'Structural Design Certificate', which confirms the stability of the proposed new structure.

Calculations are commonly required to confirm the performance of key elements such as:

■ Adequacy of foundations for the type of soil
■ Size of beams
■ Stability of walls
■ Roof structure
■ Adequacy of any works for building over or near sewers

5. Carbon compliance calculation

There is a recent Part L requirement to submit a 'carbon compliance' calculation with your building specification.

Ground levels

Elevation plans often depict buildings sitting squarely on perfectly level, table-top smooth surfaces. But in reality, ground levels are rarely perfectly flat and this should be reflected accurately on your plans. The simplest way to measure them is to hire professional electronic survey equipment. Alternatively, you could try the trusty Victorian method for accurately transferring a level over a distance. All you need is a long length of tubing (such as a hose) with a clear, graded measuring tube or a funnel inserted vertically at each end. When filled with water, the pressure will ensure that the levels at each end are perfectly aligned, no matter how far apart. This makes it possible to measure ground levels relative to a datum point.

Speeding up the process

The safest way to proceed is to tackle one stage at a time. Once planning approval has been received, your Building Regulations application is submitted. Then, once the plans approval has finally been received, a tender package is prepared and ultimately a contractor is appointed to start work. Of course, doing it this way takes a considerable amount of time, so if you're in a hurry there are ways to speed things up, albeit at a greater risk.

Legally you cannot start work without planning consent, but as we've seen, there's nothing to stop you taking the risk of starting work soon after submitting your Building Regs application.

Another time-saving option is to ask your designer to submit your Building Regs application before Planning consent has been granted so that the two overlap, saving time. The risk with this approach is that if the planners requested any significant changes, these will also need to be amended on the Building Regs drawings – causing delay and extra cost.

For self-builders looking to construct a standard kit house, it may already have 'type approval' under the 'LABC registered building types' scheme. Here, once 'registration' has been given for a design through one local authority, it will subsequently be recognised by all Council Building Control Departments, speeding up your application. Whilst there may be some local variations to foundations and drainage details, the main design details and the structure will remain unchanged.

Approval or rejection

There are basically three possible outcomes to your Building Regs application: approval, conditional approval, or rejection.

Approval

If you receive a 'plans approval notice', congratulations! But even drawings submitted by professional designers are rarely perfect first time, so it's quite likely that Building Control will first write back to you with an 'amendments letter', asking for clarification on various points or requiring additional information, such as extra engineer's calculations.

Being bombarded with highly detailed questions about your design at this stage may seem a trifle irritating, but it's a whole lot easier sorting out problems on paper now than it would be trying to rebuild walls later on site. And you can be

When can I start work?

INITIAL NOTICE ACCEPTED ✓

pretty sure that if Building Control aren't entirely clear about what you mean, the builders on site will also be scratching their heads. If you haven't got a clue what they're on about, just ask your Building Control Surveyor, who will normally be happy to explain matters in plain English.

Conditional approval

Conditional approvals normally have a number of conditions listed in an attached document (*eg* stating required modifications), or they may issue approval and request that further information is submitted at a later date. It is important that your designer resolves these conditions in good time before the start-on-site date.

Rejection

A rejection notice needn't be as bad as it sounds. It can usually be overcome by resubmitting amended plans, for which no additional fee should be charged. Most of the checking will have already been done, and the new approval may come through in a matter of a few days.

The broadening scope of Building Regulations

In recent years, the scope of Building Regulations has ballooned to major proportions. In Chapter 6 we looked at the importance of meeting increasingly stringent energy conservation targets. This means your design will need to satisfy Building Control on a number of related matters, such as insulation, ventilation and airtightness.

But the Building Regulations now additionally include a number of specialist areas, such as electrics and the installation of heat-producing appliances. Hard-pressed Building Control Surveyors haven't got time to test and check everything, so approval in some specialist areas is now delegated to approved 'competent persons' – see website for a full 'who can do what' list.

We now take a closer look at some key parts of the Building Regs that will affect your design: Part L (conservation of fuel and energy), Part F (ventilation and condensation), and Part M (access and use of buildings). Part P (electrical safety) is discussed in Chapter 13.

Part L: Conserving fuel and energy

If there's one part of the Building Regs that drives everyone potty, this is it. Thermal insulation standards and energy targets are becoming ever more demanding for new dwellings as we approach zero carbon year 2016. So to ensure compliance, and get your DERs lower than your TERs (see Chapter 6), special attention must be paid to air-tightness and ventilation.

Excessive air leakage results in increased energy consumption and a draughty, cold building. The Building

Photo: Villavent

Regs therefore require that new homes should be airtight, so the building is resistant to both inward and outward air leakage. To demonstrate compliance with air leakage targets, a physical test is carried out which involves hiring a massive fan and pressurising the house to demonstrate that there is no air leakage. See Chapter 14. (N.B. small developments of 1 or 2 homes can be exempt if a dwelling of the same type built by the same builder in the last 12 months previously passed the test).

Part F: Room ventilation

When fresh air enters our homes it's traditionally perceived as an unpleasant cold draught. As a result, in many homes ventilation is actively discouraged. Hence the enduring popularity in Britain of bizarre products such as door snakes and chimney balloons.

Photo: Velux

But ventilation is essential to provide fresh air and remove old, stale air laden with moisture, smells, and pollutants such as smoke, poisonous carbon monoxide, and CO_2. It's a fact that in most modern houses the indoor air is actually more polluted than the air outside, even in cities. This is because airborne toxins can be introduced from such things as carpets, furnishings, and certain building materials (see Chapter 6).

Moisture generated by cooking and bathing needs to be expelled because excess humidity that can't escape will cause condensation and mould growth. But that's not the only problem. Gas cookers, stoves, and open fires release the products of combustion, including water vapour and CO_2, and it's essential for them to have an adequate supply of oxygen, normally provided via air vents in the walls. But the problem with uncontrolled ventilation in older, draughty properties is that it can result in substantial heat loss.

Today, the Building Regulations require new buildings to have high insulation levels, to keep our homes warm in winter. But as insulation is increased, air leaks become more significant as a source of heat losses. Hence the current requirement for new houses to minimise air leakage. The trouble is, as buildings become more airtight it can result in poor indoor air quality (IAQ). The solution is to provide ventilation, but unless controlled there's a danger that the warmth from your rooms will disappear straight out of the air vents. So ventilation systems need to be controllable.

Part F of the Building Regs lays down requirements for three types of ventilation – 'rapid', 'background', and 'extract' – see the boxout. For some years, the simplest way to comply with the Part F has been to fit electric extractor fans in 'wet rooms' such as kitchens, bathrooms, and utility rooms. As the fans extract moist stale air, fresh replacement air is sucked in through 'trickle vent' inlets in the windows in each habitable room (as well as through any air leaks in the structure). This system is economical but not very controllable, unless you fit fans with built-in humidity sensors that automatically control the rate of extraction so that more fresh air is admitted when the air in the room is moist (ie when it's occupied).

But house ventilation doesn't have to be organised on a room by individual room basis. When building from scratch it makes a lot of sense to design one master system that controls ventilation to the whole house. This way you can even capture the heat from stale air to warm up the incoming flow of fresh air. Today, such mechanical ventilation systems are an acceptable alternative to 'fan and vent' systems. There are 3 main alternatives...

Photo: Total Home Environment

PASSIVE STACK VENTILATION

Passive stack ventilation (PSV), employs large-diameter vertical ducts that draw stale air out of the house without the need for electric fans. In a primitive form PSV has been used successfully for many centuries – as chimney flues! Working on the principle that warm air naturally rises, PSV systems use no energy and hence cost nothing to run. In addition, suction is provided in the form of an updraught generated by air flowing over the roof. So PSV works simply by exploiting the natural forces of wind, temperature, and pressure differentials.

In a typical system, moist, stale air from kitchens and bathrooms is drawn upwards and expelled at roof level. This causes fresh replacement air to be drawn in via permanent vents in the habitable rooms (eg from trickle vents in windows) or alternatively from special fresh air ducts leading

Room ventilation

Ventilation requirements will depend on the overall design (check with Building Control), but as a rough guide new homes without mechanical ventilation should achieve the following:

BACKGROUND VENTILATION

Habitable rooms require a free airflow of 10,000mm² (5,000mm² to kitchens, bathrooms, utilities, and WCs). This is normally provided by trickle vents in the form of slots at the heads of windows, or airbricks sleeved through the cavity wall to an internal grille. Additional permanent ventilation is required for rooms containing heat-producing appliances, such as open-flued gas fires of rated input starting at 500mm per kW and most 'living flame' effect fires. Finished gaps under internal doors should be no more than 10mm to allow air movement between rooms.

RAPID/PURGE VENTILATION

Rapid ventilation (aka 'purge' ventilation) is provided by openable windows that allow the occupants to rapidly clear the air of paint fumes, foul toilet stenches, cooking smells etc. So the windows in all habitable rooms must be openable with a clear opening area equivalent to 1/20th of the floor area unless an approved mechanical ventilation system is installed.

EXTRACT VENTILATION

The Building Regs require extractor fans fitted in kitchens to be capable of shifting 60 litres of air per second, and those in bathrooms and utility rooms 15 and 30 litres per second respectively. The best type to fit are the relatively quiet humidity-sensing fans that only come on when humidity rises to a preset level, or heat-recovery fans that recycle heat from extracted air.

Alternatively, PSV or continuous mechanical systems are an acceptable form of extract ventilation where certified by an approved body. The internal diameter of PSV stacks is 125mm.

Note that solid-fuel appliances like Agas and wood-burning stoves aren't necessarily compatible with extractor fans. Open fires need to take their combustion air directly from inside the room. If this air supply is stolen by extractor fans, the consequential air starvation can leave occupants gasping. Hence the requirement for additional wall vents.

Passive Stack Ventilation

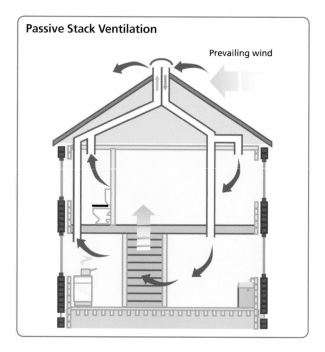

Prevailing wind

down from the roof. PSV systems are silent and since there are no fans they require no electric power to operate and need zero maintenance. The rate of extraction can be controlled by energy-saving technology that uses special dampers to help control air flow, which tends to be greater on windy days. Also, humidity sensors can be used to boost performance when rooms are occupied or when steam is produced from cooking or bathing.

PSV systems provide good internal air quality thanks to the permanent background ventilation in habitable rooms. The downside, of course, is that you lose all the room heat contained in the stale air being expelled. And because wind conditions can affect performance, it may not always be possible to deliver sufficient ventilation unless mechanically assisted.

MECHANICAL EXTRACT VENTILATION

Mechanical extract ventilation (MEV) provides a more high-tech, controlled solution. Using a system of fans, fresh air is supplied to rooms via ductwork whilst the old stale air is

Photo: Total Home Environment

Mechanical Extract Ventilation

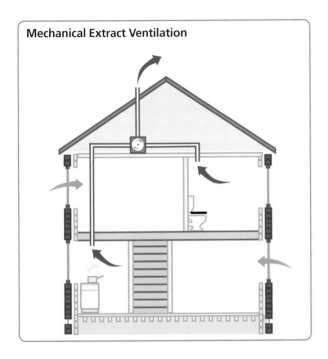

Heat Recovery

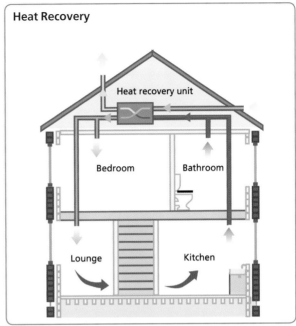

simultaneously extracted. With MEV systems the ventilation ducts from the wet rooms are combined so that stale air is expelled through a single vent. Fans work continuously at low speed to deliver a consistent flow of fresh air day and night. When extra ventilation is required, the system automatically speeds up. Filters can be fitted to improve incoming air quality by removing dust and pollen, but need to be changed each year. Similarly, 'positive input systems' (PIV) use a fan to push air into the building sourced from within the roof space (where the air may be slightly warmer and more moist).

However, for mechanical ventilation to work properly the house must be well sealed at the time of construction. So there's no need for conventional trickle vents or extractor fans (which has the additional benefit of reducing noise intrusion in busy locations). Unfortunately, this also means that open fireplaces are incompatible (although woodburning stoves can be compatible). The other drawback is the running costs, which may cancel some of the energy savings, making them not especially energy efficient. But one of the major attractions of MEV is the option of fitting a heat recovery system to conserve most of the warmth from the stale air being extracted.

MECHANICAL VENTILATION WITH HEAT RECOVERY

Heat-recovery technology can deliver the best of both worlds – fully controlled fresh air at comfortable temperatures, and ventilation with minimal heat loss. Heat recovery can be achieved by means of a simple heat exchanger that captures most of the warmth from outgoing stale air (typically about 70%) and transfers it to heat up the incoming fresh air, so the two air streams don't mix.

Such MVHR systems are most effective when designed for a *whole house*, so that warm stale air can be extracted

from the rooms that generate the most heat and moisture (normally kitchens and bathrooms). This is then used to warm up the incoming fresh air, which is fed via ducts to habitable rooms (living-rooms and bedrooms).

Whole-house mechanical extract systems require only one set of ducts. Even so, the ductwork that needs to be accommodated in the structure of the building can be quite substantial. From an energy-efficiency perspective you need a system with large-diameter ducts and low-loss heat exchangers with efficient, low voltage DC fans (otherwise the energy saved from reclaiming heat is lost powering the electrical fan).

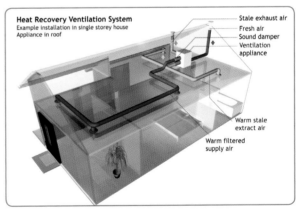

Heat Recovery Ventilation System
Example installation in single storey house
Appliance in roof

Stale exhaust air
Fresh air
Sound damper
Ventilation appliance

Warm stale extract air

Warm filtered supply air

21°C — Extract air from house
34°C — Supply air into house
-1°C — Fresh air from atmosphere
-7°C — Exhaust air from house

Genvex Premium
figures based @ 300m³/hr

However, MVHR systems are relatively expensive, and to work effectively buildings must be very airtight (air leakage of 3m³/hr/m² or less) so that the outgoing warm air is forced to pass through the heat exchanger, rather

Photo : Total Home Environment

than leaking away. However bear in mind that Part F of the Building Regs recommends that adequate natural ventilation is provided for use in summer to reduce energy consumption.

Part M: Access all areas

When designing a new dwelling, you must ensure that it's accessible for everyone, including people with disabilities. This

SECTION A-A
1:100

DENOTES PROFILE OF PROPOSED GROUND LEVEL

DENOTES PROFILE OF EXISTING GROUND LEVEL

doesn't just mean making the main entrance door a bit wider. You need to pay special attention to cloakrooms/WCs and to all habitable rooms on the 'entrance storey' (normally the ground floor). The key areas where special design rules apply are:

EXTERNALLY

The Building Regulations require 'reasonable provision for disabled people to gain access to a building'. Normally this involves providing shallow access ramps to the main entrance doors. For slopes up to 1:12 these ramps must be no longer than 5m, but for gradients up to 1:15 a maximum length of 10m is acceptable. This can pose problems on steeper sites where it's simply not feasible to provide ramps.

In the absence of a ramp, you can use steps at least 900mm wide with a rise of no more than 150mm, and a distance between landings of no more than 1,800mm. If there are more than three risers, handrails must be provided to at least one side.

On sites with very steep slopes, strict compliance could be impossible. But Building Control Surveyors are pragmatic people and may accept an alternative interpretation, such as a ramp in a spiral shape or perhaps some form of lift. Steep sites are always going to be challenging for disabled access,

and if there's no practical solution it may be possible to agree a relaxation of some rules.

INTERNALLY
Main entrance

The main entrance door should have a clear opening width of at least 775mm. If for some reason this is a problem, it should be possible to nominate an alternative side or back door rather than the front door. The main entrance should also have a flush and level sill. However, where it's unavoidable to have a step up into the building, the rise should be no more than 150mm.

Circulation space

There needs to be free access and circulation within the whole ground floor (*ie* the entrance storey). Doorways and corridors should be wide enough to allow wheelchair users to manoeuvre in and out of rooms. To help achieve this, there are rules that govern the relationship between doorway widths and the width of adjoining passageways – see table below. As a rule, where doorways are narrower the passageways approaching them need to be wider. But much depends on whether the doorway is approached head-on or sideways. See website for full Part M design details.

Internal doorways and corridors should relate as follows:

Doorway clear opening width	Hall or passageway minimum width
750mm or wider	900mm (when approach is head-on)
750mm	1,200mm (when approach is *not* head-on)
775mm	1,050mm (when approach is *not* head-on)
900mm	900mm (when approach is *not* head-on)

Photo : OPUS.eu

Photo: Roca baths

Cloakrooms
There must be a WC on the ground floor (or other entrance storey) large enough to accommodate a wheelchair. This normally means the door needs to open outwards and the washbasin is located somewhere out of the way. The end result is an attractive spacious cloakroom for all.

Sockets and switches
Power sockets must be no lower than 450mm from the finished floor surface and light switches no higher than 1,200mm.

The Party Wall Act 1996

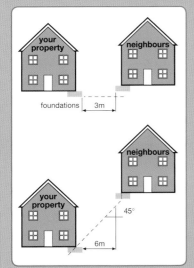

The Party Wall Act 1996 may sound like a lot of unnecessary hassle, but its objective is to prevent serious disputes arising between neighbours. It's nothing to do with Planning or Building Regulations, but is a totally separate piece of legislation. Even if you're building a detached house without a conventional party wall, the Party Wall Act can still apply as it covers excavations close to adjoining buildings and garden boundary walls.

So where any construction work needs to be carried out within 6m of your neighbour's property, you may be legally required to notify them in advance by giving them at least two months' notice in writing. Although there is currently no penalty within the Act for non-compliance, ignoring it would make it much easier for a neighbour to take legal action against you and seek an injunction to halt your construction work.

WHAT WORK APPLIES?
■ Excavations within 3m of a neighbouring property, where your trenches are deeper than next-door's foundations, which they normally will be. For example, an adjoining Victorian house might only have foundations up to 450mm deep.
■ Excavations within 6m of a neighbouring property if your new foundations are deeper than a line drawn at 45° from the bottom of next-door's foundations (see diagram).

■ Building on the boundary line. If you want to build up to a boundary wall, you will need the neighbour's consent for your underground foundations to cross underneath the boundary line. But you must agree to compensate them for any damage caused to their property as a consequence of your works.
■ Work on existing party walls, such as cutting into shared walls to insert a beam or DPC.

WHAT NEEDS TO BE DONE?
Before any work starts, you'll need to serve a formal notice on the owners of the adjoining property. If the property next door happens to be a block of flats – bad luck! You will have to entreat with each occupier individually. But in any case, it's normally best to appoint a Party Wall Surveyor to manage the whole process from the outset.

■ You have to give at least two months' notice before the work commences.
■ A written notice must be served explaining what building work is proposed, giving the owners 14 days to respond.
■ Notice needn't be on an official form, but must include drawings and details of the work (see website for sample notice).

The catch is that unless the neighbours write back within 14 days the law will assume that they do *not* consent, and that a dispute has arisen. This means you then have to jointly appoint an independent 'agreed surveyor' (at your expense) to draw up a legal agreement known as an 'award'. Alternatively, each neighbour can appoint their own surveyor, in which case the award will be drawn up jointly by both surveyors. The 'award' is a document that describes your proposed works and confirms that you will pay for any damage that's caused by your digging and building activities. Specifically:

■ It sets out how and when the work will be carried out (eg not at weekends).
■ It records the condition of the next-door property before the work begins (so that any damage can later be attributed and made good).
■ It allows access for the surveyor(s) to inspect the building works.

RIGHTS OF ENTRY
Under the Act, an adjoining owner must, when necessary, allow access for your builders. But you must give the adjoining owner and occupier 14 days' notice of intention to enter. It is then an offence for the owner to refuse entry. If the adjoining premises are vacant, a police officer must accompany your workmen on entry.

Self-Build Manual

9 THE BUILDERS

With your Building Control application safely under way, you can at last focus on getting your new home built. There are basically four ways you can 'procure' the construction of your project:

- Employ a main contractor to carry out all the building works.
- A design and build package deal.
- Directly employ trades and manage them yourself, with the option for some DIY input.
- The DIY route – self-build the whole thing, with a bit of help here and there.

Whichever method suits you best, you'll need to get an accurate price for the job, so it's essential to first work out all the details of the construction. Although your Building Regulations drawings were far more detailed than those required for planning, they may still not be sufficiently detailed for a contractor to price from. So whether you're putting the job out to tender or getting competitive quotes,

you'll need to provide detailed information explaining *exactly* what you want. Even if you're considering directly employing individual trades or taking the DIY route, it's essential to have each piece of work clearly defined well before anyone turns up on site to start building.

Procurement

Who exactly is going to build your house? The 'procurement route' that's right for you will depend on a number of factors, notably the size of your budget, how quickly you need the house built, and the extent of your own skills. So let's consider the main options in more detail.

Employ a main contractor

This is the method that's most familiar both to clients and to the building industry. With a main contractor appointed to run the building works, they should take primary responsibility for managing their own subcontractors, running the site, and ordering materials. This will make it relatively straightforward to control the overall job yourself.

With this method, the contractor submits a fixed price quotation based on a full set of drawings and a specification provided by the client. The contractor then builds the house you've specified for the quoted price by the agreed date. The contractor can only ask for additional money if there are design changes, and may even have to pay you compensation if it isn't completed on time.

The main attraction of this route is the relatively low risk, but the downside is that this added security is reflected in

the price. So to ensure the best deal, the job should be tendered and tied up with a contract – see below.

Appointing a main contractor means they take most of the risk, and the client gets a pretty reliable idea of how much the finished product will cost, and the date they'll be able to move in. But a successful outcome depends to a large extent on the quality of information provided to the contractor, which is, of course, very much up to you and your designer. So it's important to take time to ensure that a full set of drawings and a detailed specification are prepared before any prices are quoted. Careful preparation will save money in the long run by avoiding belated design changes that open the door to a lot of expensive extras.

Design and build package deal

This is the lowest risk route for procuring your new house, but it's normally also the most expensive. Here a package deal firm or a single contractor will do the whole thing for you, once Outline Planning Permission has been granted. Having agreed a price for a full design and build service they will start by drawing up plans to obtain detailed planning consent. But with this method construction prices tend to be higher because the competitive element is lost.

An off-the-peg design from a leading self-build kit supplier can sometimes provide the best solution for those with limited time but no shortage of funds. The drawback for many self-builders is the loss of control, making this a less-popular option for those who want an individually designed home.

Custom-build developers

A new breed of custom build developer has emerged in recent years. 'Custom build' is where you work with a specialist developer to deliver your own home. As well as providing the plot as part of a larger development they carry out all or some of the building work. A good custom-build developer should be able to tailor the design and layout to your requirements. See www.nacsba.org.uk

Directly employ trades and manage them yourself

This is the 'pick and mix' approach that holds great appeal for many self-builders. Cutting out the main contractor and instead employing individual trades direct can offer a financially attractive compromise between the full DIY option and the 'pay someone else to do it all for you' route. But self-managing your project in this way means piling considerable responsibility on to your own shoulders – we'll take a look at what's involved in more detail later in this chapter.

With this method, you hire individual trades to do the jobs that you don't personally feel qualified to tackle, plus you take on the combined roles of project manager and main contractor. This means assuming full responsibility for finding suitable trades people, buying materials, and day-to-day site management.

The beauty of this approach is it leaves you free to personally carry out some of the less critical building work,

whilst saving shedloads of money by cutting out the main contractor's profit and overheads. It's estimated that employing trades direct can save you as much as 25% on the build cost compared to putting the whole project out to a main contractor. But this figure assumes pretty tight project management. Of course, some main contractor costs can't be avoided, such as having to pay for insurance. Also, the discounts you can negotiate on materials are likely to be less generous than for the trade.

You obviously need to invest plenty of time to organise and manage such a project, so this route tends to work best for people with flexible hours – perhaps self-employed. To succeed you must have the detailed design and specification clearly worked out for each trade before each stage is started. You must also be clear about payment. This route has the advantage of flexibility, allowing you to make some small design changes as the job progresses, without getting stung for extras by a main contractor. Nonetheless, to minimise risk it's usually best to avoid complex designs that utilise unusual materials.

Full DIY self-build

A lot of self-builders relish the prospect of applying their DIY skills, working up a sweat and making a serious contribution on site. But very few aim to tackle all of the work on site. Most limit themselves to trades where they've got the right skills. Taking on most of the building work yourself is an option only for the most determined and knowledgeable self-builders. No matter how much of a red-blooded DIY supremo you are, there will always be some jobs where the law dictates that only certified installers can do the work, such as gas appliance installation and much of the electrics.

Photo: Helifix

For most self-builders, it's the substantial financial savings you can make on labour costs and a main contractor's profit margin that make the full DIY route so appealing. But when it comes to buying materials self-builders can lose out. Builders' merchants may be reluctant to offer the same level of discounts available to their long-term customers. However, some builders' merchants now offer discount schemes for self-builders to join, helping to bring prices down.

The other drawback with this route is the time it takes to complete the build – generally two or three times longer than it would if you employed a main contractor. Progress will inevitably be slower where new skills have to be learned. Then there's all the time-consuming project management tasks, such as ordering materials, organising site safety and insurance, and liaising with Building Control.

A good compromise solution that can get round some of these difficulties is to have the outer shell built by a contractor, and leave the rest as a DIY self-build project to tackle at your leisure.

DIY decisions

Some overconfident self-builders have come a cropper by mistakenly assuming they had the ability to take on a major role on site, so it's important to be realistic about how much time and energy you're genuinely able to commit.

WHAT WORK CAN I DO MYSELF?

Only you will know what tasks you have the right skills to accomplish and those you feel confident enough to take on. Being quite good at the odd DIY project doesn't automatically qualify you to hack it under pressure on site. This is not the time to be learning new skills on the job. But if you've already got some solid experience under your belt, such as building a home extension, it should stand you in good stead.

Many hands-on self-builders opt to tackle the below-ground construction works – tasks such as setting out, ground preparation,

laying concrete foundations, and laying the brick and blockwork up to DPC. This is because as long as the work is structurally sound, it doesn't have to look pretty; unlike above-ground work this will all be safely concealed.

TIME

Trying to run a major building project whilst simultaneously holding down a demanding full-time job can be incredibly stressful. It's your day job that is usually the source of the funds that are making the build possible and there's usually a limit as to how much time you can take off work without sabotaging your career prospects. You may actually be better off earning more money in your job to afford the luxury of a professional doing most of the build for you.

SLOTTING IN

It's vital to understand how one trade relates to another if you plan to slot yourself into the building process. You need to know which trade follows the other, and where they can overlap.

In some cases two or more trades will be working alongside each other, such as the electrician and plumber earthing the pipework or firing up the boiler. Plumbers, roofers, and brickies cooperate closely on tasks like fitting lead flashings to roofs, stacks,

and dormer windows.

Working on a single new house isn't going to keep one individual continuously employed for long. All trades need to be able to go from job to job, meshing in at the right time until the next one takes over. So if you've decided to carry out a particular task, consider how you're going to keep up with the programme. The professionals will normally want to get on with the job and

then swiftly move on to the next one. Problems can be created where a relatively slow DIYer is buggering about, thinking it's all jolly fun, and causing delays to following tradesmen. By losing a 'slot' it could take weeks before a trade can come back.

To avoid such problems, many self-builders play safe and take on the decorator's role towards the end, when there's less time pressure. Another easy but very useful contribution you can make is to keep the site tidy. Trades often think it's someone else's job to tidy up. Broken blocks, bits of tiles and old packaging seems to get littered everywhere. So make it your job every evening to sweep up, cover the mounds of sand, make sure bags of cement are stored safely away, and turn back the scaffold boards nearest to the brickwork. Check that tools are safely packed away. Clear paths and make sure stacks of bricks and blocks are stable. Your site will then run more smoothly and productively.

CAN YOU HACK IT?

Physical stamina isn't something you develop staring at a computer screen all day, so don't expect to be able to compete with people who've worked on building sites since they left school. Prolonged physical exertion builds tough

muscles and resilience to heavy lifting that could give lesser mortals back strain, hernias, and all round collywobbles.

PROJECT MANAGEMENT

Of all the DIY skills that you could contribute, project management and financial control is, for many of us, the most obvious. If you think about it, the main reward for helping out with the physical building work is to save money. But if by doing so you take your eye off the ball and let the management drift, it may not actually be time well spent.

Good project management can make a massive difference. It means thinking things through in advance and planning ahead to bring a project in on time and on budget. Conversely, those who allow things to stagger from crisis to crisis will inevitably suffer serious delays and spiralling costs. So use your skills to keep a tight rein on expenditure and a sharp eye on the cash flow. Plan ahead and get your labour organised well in advance. Order your materials in good time so that no trades are held up by late deliveries.

TOOLS

Your trusty old DIY tools may not be tough enough for heavy duty building work. Tools can, of course, be hired, but some can be devilishly complicated to get the hang of, and others – like nail guns and angle grinders – potentially dangerous in inexperienced hands.

Part of the attraction of the hands-on DIY route is being part of the craic and banter of the building site. This is an exciting new world, far removed from the tedious routine of office life or the factory floor. For many of us, part of the appeal of self-build is the sense of achievement in being able to claim 'I built it myself'. You may even dine out for a good while afterwards on the strength of heroic tales sourced from the building site. But for others there is sufficient challenge alone in finding a plot and designing a mini-masterpiece, without having to satisfy any primeval urges to wield tools.

If you do decide to take the pure DIY route, it's important that the design is planned from the outset with ease of construction in mind, employing materials that don't demand particularly high skill levels. The build should be kept as simple as possible with straightforward roof shapes, for example gables rather than hips and interlocking tiles rather than, harder to fit slates.

Finding suitable builders

Good small firms of builders tend to get most of their work via local word-of-mouth and don't have to advertise too much. However, building contractors looking to grow their business often send out mail-shots when you submit your planning application, which can be a useful starting point.

Although tracking down individual trades people involves a similar approach to finding a main contractor, there are some important differences which are discussed later in this chapter – see 'Employing trades direct' on page 149.

Here we look at how to draw up an initial shortlist of main contractors and then how to pick a winner.

Recommendations

The ideal way to find a builder is through personal recommendation. So be sure to ask friends, neighbours, and work colleagues. However, the people best placed to make recommendations are obviously other self-builders (who are normally only too happy to chat about their experiences). Building Control Surveyors are also in a good position to judge the performance of builders. Although, strictly speaking, they're not allowed to voice opinions, there may be discreet ways of letting you know. And if a Building Control Surveyor has had problems in the past with a particular contractor, he certainly won't want to encourage their appointment.

When you register your property for an LABC or NHBC warranty, there's no harm in asking them for their opinion of the firm your thinking

of appointing. Ask architects, package deal companies, builders' merchants, and tool hire companies – no one's going to recommend a contractor or tradesman who causes trouble.

Approved lists and websites

There are various self-build websites that offer listings of trades and contractors. However, in some geographical areas the numbers listed can be a bit on the thin side. Buildstore offers a 'Trades Referral Service' and Trustmark has a joint service with the Federation of Master Builders. Homepro lists contractors and tradespeople who pay an annual fee. You can search by trade and postcode, and check ratings from past jobs. See website for links. It may also be worth approaching local Housing Associations, who may publish lists of approved building contractors.

> ### GREEN CHOICE
>
> ## Green builders and designers
> The Association of Environment Conscious Building (AECB) has an online register of designers, builders, and suppliers with experience of green building.

Selection criteria

Once you've got a list of names the detective work can start in earnest. Sorting the good guys from the bad requires some careful research. Some excellent builders with lots of happy customers might be hard to track down because they don't do much in the way of marketing. Conversely, flashy brochures and all-singing-all-dancing websites don't necessarily translate into quality building work, although this does demonstrate a degree of commitment.

Research stage 1

A little preliminary delving about should give you a feel as to whether a firm is reasonably businesslike.

CONTACT DETAILS

Avoid firms who rely solely on mobile phone numbers with no permanent address. Whilst you might expect individual trades to work from home, any decent building contractor needs an office and a yard.

STAFF AND COMMUNICATION

How many permanent staff do they employ? Do they listen, or are they evasive when asked straight questions? If they make you feel uneasy during one phone call, think how you'd feel after months on site.

AVAILABILITY

Most good contractors are booked up several months ahead, so if they're available next week ask them why. A genuine reason might be where an anticipated contract has fallen through, leaving a sudden gap in their schedule.

READINESS TO QUOTE

Mention that you'll be using a standard contract and will require a fixed price. If the only contract they accept is their own, it's not a good sign, as these tend to be written in tiny grey print on the backs of quotes and are heavily biased against the customer.

TRADE ASSOCIATION MEMBERSHIP

Builders sometimes belong to trade federations, but these exist primarily to represent the interests of their members, not the customers. It's also true to say that the requirements for joining are not always terribly demanding. There's the old joke about the elaborateness of the logo being inversely proportionate to the value of the organisation. Whether this is true

or not, displaying a trade association logo is no guarantee of a quality service. However, several years' membership of a well-established organisation such as FMB is likely to indicate a reasonable standard of workmanship. This is particularly true with major warranty providers such as NHBC, because they undertake to police their members' work.

Research stage 2

Armed with your shortlist, you now need to dig a little deeper. The most important check is to talk to previous customers. You might also want to visit their offices, but don't expect a palace.

PREVIOUS WORK AND REFERENCES

Ask to see examples of their work, and ask for references.

Talk to former clients to see if they did what they promised and whether they came in on budget or made outrageous charges for extras. Go and take a look at a

---

How to spot a good builder

Despite TV schedules packed with 'builders from hell' stories, there are many who take great pride in their work and deal honestly with customers. These are some of the signs that demonstrate a businesslike attitude:

- **Keeping jobs running** – Few builders survive on just one job at a time. The ability to fluently juggle jobs on more than one site at the same time is a mark of professionalism.
- **Forward thinking** – For main contractors, the key to successful management on site is the ability to think ahead and anticipate events.
- **Helpfulness to self-builders** – Self-builders have an endearing tendency to query what's being done and to take an active interest in the building work. Some even want to work alongside trades on site. But not all builders can happily accommodate this inquisitive mentality.
- **Tidiness** – Tidy sites run smoothly and are more likely to come in on budget. On a messy site progress is held up as accidents occur and materials get spoilt or lost.
- **Progressive attitude** – The building trade is generally pretty conservative. Good builders are problem-solving, hands-on people, and tend to be wary of being told what to do by rosy-cheeked professionals fresh out of university who would struggle to drive a screw in straight. Over the years they've seen many new products come and go, and quite understandably may be a tad cynical.

Many builders are so busy working that they have little time to familiarise themselves with new innovations. Self-builders, on the other hand, with just the one project, have the time to explore new products. So if your builder is immediately dismissive of the materials or fittings that you're considering using, move on to someone else. That doesn't mean they have to swoon with joy at your every suggestion. From experience they may have some valid criticisms that haven't occurred to you, potentially saving you from a lot of unnecessary expense. But when it comes to satisfying new regulations, you need a builder with a general openness to new technology. It's not your job to play the role of teacher on site.

---

completed project and if possible speak to the owner – most people will be happy to help. Ask some fairly detailed questions, for example:

- Did the builders turn up when agreed?
- Did they clear away their rubbish?
- What was the quality of work like?
- How was their attention to detail when finishing work?
- Did they price extra work reasonably?
- Did the job start and finish on time?
- Were they helpful and considerate?
- Would you employ them again?

If all the referees sound the same and can't give much detail except to say the firm is 'really great', forget it. Try to visit one of the builders' current sites to see work in progress. Note how tidy the site is, and whether materials are protected and stored neatly, or just littered about chaotically. Is it a rushed job that's been badly planned? Don't employ builders who cut corners and ignore legislation.

It's not a bad idea to run their name past Trading Standards, and check the OFT website (www.oft.gov.uk) to see whether there have been any complaints.

INSURANCE

Suppose someone accidentally drops some roof tiles off the scaffolding, injuring or killing a passer-by. If it turns out your builder isn't fully insured, as the owner of the property you could be held jointly liable to pay compensation. More likely some irritating minor damage will occur, like next-door's fence getting damaged, or their phone cable getting severed. So it's important to cover yourself by asking the builder to produce his current certificate of public liability insurance, which must provide cover to a minimum of £1 million. Any bona fide builder will have no trouble cooperating. Always ask for proof before signing contracts. Your designer should also have full professional indemnity liability insurance. See page 160.

Quotes and tendering

Once you've got your shortlist of suitable builders, the next step is get a firm price for the job, by putting it out to tender or getting competitive quotes. But first you need to refine the build costs you've budgeted for the construction of the house, which are probably still based on an approximate price per square metre.

Cost analysis

Translating the drawings of your dream home into a list of all the necessary materials and labour needed to actually build it can be one of the most challenging tasks for self-builders. Unless you're blessed with the skills of a quantity surveyor, it

can be tempting to brush this under the carpet and leave it for the contractor to work out. But keeping control of costs is important.

One solution that's popular with self-builders is to send the plans off to an estimating company to produce a print-out of the quantities of labour and materials with estimated costs, which can be done for a relatively small fee. These are usually pretty accurate, although the sums they allow for bathrooms and kitchens may only be based on average-quality fittings – and as a self-builder you may want something superior. These estimated costings can provide a useful target to beat. VAT isn't normally included on items where it's recoverable – *ie* most of the build. But these figures don't tell the full story, so you'll need to add certain additional costs such as:

- Architect's/designer's fees
- Project management fees
- Finance costs
- Site preparation
- New services (depending on ground conditions and the distance supplies need to be run)
- Carpets
- Fitted furniture
- Landscaping
- Garden walls and fences

Competitive quotes

The best way to get a major development priced is to tender it (see 'Tendering' below), but a competitive quote is the next best thing. A quote is a fixed sum for a fixed amount of work, based on your description of the work you want carried out. This usually takes the form of a letter and a set of detailed drawings, ideally with a brief schedule listing any details of the works that aren't immediately obvious from the plans. It's worth confirming in your letter that you've already got Planning consent, otherwise they may assume you haven't and judge it as a waste of their time replying.

To judge how good their quotation is, consider the amount of care and detail that's gone into it (*eg* references to your drawing numbers etc). Also, check that they actually went to the trouble of visiting the site before quoting. Most contractors will add something to their price to cover the risk

that the job may turn out to be more complicated than it seems. But having provided a firm price your builder must then stick to it.

The specification

Most disputes with builders are (surprise, surprise) about money. This is often down to misunderstandings about what work was meant to be done for the quoted price. This may be because the client or architect didn't make it clear at the outset exactly what work was required. Builders are not psychic. The key to a smoothly run scheme is, above all, to specify precisely what you want, so you know what you'll be paying for.

An architect or surveyor can write a detailed specification, often referred to as 'the spec' (pronounced 'spess'). This is basically a long, detailed list stating each separate task required. Of course, a house designed to your own individual requirements can be a pretty complex thing to describe in detail. Fortunately, much of this information will already have been written down on the drawings submitted for your Building Regs application, although this won't cover everything (just the stuff that Building Control are concerned with).

A detailed specification is an essential part of getting a job priced, especially by tender. Trying to save money by not preparing one is a false economy. If you just send out a set of drawings it will leave gaps, so there's a good chance some of the things you want won't get priced. Only a detailed specification will clearly list each component of the job so it can be individually priced. This process is also important because it helps you focus on exactly what you want and reduces the risk of misunderstandings later.

The full specification will be sent out to contractors, together with a set of drawings, to work out how much the job will cost. This should make it perfectly clear exactly how many square metres of plastering, roof slates, tiling etc are needed, so the contractor can simply write a price next to each part of the job. The whole lot is then totted up to produce a grand total. This is especially useful when it comes to paying the builders on site, since you can see which parts of the job they've completed and pay them the price they've quoted for each piece of work. It also helps in avoiding overpaying for any extras, as you can base the price on the figures quoted here for similar work.

Devil in the detail

Once you start to look in details at all the things that go into a building, you realise how many intricate design decisions need to be made. This in turn requires research, and you

Photo: oakwrights.co.uk

soon find yourself buried under stacks of product literature, absorbed in magazine articles, and endlessly trawling websites. Being lazy and skipping this stage is a recipe for trouble, as misunderstandings on site over tiny details have a habit of growing into disputes. For example, you might have visualised traditional Victorian 'ogee' skirting and architraves, but unless instructed otherwise the builder may just stick on the cheapest bit of timber he can find.

So it's important to think carefully in advance about the details, such as the kind of light switches, sockets, taps, and basins that you want, or you'll inevitably find they've fitted ones you don't like, perhaps left over from the last job.

Visiting exhibitions can allow you to see products and materials in the flesh, and to talk directly to manufacturers to confirm prices and availability. It's best to visit exhibitions on weekdays, if possible on the first day of the show, to avoid the crowds so that people on stands have time to discuss your project. Take along some spare sets of plans to help obtain accurate prices.

There will be less room for mistakes if you quote precise catalogue numbers. If you intend to supply some fittings or materials yourself, make sure that the specification clearly states that the contractor is to 'allow for fitting only'. This means that you are solely responsible for ensuring these items are available on site exactly when the fitter needs them. If they're delivered too soon there will be more risk of theft or damage – too late and the plumber or chippie may not make an appearance on site again for another month.

Who's responsibility it is to supply materials and fittings must be made very clear at the start, or it can very easily flare up into a major issue later.

> **TIP**
> If you're planning to use unconventional products, the manufacturers will often be able to help with design and specification. Some even provide on-site training for builders at no extra cost.

Prime cost sums (PC sums)

Most of us like to take our time mulling over the precise choice of visually important things like taps, worktops, and tiles. After all, rushing such key design decisions could result

Photo: William Ball Kitchens

in catastrophic style errors that come back to haunt us for the rest of our days ('I swear the colour looked different in the brochure...'). On the other hand, you don't want to hold up the entire project whilst umming and ahhing over samples and colour swatches for months on end. And you certainly don't want to keep changing your mind as work progresses on site.

Photo: Roper

Thankfully there's a solution to this dilemna. Using 'prime cost sums' in your specification allows you to have your cake and eat it, so you can get on with the job whilst postponing such tough decisions. Prime cost sums allow clients the freedom to postpone the selection of the precise product until a later date. So you might write *'Allow the sum of £4,000 for supply only of bathroom suite'* as an estimate for what you think it's likely to end up costing. The contractor simply has to include this amount in his quotation or tender, but the actual figure you pay will be down to which suite you ultimately choose. The builder quotes only for the labour to fit it and adopts your stated 'guide price' for the materials.

Similarly, 'provisional sums' relate to specific trades and are 'an allowance for unknown items of work or goods, including labour costs, overheads and profit'. They're useful where it hasn't yet been decided what exactly is required, or where there's doubt about the likely extent of work to be carried out (*eg* for some underground drainage works). These are only estimates, not firm prices, so the actual costs (which could be higher or lower) will need to be sorted out later. Adjustments will normally be made in the final bill. The downside is they introduce an element of uncertainty that can potentially lead to disputes, so ideally their use should be restricted.

Workmanship

To accurately describe the job, the standard of workmanship needs to be clearly defined, as well as materials. Under the terms of the contract, the builder will normally be required to 'use reasonable care and skill', which is obviously open to some interpretation. One solution is to specify 'British Standard BS8000' on the plans, and to make sure they're included as contract documents so that, if push comes to shove, they're legally enforceable. This can be achieved by stapling a set of plans to the contract and having each party jointly sign all pages and drawings. Then if the workmanship becomes unsatisfactory at any stage, your builder will be in

Thoughtful contractor provides ladder to site loo

breach of contract and you'll be able to give him notice to correct the defective work, or even terminate the contract.

Preliminaries

The 'prelims' are your way of explaining to the contractor the arrangements for all those important little things that make a job run smoothly, such as carting away rubbish, water supplies, temporary power, and arrangements for storage of materials. They'll also confirm that you agree to provide the builders with free access to your site for the duration of the contract.

'Prelims' are important because there are a lot of easily overlooked potential costs involved – things like hiring port-a-loos, security fencing, and storage containers. If things aren't spelled out clearly at the outset, the neighbours might wake up one fine day to find themselves temporarily entombed behind mountains of concrete blocks deposited in their garden.

The contract

A building contract is simply an agreement between you and your builder, to undertake a stated amount of work, to a certain standard, for an agreed sum of money. Although, strictly speaking, accepting a verbal offer could form a contract in law, to run the job properly it's advisable to use a written contract, signed by both parties. This shows that both sides are serious, and gives you both certain rights and duties that are enforceable in court. If things go horribly wrong later, a contract will allow you (the innocent party) to seek compensation ('damages') for any losses incurred. You're also allowed to terminate the contract and employ someone else.

Although some self-builders are prepared to take the risk of not using a contract, it's very easy to arrange and affords you valuable protection. There are several standard contracts readily available, so there's no need to involve solicitors. If you are using an architect or surveyor, ask their advice. Even small builders will accept a simple four- or five-page contract written in plain English. The advantage of professional ready-made contracts is that they're designed to be fair to both sides.

But which contract to use? There's nothing to stop you writing your own if you want, but it's a lot easier to use a ready-made 'off the shelf' one. Perhaps the best known are the JCT (Joint Contracts Tribunal) contracts such as the JCT *Building Contract for a Home Owner*. Alternatively, the Federation of Master Builders and 'Homepro' have free ones you can download from our website.

Don't be too gobsmacked when you first read one through. It's basically a collection of all the things that could ever go wrong with a property development, based on other people's bad experiences over many years. It clarifies who's responsible for what, thereby reducing the risk of a major dispute messing up the job. All the important stuff is there – like how frequently you pay the builders, and who's responsible if the job is only half finished by the completion date.

There's nothing to stop you adding your own specific conditions if you wish. For example, you might want the builder not to start work until after 8:00am, not to work at weekends, and not to entertain the neighbours by blasting out *Radio GaGa* all day long.

Finally, before signing the contract remember that this is your last chance to make sure that you'll be getting exactly what you want – any changes from now on could start to cost serious money.

Tendering

A tender is a sort of super-detailed competitive quotation. The tender documents that you send to your list of contractors should explain exactly what you want them to build, against which a firm price can be offered. Each contractor must also be asked to confirm when they could start the job and how long they'll take to complete it.

There are two good reasons for tendering a major construction job. Firstly, it immediately weeds out lazy builders, who won't want to go to all the trouble of completing and replying to tenders. Secondly, it allows you to easily compare rival firms' prices for the same job.

Start by phoning or emailing your shortlist of building firms to ask whether they would like to tender for the job of building your new home. This saves wasting time sending out documents to firms who are either too busy or just not interested.

What's in a building contract?

These are the key points you need to fill in:

- **The parties** – You and your builder (be sure to write their full name, so that there's no wriggling out later).
- **The works** – A brief description of the project.
- **The contract documents** – List all the specific drawings by number and the tender documents specifying the agreed work.
- **The professionals** – The names of any architect or surveyor who'll be managing the contract for you.
- **The tender sum** – This is the contractor's agreed price, unless any post-tender changes have been made.
- **Project duration** – Clearly state the agreed start and finish dates (many disputes relate to time overruns).
- **Liquidated damages** – If a project significantly overruns, any 'unwarranted delays' will give you the right to make deductions from the money due to the builder. Normally a weekly sum known as 'liquidated damages' is filled in.
- **Payment terms** – Contractors are normally paid every four weeks or at agreed stages in the construction work. A small 'snagging' retention (typically 2.5%) is normally held back for six months after the completion date.
- **Insurance** – The contractor must confirm that they have adequate insurance throughout the build period.
- **Solving disputes** – A description of what action the parties can take if there is a dispute.

A typical tender will contain:

- **Drawings** – The latest version of the drawings used for Building Control, plus any additional 'builders' drawings' illustrating key construction details.
  – Plan and elevation drawings to 1:50 scale.
  – Section drawings to 1:20 scale.
  – Drawings of key details to 1:10 scale.
- **A detailed specification** – Typically 20 or more A4 pages.
- **Preliminaries** – Alternatively you could list these in your covering letter.
- **A completed form of tender** for replying (see website for forms).
- **A photocopy of any contract** you propose to use, left blank (otherwise the appointed contractor could later claim the right to renegotiate the price).
- **A covering letter** clearly stating the precise deadline – a time and date – for completed tenders to be submitted to you.
- **An addressed envelope** for the returned tender.

A tender package should be as complete a description of your new house as you and your designer can manage. The objective is to leave as little doubt as possible, because if you forget something it probably won't be priced, and could become an expensive 'extra' when you have to ask for it later. So it's essential to make sure the drawings and specification are clear and thorough.

However, there will inevitably be some unknowns at this stage. Ground conditions are never 100% certain until you start digging. And as we saw earlier, if you really can't decide on a particular fitting until later it's possible to leave it out – provided you make it clear in the tender documents (*eg* by including PC sums) and allow for it in your budget. If you do leave certain items out, at least describe the labour element. For example, you can specify that a particular fireplace is to be supplied by the client, but fitted by the contractor. This allows you to shop for it and buy it yourself later, when you've got more of a feel for how a particular room is going to look.

It's always a good idea to give your specification a thorough read-through before sending it out, and make a note of any areas ripe for making economies. This way, if the returned tenders turn out to be way over budget you'll have a 'Plan B' waiting in the wings.

HOW MANY FIRMS?

A contractor's decision on which jobs they'll tender for is primarily influenced by their workload, which of course can change overnight as they win a new contract. As a result, a firm may agree to tender one day and change their mind the next, returning your uncompleted documents, or not bothering to reply at all. Or they may inflate their tender to an extortionately high price, so that if by some twist of fate they ended up getting the job it would have a generous profit margin to compensate for the extra trouble. To allow for 'drop-outs' and overpricing it's recommended you send tenders to five or six firms. But pricing lengthy tenders accurately means a lot of work for contractors, so if they learn through the grapevine (via builders' merchants or subbies) that numerous others are also being asked to tender, they may opt out.

When contractors come to price a job, their calculations are primarily based on how they interpret your description of each task and convert it into the cost of labour and materials. But the pricing will also depend on how keen they are to win the contract, which is influenced not only by how busy they are, but by how easy it is for them to get the right trades, how profitable the job looks, and even how convenient the location is.

SENDING OUT TENDERS

At least four weeks should be allowed for fully priced tenders to be completed and returned. Some contractors will need to get binding prices from their subcontractors and suppliers, so if you don't allow enough time they may have to 'guesstimate' these with inflated prices to allow for risk.

It's clearly essential that each firm receives exactly the same information,

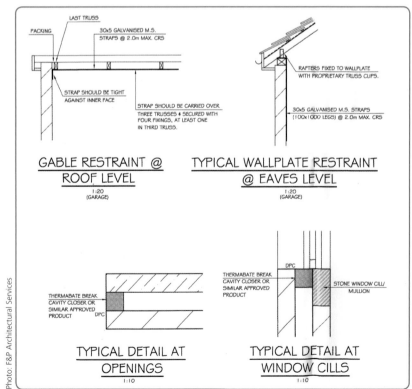

GABLE RESTRAINT @ ROOF LEVEL
1:20 (GARAGE)

TYPICAL WALLPLATE RESTRAINT @ EAVES LEVEL
1:20 (GARAGE)

TYPICAL DETAIL AT OPENINGS
1:10

TYPICAL DETAIL AT WINDOW CILLS
1:10

so you can accurately compare their submitted prices. It's also important that firms are not told who else is tendering for the same job.

If a firm gets in touch to ask questions before submitting their quote it's a good sign, suggesting they're being thorough. But to be fair to all tenderers, you'll then need to issue a 'tender update' clarifying the same points to all the competing firms. Similarly, if you grant a time extension to one firm, allow the same for all of them. By putting a simple code on the return envelope you can identify who it's from without opening it, so if one's a bit late you'll be able to work out who it is and call them to check what's happened.

COMPARING OFFERS

Come the Big Day, you should have at least three completed tenders to choose from. But no matter how much time you've allowed, you can be fairly sure there'll be one arriving at the last minute in the grip of a breathless motorcycle messenger. Or there'll be an urgent phone call asking for more time.

When you're ready, open the envelopes and check the bottom-line prices. If all the prices submitted significantly exceed your budget, don't despair. Below we look at some relatively painless ways that costs can be reduced.

Selecting the best tender isn't as simple as just choosing the lowest bid. One builder may have undercut the others because he made a mistake or is inexperienced. If you accept an uneconomic price it will be you who loses out in the end, when they fail to finish or go bust! So beware picking the cheapest price from a builder who can start tomorrow. An unscrupulous firm may be quite skilled in spotting loopholes in your specification, knowing they can later charge you premium prices for extras. It's often better to go for the middle ground, and check everything carefully before signing the contract.

Comparing all your tenders, section by section, may reveal some surprising differences in quoted prices for the same works. Naturally, the cheapest price will always appear very attractive. But take a minute to consider why it's that cheap. Is it because they're desperate for the work? Or it could be that, for example, roofing work is that firm's area of expertise. If it's suspiciously cheap compared to the others it may be that they've forgotten to include some materials or labour – in which case you'll probably end up with a badly done job or having to pay for lots of expensive extras. Unless the firm has come highly recommended, it may be cheaper in the long run to select another, rather than risk them skimping on materials and labour to recoup losses. If there are two contractors who are very close, treat them equally, and give them the same chance to reduce their costs.

Then there's the important matter of the start-on-site date. It's unlikely that a good small firm would be so desperate for work that they could start tomorrow. But on the other hand, a firm that can't begin for another year may not be ideal either. Finally, use a little intuition about the

Where to trim your costs

The main reason projects go over budget is because the specification is too ambitious. Choosing too many nice things before you know their full cost is a common mistake. The obvious solution is to shave off some of the more ambitious design ideas, whilst maintaining the spirit of the building. The thing you don't want to compromise is the fabric of the building, such as the bricks and roof tiles. Only cut things that can easily be replaced with better ones at a later date when funds permit.

- **Decorations** – Opt for plain emulsion rather than elaborate wallpapers and expensive tiling.
- **Finishes** – Stick with carpet and vinyl for the time being. Expensive wood floors and tiles can be fitted later.
- **Internal doors** – Cheaper doors can easily be replaced in future. Or fit quality ones only where they're most visible, such as in hallways.
- **Kitchens** – Large sums can be saved by fitting standard units, and upgrading doors and worktops later.
- **Joinery** – Use softwood or MDF rather than hardwood. Rather than built-in cupboards, buy free-standing furniture at a later date.
- **Electrics** – Reduce excessive numbers of sockets, and use cheaper white plastic switches and sockets, and pendant lights, which can easily be upgraded later.
- **Sanitary ware** – Forget premium designer fittings. Good quality, cheaper models can provide big savings.
- **Special features** – Reassess costly, non-essential items like swimming pools, expensive fire surrounds, solar panels, multi-media cabling, custom glazing, and central vacuum cleaning.
- **Landscaping** – Much landscaping can be postponed. Forget about ponds, grandiose water features, and dry-stone walls. Gravel drives are cheaper than fancy paving and can still look stylish.

people themselves. Talk to the contractor. Is there something about their attitude that makes you feel uneasy? If in doubt, go for the good communicator or the one you feel you can best do business with.

WHAT IF IT COMES IN OVER BUDGET?

If all the prices come back well above the level you anticipated, don't panic! It's often possible to reduce costs by as much as 20% by going through a cost-reduction exercise with the lowest tendering contractor. Most builders know their stuff, having seen it all before, and may come up with helpful cost-cutting ideas.

Sometimes a high price can be the result of too few tenders being returned. Or they may be from firms who priced very high because they weren't particularly keen to win the job. Big building firms have hefty overheads, such as large offices and fleets of vans, and can rarely be competitive against a small builder working from a home-office and yard. In some cases, it may then be best to re-tender the job to a fresh list of contractors. All you stand to lose is a few weeks.

If it turns out that the building is too large for your budget, you could postpone construction of any 'non-essential' parts such as garages, conservatories, or lean-to

utility rooms. These could be added later when funds are available. It's always wise to keep this option in mind at the design stage.

AND THE WINNER IS...

Having made your choice, you'll then need to set up a meeting to go through your drawings and specification in some detail prior to signing the contract. If you're employing an architect, they should be present so that any complex parts of the design can be discussed in detail.

Run through key things like the depth of foundations, the method of construction, the type of bricks (or stone, render, cladding etc), the sizes of roof timbers, and the types of windows and doors.

If the builder's face turns a peculiar shade of ghostly white at any stage it means he's just realised something's been seriously underpriced. If his pricing is very low for one particular part of the job compared to the others, mention this by saying 'your price for the [eg window] work seems to be less than the average for this kind of work – would you like to check your price before signing the contract?'

It's always better to agree a revised price at this stage than to have a dispute later on site. But it must be agreed in writing, as it changes the tendered sum. Legally, you could hold them to their tendered price, but this would only encourage a cheap job to save money.

New detached house, 'Pylon View', Crack Street		
PROGRAMME OF WORKS		**Date**
1	Start on site	3 March
2	Demolish old bungalow	7 March
3	Groundwork: excavate foundations & pour concrete	14 March
4	Excavate and lay drainage etc	
5	Ground-floor beams	
6	Construct main walls up to DPC	
7	Build main walls to first lift	
8	Complete main walls	
9	Roof structure	
10	Roof coverings	
11	Windows & doors	
12	Plumbing, heating & electrics first fix	
13	Floor screed and plastering	
14	Internal joinery	
15	Kitchen and sanitary fittings, connect drainage	
16	Plumbing, heating & electrics second fix	
17	Decorations	
18	Practical completion	

Bodgit & Scarper Building Contractors,
Costa Del Boy, RSJ Boulevard, Espana

This is the honeymoon stage. Everything is sweetness and light, as the contractor rejoices in the warm afterglow of winning the tender. So it's as well to take the opportunity to bring things sharply down to earth and remind them about all the important issues – the agreed start date and completion date, arrangements for security and storage of materials on site, the welfare facilities (loos etc), health and safety matters, frequency of site meetings, and the agreed method for dealing with any changes you might later require to the work (ie written instructions). Don't mention to the contractors that you've got a contingency fund in reserve, to cover any unexpected items that crop up later on site. If anything, try to convey that you're already at your financial limit and even the smallest extra cost would be painful.

Although as a rule they don't like committing to this, ask the contractor to provide a week-by-week programme of works, so that you'll be able to monitor progress. Even a simple 'milestone programme' would be useful.

The next step is to fill in the contract with the agreed start and completion dates, the amount of retention you plan to keep, how frequently you propose to make interim payments and the penalty (if any) for overrunning.

As far as timescales are concerned, it's not unknown for contractors to be prone to bouts of over-optimism at this stage, which will only cause problems later if you have to apply damages for non-completion. So it may be worth asking them to think again about the projected time to finish the job. They might want to add an extra week or two, just in case. They are, of course, perfectly at liberty to finish the job earlier if they wish (but they won't). All contract documents, including the drawings and the form of tender, now need to be signed by both parties and stapled together with the contract itself. As the client, you hang on to the originals.

If you're not using a ready-printed legal contract you can complete the deal by writing to the contractor to formally accept his offer. Your 'letter of acceptance' is a legally binding contract document and should include a list of the drawings and documents on which he based his quote, together with confirmation of the agreed price (the 'contract sum'), the agreed start and completion dates, and the stages of payment etc. You must also request that the builder formally replies and acknowledges receipt of your letter.

Employing trades direct

If you've always wanted to manage your own team, now's your chance. Rather than deal with a single main contractor, you may prefer the idea of directly employing individual trades (aka 'subbies') and organising your own site. Strictly speaking, tradesmen employed by a main contractor are subcontractors. So if they work directly for you then they aren't actually subcontractors at all, although everyone still calls them 'subbies'.

This arrangement often suits self-builders who plan to do some of the building work themselves. But there are some important differences between employing a main contractor and running your own show. Subbies tend to operate either

as one-man bands or as small firms and expect to be paid weekly (in cash) rather than in stages like main contractors. Their true loyalty tends to be to larger firms of contractors who can provide them with regular employment as part of a building team. They can't afford to let down a main contractor who consistently provides them with work, but they can afford to lose a one-off client.

This means that potentially there's a greater chance of

them not turning up on an appointed day, or even leaving a job only part-complete.

As a client who has chosen to employ some trades direct, perhaps doing some of the building work yourself, you'll need to take on many of the time-consuming responsibilities that you'd otherwise pay a main contractor to do. This means budgeting for increased overheads, especially phone bills and mileage. This project management role is virtually a full-time job, and the buck stops with you, so it's only advisable to take on this managerial role if you know your joists from your rafters. Otherwise there's always the risk that less scrupulous sub-contractors who think they're working for a novice could be tempted to cut corners.

Alternatively, you could arrange things so that your architect employs a number of subbies on your behalf. But this would significantly add to the architect's fees, because you're extending their role to 'contract administrator', where they monitor progress, value completed work for interim payments, and sign the completion certificate.

As client and main contractor rolled into one, your managerial responsibilities should not include policing your subbies' tax affairs. A one-off building project should not be construed by the Taxman as being connected to running a business or trade. Only if they deemed you to be a 'professional developer' would you get lumbered with all the main contractor tax liabilities.

Finding trades

Good tradespeople can afford to be picky. For the same reasons that clients often play safe and only appoint builders by recommendation, many subbies will be reluctant to accept a job unless it comes via a trusted contact.

The fact is, even major house-builders sometimes struggle to find reliable trades when the economy's booming. At worst you may have to take a calculated risk and employ someone and be prepared to sack them if you're not happy.

As always, the best way to find good tradespeople is through other self-builders. But it's also worth asking around at builders' merchants and tool-hire shops – they're unlikely to recommend people who don't return tools or fail to pay bills. Strangely, it's rarely worth asking tradesmen to recommend someone from within their own trade, as they

Photo: Wavin Hepworth

tend to be critical of competitors' work. However, they're usually happy to recommend trades that go before or alongside them, as they don't like following a bad one.

It's also worth checking trade associations that list local members – some offer insurance-backed guarantees. Simply being a member doesn't tell you how good they are, so ask them who they've worked for recently and be prepared to go knocking on doors of former clients.

When someone's dead keen to win a job, they may suddenly act like your new best mate. But what will they be like when things go wrong, or at the fag-end of the job? The only people who really know this are clients for whom they've worked before. This means plucking up the courage to ask complete strangers questions about someone who carried out a job for them in the past. If possible, take a look at the finished building to judge the quality for yourself.

When employing individual trades, a written contract of the type you'd use with a main contractor is not realistic. Many will run a mile at the thought, and in any case contracts can sometimes be difficult to enforce against an individual. The best arrangement is to get written quotes and confirm acceptance in writing, making reference to the relevant drawings (by plan number and date issued).

An exchange of letters and plans approved by Building Control is often sufficient. If your design incorporates a lot of newly developed products or alternative methods, avoid employing tradesman who are unfamiliar with them. Having researched building innovations, your knowledge may be way ahead of some trades, and you don't want to end up paying them to learn on your job.

It can sometimes be tempting to cut costs by arranging for 'a friend of a friend' to undertake jobs such as electrics or plastering. But with these kind of informal arrangements it's not unusual to find yourself left in the lurch because something else has come up at the weekend when they promised to finish your job. And if an important piece of work hasn't been done, the following trades will be held up, making this a false economy. Similarly, employing non-English-speaking trades invites all kinds of communication problems. A roofer who has just arrived from Bulgaria may have a totally different way of working that doesn't take account of UK Building Regs. Plus there is a tendency for illegal migrant workers to grab their tools and leg it out the back door as soon as the Building Control Surveyor makes an appearance! So for the more complex parts of the job, it's normally safer to stick with experienced, home-grown trades folk.

Pricing the job

To get each individual job priced, you need to obtain quotes from each trade in turn. But first, it's worth doing a little homework to make yourself aware of the 'going rate', so

Quirks of the trade

Each trade has its own quirks and particular ways of working:

- **Brickies** – Often work in gangs of two, plus a labourer. They don't usually supply materials. When calculating a quote, they may simplify things by 'measuring through openings' on the walls, not deducting anything for the air spaces that comprise the window and door openings (perhaps 25% of the surface area). This compensates for all the 'fiddly bits' at reveals, and window templates etc.
- **Carpenters and joiners** – Usually work as individuals and don't supply materials.

Trades that supply materials as well as labour ('supply and fix') typically include roofers, plumbers, electricians, plasterers, and some kitchen fitters. They make a mark-up on the materials they buy on your behalf.

- **Roofers** – May be hired in gangs, sometimes also organising scaffolding.
- **Plumbers and electricians** – Usually one-man bands or small firms.
- **Plasterers** – Often work in pairs or small gangs.

that it's easier to judge whether a quoted price is reasonable. But should you get quotations or estimates? A *quotation* is a firm price that is legally binding. It's a fixed sum for a fixed amount of work. If you're presented with a larger final bill, you're only obliged to pay the agreed quoted price, unless you've requested 'extras' or you've agreed to 'changes'. An *estimate*, on the other hand, is the builder's best guess as to what the cost might eventually be. It's not legally binding, and allows the builder to present a higher (or lower) final bill, and is consequently far too risky for major works.

If you're clear about what you want and provide sufficient information, you should in return receive accurate quotes. This means there's less chance of misunderstandings arising later on site and being charged expensive extras for things you forgot to include. But whilst a surveyor or architect might reasonably be expected to nail down every last detail, there's always a risk that you could forget something. So it's best to keep the wording fairly general, for example writing *'to include all necessary roofing work'* and *'all first and second fix work'*. This should prevent them claiming that a particular task wasn't included and charging it as an extra.

If possible, get at least two quotes for the same job. These should be submitted in writing as a firm price for the completed job. If you can't get a fixed price for the job, individual tasks may be quoted in terms of a 'metreage rate', known as 'pricework'. This is where you're given a price expressed as a 'measured rate', either *per square metre* (tiling, brickwork, plasterwork etc) or *per metre run* (for laying pipes etc).

If someone claims there are too many uncertainties to know in advance what the job entails, it usually means they're too lazy to read the plans. Such a claim may be true for some refurbishment jobs, but not for new build. So unless you trust someone implicitly, it's normally best to avoid quotes based on time taken, such as 'day rate' (a price per day), because of the obvious temptation to sit around

Tips for appointing and working with trades

- Go by personal recommendation where possible.
- Get more than one quote for each job.
- Take up references and talk to previous clients.
- Check that membership of any trade bodies is up to date, and especially that electricians and gas fitters are suitably qualified and registered (eg *Gas* Safe registered).
- Be careful about beating them down on price – if they feel 'robbed' they may claw it back later.
- Make sure all insurances are in place.
- Be very clear about what you want, and try not to change your mind.
- Remember, if you leave finding a trade to the last minute you'll be charged more.
- Agree a price rate in advance for any extras.
- Guard against poor quality work by confirming everything in writing, with reference to drawings.

and string the job out indefinitely. Daywork may be acceptable for small parts of the job, at agreed rates, but not for the whole job.

If any prices come in a bit on the high side compared to the competition, be prepared to negotiate. But knocking the price down too far can backfire. By reluctantly accepting a lower price, a tradesman may be tempted to skimp on the job or go out of their way to charge for extras. So it's important to be clear about what exactly is included in each individual's price.

Like any team, yours will only be as strong as its weakest link. So you need to ensure that each person knows exactly what's expected of them. Otherwise the job will get done the way that's easiest for them rather than the way you want it.

10 PROJECT MANAGEMENT

Photo: buildstore.co.uk

One thing is certain about any building site – at some point things will go wrong. No matter how well you've planned and organised your project, with a complex operation like building a one-off house you can never cover *every* eventuality. So you need to be ready to take effective action to prevent any problems that arise from getting out of hand.

Most of the advice in this chapter is equally relevant whether you've chosen to appoint a main contractor or to directly employ individual tradespeople and 'cut out the middleman'. However, some of the issues discussed later in the chapter ('Extras and changes', 'Penalties', 'Disputes' etc) are applicable for projects where a main contractor is primarily responsible for all the day-to-day hiring and firing of subbies as well as coordinating materials and trades on site. Here, the project manager's role should essentially be limited to overseeing the main contractor.

In contrast, where you directly employ your own workforce, the management role will be considerably more demanding and time-consuming. As we saw in the last chapter, this will mean taking personal responsibility for the smooth running of the entire project on a daily basis.

The importance of good project management cannot be overestimated. Achieving the holy grail of completing the job on time, whilst simultaneously meeting cost and quality targets, doesn't tell the full story. Project managers are often the unsung heroes who prevent problems arising in the first place. For example, preventing someone getting hurt on site means you haven't got to deal with all the stress and expense of possible criminal charges and ensuing claims for damages.

Organisational skills are often something that self-builders can contribute personally, so it's rare for this role to be delegated. In any case, it's not always easy to find

The project manager's role includes some or all of these tasks

- Refining the budget costings.
- Listing quantities of materials.
- Drafting a programme of works.
- Monitoring costs as work progresses.
- Negotiating labour prices and appointing trades and sub-contractors.
- Quality control on site.
- Hiring, firing and managing site personnel.
- Authorising payments and arranging progress certificates.
- Ensuring the build complies with approved drawings and the Building Regs.
- Ensuring insurance is in place.
- Complying with health and safety legislation

suitable professionals willing to take on a relatively small development. If you've employed your architect on a 'full service' package, they will carry out an inspection once every fortnight or so, instruct any necessary alterations, and issue extra information if required. If work is unsatisfactory, they can require the contractor to correct or rebuild it. However, architects aren't always as good at site management as they are at design, and paying more for the 'full service' can increase the total cost to more than if you'd employed a main contractor. And as the client you'll still carry some of the risk if things go wrong.

Even with package-deal self-builds where everything's done for you, the project manager's first loyalty is likely to be to the kit company who employs them. If you decide to source your own manager, try to find someone with good experience of one-off housing projects who's up to speed with current site regulations. The ideal person would be a retired Building Control Surveyor. Of course, their fees will be higher the more they take on, so it's important to define exactly which tasks you're paying them to carry out.

Five golden rules to managing a build

- Never pay in advance.
- Agree payment dates with the builder before work starts.
- Agree a completion date with the builder before work starts.
- Make regular site inspections.
- Record key events in writing.

Management style

So you think you can project manage? Having the right temperament and the ability to handle people in challenging situations is just as important as knowledge of the building process. But management attitudes in offices and factories don't always translate easily to the building site. For example, arrogantly boasting about 'showing these guys how to do their job' probably isn't the best approach.

Fortunately, there are some simple common-sense rules to getting the best out of people on site:

Attitude

Adopting the right attitude is key. Think of yourself as the hub of a large wheel, where it's your job to coordinate all the various trades on site with deliveries of materials, plant hire, stage payments, and Building Control inspections. Although you may be the person least experienced in the subject of building, the ability to manage people and situations is more important than detailed technical knowledge (which you can buy-in from professional advisers).

When it comes to managing trades, remember that these are self-employed people who have chosen to opt out of the mainstream and instead endure the harsh 'hire and fire' world of the building site. Each represents their own business, and should not be treated as employees or servants. But, at the same time, it's important to remember that this is your project, not theirs, and that this is not an equal partnership. To them it's just another job.

Life on building sites is often uncomfortable and dangerous, which may explain the pervading dark sense of humour. A site with laughter generally runs better than one where there's an air of tension and pent-up animosity. Building sites are not formal places, so adopting a formal attitude to relationships can sometimes incite feelings of hostility. On the other hand being too matey could encourage some individuals to take advantage of your inexperience. The best approach is to be reasonably friendly, but to maintain that little bit of distance.

Be prepared to compromise

You rarely get 100% of what you want on any building project, so there will be times when compromise is necessary. If you genuinely believe that changing something and deviating from your masterplan will damage the appeal of the finished product, by all means dig your heels in. But because this is your baby, it's very easy to become totally attached to every tiny design detail, and get overly 'precious' about something that isn't ultimately going to make much difference.

A failure to compromise can be damaging to the progress of the entire project. So if, for example, the bathroom washbasin that you've set your heart on is no longer available, be prepared to consider alternatives, even if it means spending some of your contingency fund – that's

what it's there for. Details that seem crucial will, in time, more than likely pale into insignificance.

A self-build project can take over your life, so it's vital to stand back and look at things in perspective, and not jeopardise the success of the entire project by arguing about something inconsequential.

Progress and timing

Should you set yourself a strict timetable that must be adhered to at all costs? Or would it be better to take a more relaxed approach, focussing more on the quality of the end product and enjoying the experience? Most self-build projects fall somewhere between these extremes, typically taking a couple of years to complete when directly employing individual trades.

Even where there's a main contractor running things, it's unrealistic to expect everything to run like clockwork all the time – the best firm in the world can be the victim of last-minute 'no show' deliveries, or subbies going AWOL. So the occasional day with no one on site is to be expected, and a bit of slack needs to be built into the programme to take account of this.

Larger firms will normally be juggling manpower between several simultaneous jobs, and occasionally an electrician or plumber will be urgently needed at two sites at the same time. So you need to allow your contractor some breathing space to manage their workload.

However, as project manager you need to regularly compare progress on site to your original programme, and a site that remains spookily silent for days on end will soon mess things up. If this is the case, you'll need to formally request that progress is restored. But you need to be reasonable, as some things are not foreseeable.

On site there are two major influences on progress – weather and cash flow. Only the latter is directly within your control. Thousands of man-hours are lost each year due to inclement weather – as builders sit in huts whilst trenches flood or freeze. The weather isn't the contractor's fault (but unless it's exceptionally bad for the season, this isn't necessarily an acceptable contractual reason for projects overrunning). Where you encounter prolonged severe conditions, the completion date may need to be extended. So to minimise disruption, try to avoid attempting groundwork in winter, as the plot turns into a sea of mud. Site excavation and pouring foundation concrete is best done during long periods of dry weather when the ground is firm. Ideally, aim for a weathertight shell by the end of autumn, so work can continue indoors.

Understand the other guy

Quotes usually include a bit of 'fat' to allow for any minor changes. But if major unforeseen problems arise, the quote probably won't have taken account of it. You may be within your legal rights to have this work done contractually, but forcing people to do what they may regard as unpaid extra work will very likely result in them skimping the job or walking away. Ultimately this would be more expensive than renegotiating. Your objective is to get your job finished, and you want the work done properly. The trick is to spot the difference between someone trying it on, and genuine misfortune. If in doubt ask someone with site experience, such as a surveyor or architect.

Communication

One of the hardest parts of any successful build is achieving good communication. Holding brief site meetings with the main contractor perhaps every couple of weeks is essential. This means all parties, including yourself, can be regularly updated on progress. Much as builders instinctively dislike 'wasting time' turning up to tedious meetings, in the long run this will save time by minimising the risk of costly mistakes and should therefore prevent minor disputes from turning into major ones.

Photo: Wavin Hepworth

If you've appointed a project manager, such as your architect, to run the job on site, confusion can arise where clients issue directions directly to individual subbies without informing the project manager. Where a client wants to alter the design or request some extra work after the job has started, they should first raise the matter with either the project manager or the main contractor. Unless you're employing trades direct, avoid making arrangements with individuals on site. This is outside the terms of the contract, and will seriously dismay your main contractor, resulting in a bad atmosphere and a lack of future cooperation.

Listen and learn

Here you are, on a building site with lots of blokes who've been doing this for most of their lives. Many will know considerably more than the average designer about how buildings actually fit together. Yet they're asking you for guidance! Some novice self-builders don't know one end of a brick from another, but suddenly find themselves in the deep end having to make a lot of key decisions, which they then feel obliged to defend to the bitter end. But building a house shouldn't be a battle of pride – it's a team game.

There's an unfortunate attitude prevalent in British management that somehow it's a sign of weakness to consult with anyone less senior, much less the person who actually does the job in question. It isn't. It's actually a sign of confidence and intelligence.

Most experienced builders will come up with useful suggestions as the project progresses, which in some cases may simplify the design and save you money. They have an important contribution to make, so don't be too proud or suspicious and reject their ideas out of hand. Listen to what they suggest, but be careful to sift out advice that's simply aimed at making their job easier. Also be aware that some suggestions come with a price tag. If your friendly builder suggests that something 'might as well be done', don't necessarily assume he'll do it out of the goodness of his heart.

There are times when you'll be put on the spot and may not immediately know all the answers. When a tradesman asks if you want them to do something a certain way, they may just be demonstrating how knowledgeable they are. If you reply by saying 'What would *you* normally do?' they'll probably tell you what they see as the best solution. If this involves any significant extra cost, seek the opinion of others, such as your architect or warranty inspector, and do some research.

Don't get in the way

Many self-builders have good DIY skills and plan to undertake one or more of the trades themselves, employing subbies to do the rest. But professional tradesmen are busy people, experienced at judging how much time to allocate to each individual task. So it's important not to let your endeavours hold them up and delay progress. It's unlikely that you'll be able to work as fast as an experienced tradesman, and you can't expect them to hang around waiting for you to finish a job. Time is money. So it's sometimes better to stand aside and concentrate just on the project management.

Photo: Wavin Hepworth

But even in this role you have to guard against getting in the way. There's an art to judging the quality of workmanship without standing intimidatingly behind people, questioning their every move!

Don't keep changing your mind

When things don't turn out the way the client imagined, or a job runs over time, the builders usually get the blame. But probably the single biggest reason for delays is clients who constantly change their minds, requiring work to be taken down or done again. OK, you might occasionally need to change something like a door that would be better hung from the other side. But don't do it too often or the relationship will swiftly deteriorate and the project will lose momentum, as well as clocking up additional charges for all kinds of 'extras'. So it's essential to think things through in your head in advance, rather than working on a 'suck it and see' basis. If having the freedom to change your mind as you go along is important to you, then the best approach is to employ individual trades direct rather than tender the whole job in advance.

Know when to terminate

If you do make the wrong choice of builders, it normally becomes apparent fairly soon. If you're employing trades direct and they seem consistently incapable of doing a good job, then you need to move swiftly to terminate their employment. Don't be persuaded to give them a second chance. A bad subbie will be used to having a job terminated, so just pay them for the work done, accept that you made a mistake choosing the wrong person, and move on.

Running the project

If there's one key secret to a successful outcome it's to first 'build the house in your head'. It is claimed that every hour spent planning the job can save at least three hours sorting out problems later on site.

Planning ahead means identifying in advance who's going to be doing what, and when they're going to be doing it. This can then be coordinated with supplies of plant and materials, with the whole process lubricated by big dollops of cash injected at key stages. But for a project to run smoothly, before any work starts on site you need to have systems set up to help you manage it.

Programming

To successfully manage any build it's important to understand the sequence of work on site. But where you're personally managing your own team, it's essential to prepare a written programme showing what's supposed to be happening week by week. This will enable you to coordinate the delivery of materials with the arrival of different trades on site, and to ensure that sufficient funds are in place at the right time to pay for labour and materials. Get this wrong and you could end up halfway through the project with the nightmare scenario where some trades are refusing to work for you because they haven't been paid, and the mortgage company is refusing to release any more funds until the next stage is complete.

At its simplest, a programme shows when each part of the job is due to be carried out and by whom. It should flag up dates when key decisions need to be made, such as ordering materials, and will help you give trades plenty of advance notice.

To draw up a programme, create a spreadsheet listing all the key tasks down the left-hand side of the page and the weeks along the top. Next to each task, mark any that you plan to carry out yourself, and which specialist trades are needed. It's vital to check the availability of stocks well in advance, particularly materials like hand-made bricks and roof tiles that can be subject to shortages. Check the availability of labour and ask trades how long each task is likely to take. Then all the various times and dates can be completed, allowing some flexibility for bad weather and the inevitable delays where, for example, tradesmen may need to go off to other sites. But remember, these dates and times are only your best guess – they're not set in stone and you may need to fine-tune things as the project develops.

Site meetings

Where a project is primarily managed by a main contractor, you normally discuss the key issues at the contract-signing meeting. This is also the time to stress that the site must be run as safely as possible, with effective security, whilst minimising inconvenience to neighbours.

As we saw earlier, successful projects depend on effective communications between all parties, so this is a good time to book dates for regular site meetings. This also gives the contractor an opportunity to flag up any concerns and to request additional detailed drawings.

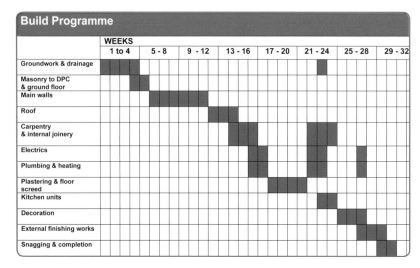

Build Programme

	WEEKS							
	1 to 4	5 - 8	9 - 12	13 - 16	17 - 20	21 - 24	25 - 28	29 - 32
Groundwork & drainage	██					█		
Masonry to DPC & ground floor		██						
Main walls			██					
Roof				██				
Carpentry & internal joinery					██			
Electrics					█	█		
Plumbing & heating					██			
Plastering & floor screed						██		
Kitchen units						█		
Decoration							██	
External finishing works							██	
Snagging & completion								██

Improvised 'site office'.

Certificates

There's often confusion about precisely what type of certificates you need for a newly built house. In particular, the differences between *building warranties* and *architects' certificates* often get blurred.

This is important, because in order to borrow money to finance the construction mortgage lenders normally require professional certification before releasing funds. Once the house is complete, if you need to sell it or remortgage within at least the first six years the buyer's solicitor will normally insist on seeing certificates. But which ones?

BUILDING WARRANTY

People buying a new house expect a building warranty. Likewise, most banks will only advance mortgages on newly-built properties where they are covered by a building warranty (although some will accept a PCC Certificate – see below). Although NHBC are perhaps the best-known provider of ten-year insurance-backed warranties, LABC is now the major provider of self-build warranties in England and Wales.

Most self-builders pay for a building warranty to provide a safety net against catastrophes such as your builder going bust or structural failure. Cover should be arranged before starting construction – once the foundations have been poured it's usually too late. However, the inspections undertaken on site are fairly limited, and the warranties don't cover problems they regard as minor, such as flaking paintwork.

No certificate will guarantee that your building has been perfectly built, but the safest route is to have both a building warranty and a PCC certificate.

CML PROFESSIONAL CONSULTANTS' CERTIFICATE (PCC)

The main type of certificate issued by architects/contract administrators is often referred to as an 'architects' certificate'. However, it's correctly known as a 'Professional Consultants' Certificate', published by the Council of Mortgage Lenders (CML). Issued towards the end of the project, this certificate can only be signed by a suitably qualified professional consultant, usually an architect or chartered surveyor. A signed certificate confirms that an experienced person has monitored construction on site at appropriate intervals to check the standard of the work and also the general conformity with the Building Regulations. The person who signs it will then remain liable to the owner, subsequent owners, and any mortgage lender for a minimum of six years.

INTERIM CERTIFICATES

Interim certificates are the other main type of certificate issued by architects/contract administrators. They confirm that parts of the building works are complete, and are issued at regular intervals throughout the build – normally every four weeks, or when key stages are reached.

Once issued, interim certificates trigger payment to the contractor for all the work they've completed to date. But because they're only 'interim' they're not proof that the work has been done 100% properly, and you retain the right to hold back payment later if anything's not up to standard or it turns out there are hidden defects. These inspections have the added benefit of showing the contractor that their work is being closely monitored and is linked to payment.

Where self-builders are reliant upon mortgage funding released in regular stage payments, interim certificates are sometimes issued at 'key stages' to trigger these payments. The usual key stages are:

- Excavation of foundations
- Walls built up to DPC level, drainage backfilled, and ground floor structure in place
- Walls up to upstairs floor joists (chamber joists)
- Pre-plaster
- Completion

However, some lenders are more concerned with the certified value of the building to date, rather than the detailed quality of the build, and will send their own surveyor to confirm the stage reached and provide an opinion of market value.

CERTIFICATE OF PRACTICAL COMPLETION (PC)

The majority of Building Contracts require the Contractor to achieve Practical Completion (normally abbreviated to PC). To confirm this stage has been reached, a Practical Completion Certificate is issued by the architect/contract administrator. 'Practical Completion' is usually understood to mean the date when you can occupy the house, without any major inconvenience from building work. The builder may still need to complete a number of outstanding items, which should be listed along with the certificate.

The usual safeguard against any minor 'snagging' items not being finished is the retention mentioned in the previous chapter (typically 5%), half of which is released at Practical Completion, which is also when the defects liability period commences, leaving 2.5% to be retained for a further six months.

FINAL CERTIFICATE

The remaining 2.5% retention is normally paid to the contractor six months after Practical Completion marks the end of the build, subject to a 'final certificate' being issued by the architect. Before the final certificate is signed, the contractor must make good any outstanding defects and snagging items. There is potential for misunderstanding between clients and architects with Final Certificates, since it is not a guarantee that everything's been done perfectly. What an architect has done is 'inspect generally the progress and quality of the work'.

BUILDING CONTROL COMPLETION CERTIFICATE

After their final inspection, Building Control should provide a completion certificate (confusingly also sometimes referred to as a 'final certificate'). Although a PCC can be issued without the Building Control certificate it is usual for the professional consultant to issue his certificate after seeing the Building Control completion certificate. This is a 'belt and bracers' approach for the consultant, since he has to confirm that the new home complies with Building Regulations.

Keeping records

Keeping accurate records doesn't happen by magic. To be properly organised you'll need to create a project file – preferably a large ring-binder – that can be produced at meetings. This should contain key documents with separate sections for each of the various parties, starting with a page listing everyone's contact details.

Project folder key sections:
- Site meetings
- Main contractor or individual trades
- Architect/designer
- Structural engineer
- Planning
- Building Control
- Utilities
- Materials suppliers
- Product information

By the end of the project everyone will have forgotten exactly what happened, or didn't happen, weeks ago on site. It's therefore worth keeping a simple daily site log of things like weather conditions (at least for the external work) and what labour and materials were present on site each day. Ideally take dated photos to prove it.

Site safety

Building sites are renowned for serious accidents. In the event of injury or death occurring on your site, it's possible that criminal law could apply, and you could be held personally responsible. Ignorance of the law is no defence.

If you're employing more than five people on your site at any one time the Management of Health & Safety at Work Regulations 1999 come into play. This means you must be able to demonstrate that you've taken reasonable precautions to prevent injury – for example, you should:

- Keep some hard hats available on site, together with luminous vests, goggles and masks, in your 'site office' or shed.
- Clearly display notices requiring hard hats and vests to be worn at all times on site.
- Provide a First Aid box and an accident register notebook.
- Ensure scaffolding is only erected by qualified firms with adequate insurance cover.
- Provide staff facilities including a WC, and if possible an enclosed space for meals, away from the building works.

Approximately half of all construction fatalities are due to falls, hence the Working at Height Regulations, which, amongst other things, restricts the use of ladders. But ladders aren't the only worry. The risks from such routine objects as power tools, sharp blades, flame guns, toxic or flammable sprays, and live electric cables are very real.

As if that wasn't dangerous enough, self-builders living on site as a family can create a potentially lethal mix with children, pets, and unwary visitors in the vicinity. Combined with bricks dropped from scaffolding and dangerous holes in the ground, this is a personal injury lawyer's dream. So if you're living on site in a caravan the main building site should be fenced off from the family area. It's not just hazards from tripping and falling, avalanches of piled up materials or being hit by flying bricks you have to worry about. Even the most familiar materials can harbour risks – for example cement dust, concrete and wet mortar are corrosive and can cause nasty burns.

The Construction (Design and Management) Regulations

But CDM regulations impose a legal obligation to protect the health and safety of site personnel. However, where as a self-builder you intend to live in the completed house, you're classed as a 'domestic client' and most of these conditions can be disregarded. However, where the intention is to sell the property at completion, you must fully comply with the CDM Regs (see website). Self-builders normally need only be directly concerned with two regulations:

- **Regulation 7** – Requires that you notify the Health and Safety Executive (HSE) about the project in advance.
- **Regulation 13** – Imposes a legal duty on designers to minimise risks, so it's a good idea to specify aspects of site safety in the 'prelims' and to encourage the builder to keep the site as tidy as possible at all times. You must also ensure that children are physically excluded from entering sites by erecting suitable child-proof barriers.

SAFETY

A lonely death

As a self-builder, you sometimes find yourself working in isolation, perhaps late into the night or at weekends. Taking some simple, common-sense precautions can make the difference between life and death should an accident happen when working alone:

- Always carry a charged mobile phone.
- Let a friend know what time you expect to finish and have them contact you at set intervals.
- Only lift stuff that you know you're capable of lifting safely, and don't overload wheelbarrows.
- Get into the habit of wearing a hard hat on site, along with steel-capped reinforced boots and a safety jacket. Wear goggles when using cutting or grinding tools etc, and don builders' gloves when necessary.
- With electrical power tools, use an RCD (residual current device) that automatically disconnects the power supply.
- When hiring plant and machinery always read the instructions.
- Use scaffolding for work at height, not ladders. Before ascending scaffolding notify the contractor.
- Be tidy! Most accidents occur because sites are cluttered.
- Look where you're stepping.
- Provide a First Aid kit on the wall in an obvious place (eg the site hut), containing plasters, bandages, antiseptic, and eye-rinse.
- Keep a fire extinguisher or a bucket of sand handy in case there's a fire or flammable liquid spillage.

Photo: Prestige Fire Protection

SAFETY

Dangerous places

- Trenches more than 1m deep may need to be temporarily shored up with timber. Never work alone in a deep trench. And if there's any risk of collapse, always deal with it from above!
- Don't get up on roofs unless you're confident at heights. Even then don't do it without scaffolding.

- Self-builders – and their visitors – regularly fall down open stairwells. A simple temporary handrail can prevent serious or fatal injuries.
- If you must use ladders, tie their tops to the floor or scaffold above. Where ladder feet could slip, fix a board across the bottom. Remove ladders at the end of play each day. Do not use old-fashioned wooden ladders.
- There's an art to walking on floor joists. Many inexperienced people slip through and get ruptured – or worse. Watch out for temporary boarding slung over joists, as it's often loose, with unsupported ends.
- Drainage manholes should always be covered up with robust, rigid boarding.
- Watch your back! One of the easiest ways to put yourself out of action for a few weeks is by carrying or offloading heavy materials or by aggressive digging. Even the scrawniest professional builders are used to handling heavy weights – your body may not be.

Insurance

Unlike main contractors, subbies don't usually carry full professional insurance. So as soon as you decide to employ individual trades, it becomes your responsibility to arrange this. The basic cover you'll need should all be provided in a self-build site insurance policy:

- **Employer's liability insurance** – This covers you for accidents to employees; for example, if one of the trades you employ has an accident and their solicitor advises them to claim against you. In this context, 'employees' includes:
  – Direct employees.
  – Labour-only subcontractors, working directly for you or via another person/firm.
  – Persons you've hired from another employer.
- **Public liability insurance** – This covers you should your building works damage or injure a third party, such as friends or family lending a hand on site. It also provides cover for anyone you invite to take a look around, for example if your architect disappeared down a hole in the ground during a site visit. Even injury to trespassers is included, such as where some kids decide to climb your scaffolding one night and one of them slips to their doom, or where a burglar injures himself breaking in and subsequently tries to sue you!
- **Contract works insurance** – This covers you for site risks, including theft, and damage caused by vandalism, fire, flood, storm etc.

However, if you personally get injured, self-build insurance is of little use. Imagine what would happen if you become incapacitated. There's no way you can manage a building project while recovering from injuries in a hospital bed. So work on site would probably grind to a halt for weeks or months unless you somehow managed to hire someone to run things in your absence, at considerable expense. To cover such an eventuality, you need to take out additional insurance for 'personal accident, death, and permanent injury'.

Dealing with suppliers

Ordering materials

Getting the right materials delivered to the right place at the right time can be a demanding task, one that clients using main contractors often take for granted. So if you plan to take responsibility for sourcing your own stuff, start by writing out a detailed shopping list of all the materials you need (known professionally as a 'bill of quantities'). This is based on your specification and plans, from which all the areas and volumes can be calculated with the help of a scale rule. Your specification should state clearly if you want any trades to supply their own materials or just 'labour only'.

Photo: buildstore.co.uk

When ordering materials you often have to 'over-order' by accepting the nearest pack-quantity size. Bricks, for example, are delivered in packs of 410, which may prove inconvenient if you don't need the full amount. You also have to remember to make an allowance for materials that get wasted or damaged in some way, perhaps to the tune of around 10%. Ordering more than you need may be expensive, but buying too little is usually more so.

Having worked out exactly how much of what you need to order, the next step is to select two or three major suppliers (usually builders' merchants) and give them your shopping list together with a copy of your drawings. Ideally show them samples of the specific bricks and tiles you require.

Discounts

A source of much gnashing of teeth amongst self-builders is the fact that builders qualify for big percentage trade discounts on materials, whilst the rest of us seem to qualify for little more than a frosty smile in the average builders' merchants. But to qualify for membership of the maximum discount club, most contractors have had to do a lot of hard bargaining and provide regular business over several years. Nonetheless, even small builders expect to pay substantially below the list prices shown in catalogues, so there's no reason you shouldn't be able to negotiate a sizeable discount for such a large order. But no one's going to offer you builders' trade prices unless you haggle.

You first need to open a trade account, which should provide you with at least a month's credit. It's sometimes claimed that to qualify for deeper discounts it can help to include the word 'Builders' in the name of your account. It also doesn't do any harm to stress the fact that you're a good payer.

For some materials it can be worth comparing prices with those at the big DIY sheds like Wickes (who are actually owned by Travis Perkins), as these can sometimes be as competitive as trade prices elsewhere.

But some specialist merchants, notably plumbing and electrical suppliers, still openly discriminate against non-trade people, even if you've got an account. Fortunately the massive combined buying power of self-builders is now recognised, with many suppliers offering attractive volume discounts and immediate access credit to the tune of £15,000 or more.

Photo: buildstore.co.uk

A single large supplier should be able to provide all of the big stuff you'll need, from cement and aggregates to roof tiles. Builders' merchants are normally most competitive when supplying timber, mass-produced joinery, sand and cement, drainage, and plastering materials.

To compare quotes from rival suppliers, ensure that you're actually comparing like with like and that the materials are the same. Also check that VAT is included, and that quantities and delivery dates are all clearly stated.

If you don't fancy your chances haggling with the stony-faced hardmen of the building supply world, there is a canny alternative. Assuming you have implicit trust in your builder and his trade connections you could offer to pay him to source the same materials at perhaps 5% to 10% below your best supplier's quote. The attraction for the builder is that if he can then obtain some of the materials for a lower price through his contacts he gets to keep the difference.

Bargain materials

Cheap doesn't always mean good. It sometimes means dodgy and expensive. If you come across goods being offered on the cheap, check their quality. Don't do business with passing tarmac salesmen with no fixed address, and avoid offers of cheap concrete – this is too important to risk gambling with the quality of the mix. Remember that if you unknowingly buy stolen goods the legal ownership of those goods doesn't pass to you, and they can be legally reclaimed by the original owner.

You can sometimes make significant savings by going directly to a foreign supplier or manufacturer, cutting out the UK agent's share of the deal. But rather like importing a new car at a bargain price on the 'grey market', there are risks

Photo: potton.co.uk

attached. It's one thing to order drinks in a café abroad, but are you up to communicating a technically complex specification in Polish? Are the materials compatible with specified British Standards, and do they comply with UK Building Regulations? Then there's the cost of long-distance deliveries, and any import taxes and a weakening pound, which may cancel out much of the saving. And what if the quality turns out to be rubbish – is there a reliable guarantee? If not, who'll pay for the cost of returning the goods?

Supply and fix

Some trades, such as electricians and plumbers, operate on a 'supply and fix' basis, providing their own materials. This isn't necessarily a problem. Roof tiling can sometimes work out cheaper on this basis due to the large trade discounts available for bulk purchase, even after allowing the roofers to keep a share of the savings they negotiate on materials. So if you supply the tiles, the price charged for labour may be dearer in order to compensate.

Photo: Marley Plumbing

Plastering is another trade that's traditionally dealt with on this basis because plaster has a limited shelf-life. If you insist on supplying the materials, there's always a risk that any bad results can be blamed on the materials you supplied. Plumbers normally supply pipework and basic materials but not sanitary fittings, unless requested. However, if your contractor's quote includes supplying and fixing high-value components such as bathroom fittings or kitchen units, it's worth asking them to state the cost of the fittings separately. The contractor's price should be pretty close to the price you'd pay if buying them yourself without a discount. They'll get a trade discount that will cover their 'margin' and so shouldn't need to bump up the price. Builders can normally buy kitchen units at discounts of 20% to 40% off the list price.

Manufacturer design services

Many suppliers of kitchens and sanitary fittings offer a free design service, as do some specialist plumbing and heating equipment suppliers. It's worth making use

Photo: William Ball Kitchens

of such services. For example, kitchen suppliers can normally provide a detailed 3D image of your proposed kitchen so you can easily visualise the new units in place before placing your order. Similarly, manufacturers of roof tiles can provide free calculations by running your roof data through their computers (to confirm total quantities of tiles, and optimum laps and gauging for different roof pitches). Manufacturers' websites can offer a mine of information about the quality of materials and after-sales service.

Lead-in times

Some materials have long lead-in times. Failure to anticipate these is a common cause of delays on site. Timber frame companies can require eight to ten weeks' lead-in time. But as a rule, anything manufactured to order is likely to require plenty of advanced warning, such as specially shaped bricks and traditional clay tiles.

Orders for purpose-made 'big feature' items such as specialist glazing, designer doors, staircases, and kitchen units need to be placed early enough to allow for longer lead-in times. Even then, delays often occur at second fix stage when custom fittings don't turn up as promised. So it's advisable to have some easy-to-source alternatives in mind.

Wherever possible, buy from well-established firms who should still be around to honour guarantees in the future, and be sure to check that custom-made items comply with Building Regs. Once your order's been placed,

Photo: velfac.co.uk

avoid changing the specification. As well as incurring extra charges this can cause your delivery to be relegated to the bottom of the list.

Most specialist suppliers require some form of deposit before commencing manufacture, so try to keep any advanced payment to a minimum and agree that any deposit would be returnable in the event of delivery being delayed.

Deliveries

Delivery dates are crucial. Get these wrong and you could end up paying trades to sit around idle on site, and then have trouble getting them to turn up again next time. But it's an unfortunate fact of life on site that materials sometimes arrive later than promised. It's not unknown for an essential delivery like the concrete wagon to let you down, despite

having been paid for in advance, which can obviously mess up your programme. So it's best to double check with a quick call to your supplier a day or so in advance to confirm the delivery date.

If something you've ordered doesn't arrive, you've got two options: continue waiting, or buy elsewhere. But even if you can source a suitable alternative at short notice, the fact that you've already paid a deposit can complicate matters. Plus the new supplier's delivery time may be considerably longer than your existing firm, who should now be working flat out to make up time.

Most suppliers give you a couple of days' notice of a delivery date, so they know there'll be someone on site. If there's no one around, the driver may make his own decision about where to dump the goods unless you've given them specific instructions. Remember that some materials may need cash on delivery.

There are a few golden rules for dealing with suppliers. First, don't pay for goods in advance, unless absolutely essential. Second, where you have to pay a deposit insist that it's kept 'ring-fenced' in a client account, where it should be protected if they went bust. And third, always check materials upon delivery. Even with an impatient lorry driver huffing and puffing you're perfectly within your rights to check the goods for damage before signing them off. If in doubt write 'unchecked' next to your signature on the delivery ticket. If you find evidence of damage, put the items back on the lorry, and contact someone fairly senior at the suppliers to resolve the problem.

If any goods prove unsatisfactory, you have the same legal rights as anyone else buying from a shop – *ie* that the goods should be *fit for purpose* and of *merchantable quality*. Any reputable supplier should change faulty goods without quibbling.

Storing materials

The key to managing incoming materials is to plan ahead. To avoid access problems, you need to allow at least 3m width for vehicular access, plus space for trucks to turn, park, and off-load. Then there's the matter of security – the longer you leave stuff piled up on site the more temptation there is for thieves to liberate it.

Photo: Top container hire

them off. Site insurance is therefore essential for self-builders. You may even be able to make a claim where goods get damaged on site.

Loss of certain goods at key stages, especially custom-made items, can seriously disrupt your build programme. So when higher value, easily transportable goods are delivered make sure they're unpacked directly, as this makes them less attractive for a thief to sell them on. Better still, try to get valuable items fitted as soon as possible. Otherwise, lock them away carefully. It can even be worth taking items like boilers and hobs home with you for safekeeping.

Wastage and reclaimed materials

It's better to have stuff left over than for work to be held up because you're a few tiles short. Reclaimed materials often have a higher wastage factor than new ones. So you may need to over-order by up to 20% on second-hand bricks, slates, and tiles. One batch of period bricks or tiles may have been sourced from a particular demolition site, and the ones in a later delivery may be very noticeably different.

A supplier goes bust

Where you're ordering materials from suppliers, paying on credit is a useful precaution. This way you're covered if the supplier goes bust before the goods are delivered. Otherwise you'd end up at the back of the queue with all the other unsecured creditors, trying to get your money back. If this does happen to you, all you can do is get on with the job and urgently source the same supplies elsewhere to prevent the build on site grinding to a halt, effectively writing off the money. Even where a supplier goes bust *after* you've received the goods, you still can't relax. You actually need to move quickly to secure the items from the attention of creditors who may want to repossess them. Once they're fixed in place they technically become your property if you've paid for them in full.

However, because builders qualify for juicy trade discounts it can sometimes make sense to have them order materials on your behalf (*eg* on a 'supply and fix' basis). But what would happen if the builder then went bust? The answer would depend on whether you'd already made payment. Suppose, for example, that your builder had placed an order for your purpose-made windows, but you hadn't yet paid him for them. In such a situation it may be possible to renegotiate the contract directly with the firm supplying the windows so that you pay them instead. However, if you've already paid the builder up front for the materials and they go bust before delivery it will be a matter for the liquidator.

Staggering delivery times can help avoid overcrowding the site, with mountains of materials blocking access routes. Try to get deliveries stored away swiftly, covered up, and distributed around the site to where they'll be needed. Make full use of lockable garages and garden sheds.

Of course, it helps greatly if you've got a generously-sized site with plenty of space for storage, so that materials requiring protection can be delivered close to the day when they're needed. Things like bricks, blocks, and roof tiles can be stored for quite a long time and are mostly supplied in packs pre-wrapped in polythene sheeting to help protect them from excessive damp and frost.

Particular care should be taken with piles of sand – the local cat and dog populations will be itching to make full use of them! Tarpaulins or plastic sheets weighed down with blocks should do the trick.

Materials like plaster and cement powder need to be delivered fresh, just prior to use. Materials made of timber are especially vulnerable to distortion and warping, especially if allowed to get damp. Roof trusses must be stacked on a level base kept clear of the ground and covered. Don't forget to leave sufficient room for builders to work, with plenty of clear space around the building. A tidy site is a safe site, so hiring a giant container can solve a lot of storage and security headaches at a stroke.

Theft

Unfortunately, items sometimes disappear from sites shortly after delivery. Your site will need to be fenced off to deter intruders, using steel fencing, which can be hired. However, thieves may be keeping an eye open for high-value, easily sold deliveries, or someone on the inside could be tipping

Plant hire

Just when you're starting to appreciate how much the main contractor does for his money, there's the curious world of plant hire ready to absorb more of your time. Just about everything can be hired, from JCBs, scaffold towers, and cement mixers to angle grinders, power breakers, and tarpaulins, normally on a daily or weekly basis. But hiring only really makes sense over short periods. Hire for more than a few weeks can mean you could have bought the tools for the same money. Be clear about whose job it is to arrange plant hire – there's a limit to how many excavators you can make use of at one time.

If you're employing subbies, it's a good idea to discuss each trade's requirements beforehand. Whereas some subbies tend to arrive fully tooled up (*eg* electricians, plumbers, and carpenters) others, such as brickies, may only possess spirit levels and trowels, leaving you to fork out for cement mixers and scaffolding.

Scaffolding is normally priced for a hire period of eight to

ten weeks, with a small surcharge levied for each extra week. The price should include scaffolders making return site visits at different stages to erect the next 'lift' (storey) when needed. Prices are quoted in terms of pounds per metre run for each lift.

Mains services providers

Builders' merchants may not always be paragons of courtesy and helpfulness, but compared with some utility companies they're customer-service champions. Many privatised gas, electricity, and water businesses offer customers two levels of service: slow and expensive. Usually both apply. First there's the challenge of getting hold of the right department. Then there's the huge amount of paperwork and the fact that you have to pay up-front with lead-in times of six weeks or more. So once you've managed to obtain a quote, be sure to make payment swiftly. And when you've finally been given a reference number, you need to act promptly to arrange a date for the supply connection work (which you can later put back if necessary).

Water is the most urgent of the services, and a building supply should be arranged as early as possible since starting work on site may depend on it. If there's a delay, a friendly neighbour may be willing for you to run a hose to their outside tap, or at worst you could organise a water bowser. Electricity isn't always essential, as trades can use petrol-powered generators, diggers, and mixers.

Extras and changes

As noted earlier, it's essential to carefully think through your plans at the design stage so that you don't keep changing

'Did I mention that we wanted a pool?'

Photo: oakwrights.co.uk

Photo: Charnwood stoves

your mind as the build progresses. Of course, it's impossible to anticipate *every* eventuality, so the need for a few small changes is inevitable. These will be charged as extras and paid for from your contingency fund. But if you find yourself having to make any major changes, such as the unexpected need for special foundations, it's best to talk to your contractor and negotiate a revised contract price, as if you were starting again. In such a situation it's a good idea to get additional quotes from specialist firms (in this case from groundwork contractors) to help compare prices.

Varying the work

In an ideal world, it shouldn't be necessary to vary the work as it progresses. But in reality there are four situations where this may happen:

YOU REQUEST ADDITIONAL WORK

As you watch your new home take shape, something will occur to you that you wish had been included. Some builders make nearly half their profits from customers requesting extra works. You may therefore notice hands being rubbed with glee when you enquire 'Could you just move the kitchen so it opens into the living room?'

Always be sure to get an estimate of the cost of any additional work before it's carried out, and then note how long the work takes and the materials used. Keep clear records, because, as time goes by, it's very easy to lose track, culminating in an unpleasant surprise when all the costs are finally totted up at completion. What can really cause trouble is issuing belated 'surprise' instructions, such as extra roof work long after the scaffolding's been taken down, or for an extra radiator once the system's already up and running.

THE BUILDERS SUGGEST EXTRA WORK

Most builders have considerable experience and may have some genuinely smart ideas about how to improve the design, potentially saving you money by doing it another way. On the other hand, many are also highly skilled at the art of 'soft selling'. Friendly, casual suggestions, such as 'While we're at it we may as well put a skylight in the bathroom', may be perfectly pitched to illicit the required

consent from the client. It's fatal to assume there'll be no charge. Always ask how much it will cost.

UNFORESEEN CIRCUMSTANCES

The builders stumble across something unexpected, such as previously unidentified underground pipes running across the site. This requires extra work that wasn't specified. But who pays the additional costs? It depends on whether such work was reasonably foreseeable and should have been built into the contractor's tender price. Or, more likely, it should have been specified by the client.

ITEMS NOT CLEARLY SPECIFIED

Basically, you expect to get what you asked for. If you specified 'guttering' but didn't specifically mention the custom-made stainless steel guttering that you really wanted, the contractor will price for fitting cheaper standard half-round black plastic guttering. If your architect leaves 'holes' in the specification that end up costing you extra cash, you may be entitled to claim a proportionate reduction in their fees.

Instructions

By the end of a project, the subject of 'who said what to whom' tends to be shrouded in mystery and collective amnesia. For example, when you query why the bathroom walls have been painted a vile shade of slime green, the builder may quite reasonably point out 'That's what it says on the drawings, boss.' You know full well that you clearly told his mate Lucky Dave to change it to magnolia some weeks ago, but this news is greeted with utter disbelief.

The moral of this tale is to notify the main contractor first, not just a subbie (unless you're employing trades direct). Keep a written record of every instruction and change agreed, with dates and details. Send a copy to the contractor, and if possible keep a folder on site that everyone can refer to. Similarly, any changes proposed by the contractor, such as substituting an alternative for your first choice of white oak flooring because it's no longer available, should also be agreed in writing.

A written 'instruction' can allow the work to proceed where the cost isn't known, without holding things up. Instead, the cost implications are kicked into the future, to the 'final account' stage. But this can obviously be risky, so if possible costs should be nailed down in the form of a 'change order':

Agreed changes		
Date	**Instruction**	**Contact**
2 Feb	Paint colour to kitchen walls changed to magnolia	Lucky Dave
1 Mar	Change main external door to 'Malton Glazed' hardwood, 78 x 33in, ref 208-243	Crazy Jack MacLad

Change orders

It's always better to hammer out a price at the time that the required extra works are requested. To eliminate any possible doubt, a form known as a 'change order' (formerly known as 'variation orders') can be submitted, listing all the agreed changes and costs.

Change orders are written instructions from the client to the builder. These authorise him either to omit work that was previously agreed, or to carry out additional work, with the agreed increase or reduction in costs. It's also very much in the builder's interest for any extras to be confirmed in writing, since there's no legal obligation for you to pay for them unless they were formally agreed, so written evidence is important.

Ref	Change order	£
JB 1 20.7.11	Omit: external lighting system	(1,250)
BG 3 30.7.11	Add: Supply and fit 1 no. shower screen (B&Q ref 12345) to bath	105.00

So that you don't get fleeced for extra works, try to ensure that the builder sticks to the same profit margin for the extra works as were tendered for the original job. Of course, it helps greatly if the builder's original quotation or tender clearly spelt out the rates charged, for example 'brickwork charged @ £X per square metre', so that additional work can be priced at the same rates. Similarly, the cost of extra materials can be based on the original prices.

An alternative method is for extra work to be priced using 'daywork' rates, which are best agreed with your builder at the outset. Labour charges can be verified with trade associations, and the prices of material researched online. You can then monitor roughly how long the extra jobs take. But bear in mind that the cost to a main contractor of bringing a subcontracted electrician back just to fit one socket will be disproportionately expensive. The total cost would need to also include materials and allow a profit margin.

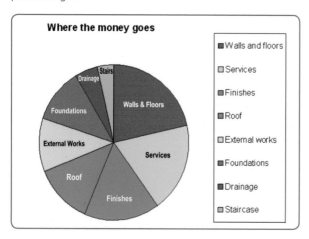

Where the money goes

Walls and floors / Services / Finishes / Roof / External works / Foundations / Drainage / Staircase

Changes to the contract period

Requesting additional work not only means a bigger bill, but is also likely to have a knock-on effect on your completion date. The job may therefore need to run over a bit, and the delays to your programme should be agreed along with the costs. Confirm the revised completion date in writing so that the builder won't be hit with any penalty clause for missing the original completion date. The builder may also incur extra hidden costs, such as hiring security fencing for a few more weeks, which he will have to pass on.

Money matters

If you're running the entire project yourself, there's more chance that your budget will overrun than where you employ a main contractor. Of course, this all very much depends on the accuracy of your original budget and how good you are at controlling costs. But on a day-to-day basis, the smooth running of a project will largely depend on how you manage your cash flow, because without sufficient cash available at key stages the entire project will simply grind to a halt.

Cash flow

As the build progresses, most self-builders find themselves spending money at a dizzying rate. Careful management of your finances is therefore essential to avoid the scenario described earlier, where work stops through a lack of funds, and funds stop because there's no work's being done. A stop-start project is a nightmare, so it's essential to know exactly when each tranche of cash will be available, *ie* at which stages funding will materialise. An 'advance-payment' mortgage can greatly help a positive cash flow by getting a life-affirming injection of funds in good time. Another way to breathe some colour into an anaemic bank account is by reducing outgoings. For example, by postponing an order for materials beyond the end of the month you can often qualify for an extra 30 days' credit at builders' merchants.

Builders similarly sometimes hit cash flow problems, where they have to pay out money for materials and labour before payment has been received for completed work – see below. Such financial difficulties tend to snowball over time and may originate from jobs done quite some time ago. A late payment from you could be the final straw, so it's important to ensure that sufficient funds are available to pay people at the right time.

Payments and retentions

The stages at which you pay your main contractor should be agreed in advance. Normally 'interim payments' are made every four weeks or so once an interim certificate has been signed off. To avoid 'double counting', each consecutive payment starts by totalling up all the work completed to date, and then deducting all previous amounts paid.

Normally the amount you pay a builder is based on the price they quoted for each piece of completed work. Where a job is only half complete, it's fair to pay a pro-rata amount of the price quoted for the full job. Valuing partially completed works is always going to be a bit of a guess, so it's not worth quibbling about the odd £50 here or there. In addition you must include the value of any substantial materials paid for by the contractor and stored on site, such as bulk deliveries of 'unfixed' bricks, blocks, tiles etc. Once you've paid the contractor for materials they'll legally be yours, so watch out that they don't later mysteriously disappear.

Next, any extras or changes that you've agreed may need to be added; if you don't have a firm price, estimate it approximately. Finally, after deducting any agreed amount of retention you'll arrive at the actual sum due to be paid.

The contractor may want to submit his invoice based on the work that he claims has been done, but to ensure you're not paying too much always carry out your own valuation. An alternative method of payment is by making simple 'stage payments'. This mirrors the way mortgage lenders advance funds to you as each major element of the work is completed, verified by their valuer. So you can then pay the builders a previously agreed amount for work completed to key stages (eg at completion of foundations, DPC level, completion of main roof etc).

Although as a rule large quantities of bank notes and building sites don't make good bedfellows, trades employed direct often expect to be paid in cash. However, main contractors can be paid by cheque or debit card. In all cases a receipt should be forthcoming.

PAYMENT RULE 1: NEVER PAY FOR WORK IN ADVANCE
Always pay in arrears, and only pay when the relevant work has been completed. If the builder goes bust halfway through the job, or they turn out to be no good and you want to terminate the contract, you'll be in a far stronger position. Also, paying for work that hasn't yet been done removes much of the incentive for good-quality workmanship to be completed on time.

If you're paying individual trades weekly, try to agree in advance how long the job is likely to take, and pay an equal amount each week (it's a good idea to agree in advance that you'll deduct 10% from weekly wages, to be added at the end of the job to the last payment).

There may be occasions when expensive materials need to be bought by builders, and they haven't got enough money, and ask for payment in advance. Don't agree to this. Instead, offer to buy the materials yourself and employ them on a labour-only basis. This way you're only paying for work done, and the ownership of the goods will be yours should they go bust.

PAYMENT RULE 2: ALWAYS PAY PROMPTLY
No one is going to do their best work if they're not paid on time. You're normally obliged to pay within 14 days of the due date (or within 30 days of Practical Completion) and failure to pay is a breach of contract. Typically, builders are at their most financially vulnerable towards the end of a job,

and paying promptly is an easy way to create a positive atmosphere of trust and cooperation.

Many subbies operate on a hand-to-mouth basis and expect to be paid promptly in cash, usually at the end of the week. If you don't have the money to pay them, they'll slip off to another job where they know they'll be paid. So if there's any doubt about funds being ready in time, don't wait until they're queuing up on a Friday afternoon with hands outstretched. Depending on your relationship, and how certain you are of having funds available in the near future, tell them in advance and leave the decision to them. At worst, if you anticipate problems consider calling a halt to the job until funds are available.

Penalties
Most contracts include a penalty clause, requiring the contractor to fork out money (known as 'liquidated damages') if the job isn't completed on time. This is agreed by both parties at the outset and stated in the contract as a  maximum sum of money payable for each week that the job overruns without a good reason. The client is in theory entitled to deduct this money from the builder's final payment. In reality, however, it can be hard to justify legally unless you can prove you've suffered real financial loss as a direct result, or that the delays were excessive and unjustifiable. Only larger firms of contractors are likely to accept such clauses. Nonetheless, this is a useful piece of ammunition in the event of delays and slow progress.

Practical Completion and final payment
Towards completion of the project, the main contractor will submit his bill, the dreaded 'final account'. It's at this point that many a client has required the assistance of a stiff drink. But the final payment should not normally be made until Practical Completion. This is the stage at which the client obtains full use of the building, although some small faults and minor finishing and works may still be outstanding and will need to be rectified during the next six months' – known as the 'defects liability period' (or 'latent defects period'). If you can schedule things so the final payment is going to be a substantial amount it means there'll be a greater incentive for the builders to fully complete the job.

Snagging

The chances are that your building work will have been regularly monitored by several parties. As well as your own inspections and those of your architect or some other project manager, there will be site visits from Building Control and quite possibly a warranty provider. But that doesn't mean every last detail will be perfect. And getting faults sorted after the builder has left the site may not be easy. So directly the contractor informs you that the work is finished, a formal 'snagging' inspection should be arranged. See Chapter 14.

Normally the contract permits the deduction of a small retention (say 5%) from each payment. Half this retained money is released to the builder at the end of the build, at Practical Completion, and the remaining half is paid after a further six months, the 'defects liability period'. However, if you're dealing with a trusted local firm you may not have needed to keep a retention, so it would be a good idea to delay the final payment until any minor snags are sorted. At the very least wait for the Building Control completion certificate before you release the final payment.

Guarantees

Many contractors guarantee construction work for a period of 12 months against material and workmanship defects. Insurance-backed guarantees are preferred, since they'll still

Photo: Wavin Hepworth

be valid if the contractor goes bust. But you also have a legal right for the contracted work to be carried out 'with reasonable care and skill'. So if a significant defect becomes apparent soon after completion, you may reasonably claim that the workmanship was negligent. If a contractor fails to honour the guarantee, you must write informing them that you intend to appoint another firm, at his expense, and give him a final opportunity to sort it out. But talk to Trading Standards or a solicitor first.

Consumer law should cover you for the quality of the various appliances such as washing machines and cooker hoods. Should your gleaming new oven promptly catch fire or the boiler conk out, legally it's the responsibility of the supplier of the equipment to deal with the problem – which may be your contractor, unless you supplied it yourself. Contractors must also provide test certificates for fitting appliances, as well as for all electrical, gas, and heating work, to prove that it all operates safely.

Latent defects

In some parts of the structure, such as foundations and roofs, it may take quite a while for defects to become apparent, often long after the 'defects liability period' has expired. It might only be after the first heavy snowfall or a severe storm that stresses cause structural cracking due to errors in the design or construction. Contracts normally allow you to take legal action for such 'latent defects' within 6 years of the appearance of problems. In some circumstances this can be extended to 12 years (*eg* where company directors sign the contract as a 'deed').

Dealing with problems

It is sometimes said that by the time a developer gets round to the last house on the estate, they've learned how to build them properly. So with a one-off 'prototype' property it's inevitable that some things will not go fully according to plan. Checking work as it progresses means that most problems can be nipped in the bud, so if you see anything that looks a bit dubious don't be shy about mentioning it to the contractor. If you think they're fobbing you off, seek the opinion of a surveyor or architect, or have a chat with your Building Control Surveyor.

If more serious problems arise, it's essential to be ready to take control of the situation. Firstly, always remain on speaking terms with all parties, and concentrate on solving the problem rather than apportioning blame. If things get serious, the contract should provide for dispute resolution, by naming an independent third party who can adjudicate (see 'Disputes' below).

A lot of difficulties are simply down to a personality clash with one individual on site, which can often be solved simply by asking a senior manager to reassign the person in question to another site. If a particular subbie is proving unreliable, bring it to the attention of the main contractor.

If you later need to withhold payment because of problems with workmanship or materials, it's essential to follow the rules set out in the contract, and notify the contractor in writing immediately, stating your reasons. If you fail to notify them and still withhold payment, you'll be in breach of contract and the builder can give you seven days' notice of his intention to stop work. If you're unhappy with the main contractor, bear in mind that it will be a major setback if they walk off the job. But as long as you haven't paid them in advance, you should have the upper hand financially. This means the contractor will also have a lot to lose and should be just as keen as you to get the project back on track.

Defining the problem

There are two basic types of problem: practical problems on site, and dissatisfaction with the builders' performance.

PRACTICAL PROBLEMS ON SITE

Occasionally an unforeseen technical difficulty will arise. The initial instinct is for everyone to immediately start blaming everyone else. To resolve matters, your first step is to accurately identify the problem. For example, it may be that the kitchen units don't fit because the available space is too short.

The next step is to consider possible solutions, such as using shorter units, or cutting back the plaster on the walls. Most builders have got a lot of experience at solving such problems, so don't automatically dismiss their suggestions. Many are good at thinking on their feet and coming up with solutions to practical problems. But this helpful 'can do' attitude can sometimes be hampered where an architect is too 'prissy' about agreeing to change something. This may be down to carefully calculated compliance with Building Regs or could just be because he's stubbornly fixated with a minor design feature.

When problems arise it may seem like the end of the world, and a threat to the entire project. But when everyone's calmed down there's always a solution. It's just a question of compromising, swallowing pride, and finding a way forward. Once a solution has been agreed it should be confirmed in writing. This should then allow work to proceed subject to reaching agreement about cost. But defining who's responsible can often be tricky. Were the plans badly drawn? Has the contractor missed a key detail? Hopefully it should be clear from the specification who was responsible. Often small problems where responsibility isn't obvious can be settled by agreeing to split the cost 50:50.

DISSATISFACTION WITH THE CONTRACTOR'S PERFORMANCE
You may be unhappy with the contractor's rate of progress or quality of work. Or both. If so, don't waste time discussing it with subbies – arrange a meeting with the main contractor. If there's no improvement write a formal, but not unfriendly, letter requesting that the matter is rectified. Most disagreements with builders boil down to one of four things:

Poor quality workmanship
Things tend to get botched when the person doing the job is tackling something unfamiliar. Specialist materials like lead, zinc, stone slates, and marble normally require specialist skills, not a 'jack of all trades' learning on the job.

Any decent builder will know very well whether work is sub-standard or not, and should rectify it when requested. If not, the obvious answer is to simply say you're not paying for poor work. You may find keeping photos a useful record. Ultimately an independent contractor may need to be paid to make good the defective work (remember some builders love nothing better than to 'slag off' other builders, so surveyors are a more impartial source of advice).

Prolonged delays
If the project is running late, the contractor will immediately give you ten good reasons why it wasn't his fault. Typically these relate to things like bad weather and undelivered materials, but a whole raft of personal problems may also be cited. Where work has not been completed, or is way behind schedule, you first need to check the contract and advise the builder that his inaction will put him in breach of contract, and that the breach must be remedied by him getting off his backside. Quote the relevant contract clause. If there's no response within a week, you may then need to go into 'dispute mode' (see below).

Charges for extra work
As noted earlier, charges for extra work should be comparable to the originally tendered work, or at least 'the going rate' in that trade. If all else fails you could refuse to pay any more than you think reasonable.

The extent of works required in the specification
You think a piece of work should have been included in the original price. The contractor does not. In order that the entire project doesn't grind to a shuddering halt, for the time being you may have to 'agree to disagree', and later go with the opinion of a mutually acceptable third party professional.

Dealing with financial difficulties
These are the warning signs: long periods when there's nothing's happening on site, or follow-on trades such as electricians and plumbers failing to turn up.

THE CONTRACTOR HAS CASH FLOW PROBLEMS
Builders in financial difficulties invariably have a hard luck story – typically someone hasn't paid them on another job.

But no matter how plausible, resist requests for money up front. The best solution is probably to agree with the main contractor that you make direct payments to subbies and suppliers to get things moving whilst still keeping control of your money.

Most builders rely on large overdrafts, juggling the money from one job to pay for another. Occasionally things get a bit stretched and one of the jobs has to stop for a while. If it's your site, this can obviously be a worry, but it doesn't necessarily mean they're going bankrupt. The first step is to find out exactly what the situation is. Talk to subcontractors. Be careful not to generate rumours of financial disaster that become a self-fulfilling prophesy as subbies and suppliers beat a hasty retreat for fear of not being paid. So don't say 'Do you think they're about to go bust?' A better approach would be 'Is everything going well between you and Brill Builders Ltd?'

If materials are failing to arrive, the builder may explain that 'There's been a bit of a cock-up.' But the reality may be that they haven't been ordered or the builders' merchants have put a stop on your man's account. To figure out what the problem is, you need to rapidly start making tactful enquiries at the builders' merchants, such as 'I'm calling to see what the situation is with the materials Brill Builders Ltd have ordered for my site.' Although they can't discuss their client's account details with you, their answer may alert you to trouble with the account. Or they may say that nothing's actually been ordered. But it's important to act calmly and not make matters worse. The last thing you want is to have to find a new contractor at this stage. Even if you haven't yet paid for the full value of the completed work to date, a new builder will charge considerably more to finish another's work. This isn't simply down to being opportunistic because you're desperate. They know that the first builder may have started skimping on the job once they knew they were getting into trouble.

There are three key things you can do if your builder is having cash-flow problems:

■ Agree to pay weekly according to an agreed schedule of work, so you never pay for more work than has been done.
■ Agree that the main contractor continues on site, but you pay subbies and suppliers directly for materials and labour.
■ Terminate the contract, and find a new builder, or deal direct with the subbies.

If you sensibly used a written contract (not everyone does) it should detail the arrangements for terminating and managing disputes.

THE CONTRACTOR GOES BUST

If your builder goes bust, you need to swing into action pretty rapidly. If the shell of the house is largely complete, change all the locks. If you're not living on site already, consider camping out, or have a friend maintain a presence. Secure the site's fencing with heavy-duty chains and locks, because a small army of hard men from the builders' merchants will very soon be dispatched to your site, intent on recovering their materials. They have the right to do this because legal ownership of the goods does not pass until paid for. Except for one killer fact – their materials are on your land, and they have no legal right of entry. So you need to prevent them gaining access.

Fix posters to the security fencing explaining that the goods were supplied to you by your builder, for which you have already paid, and that they have no legal right to enter your property. Also point out that recovering goods is a matter for the liquidator. If possible hand them a copy of this notice. This doesn't mean you can actually use the materials, but you can store them safely away. Such a situation would be considerably easier where you've not yet paid the contractor for the materials. This way you could come to an arrangement to pay the suppliers, saving both parties a great deal of trouble.

But there is another potential danger – this time from vengeful subbies, enraged at not being paid. Things can potentially get pretty unpleasant, because some may be intent on invading the site and repossessing anything of value in compensation. But again, they have no legal right to enter your property, or to cause criminal damage. Goods that are fixed in place are legally part of your property. But goods such as a boiler supplied by a subcontracted plumber but not yet fixed might legitimately be considered the property of the subbie. But where you haven't yet paid the main contractor for labour and materials supplied by the subbie, you should be in a position to renegotiate and pay them directly, thus solving the problem.

Basically, you need to make them understand that *you* are not the problem, and that you may well be the solution. So take the initiative and make contact first. You may be able to turn things to your advantage by inviting them to quote for completing the job, whilst submitting a joint claim to the liquidator for monies owed to date. Another essential step towards getting the job restarted is arranging a new self-build insurance policy in lieu of the original contractor's insurance.

The bottom line here is that you have the upper hand if you can keep these people off your site. As long as you haven't paid your builder in advance you should have the financial advantage, having only paid them up to the last agreed stage payment.

Where individual tradesmen that you employ directly go bust, it shouldn't have such major repercussions, as long as you've only paid for the work that's been done. At worst it may delay progress a little, perhaps holding up follow-on trades until a substitute can be found.

THE LIQUIDATOR

The liquidator's job is to salvage as much money as possible from the bankrupt builder's assets and then distribute them fairly to the creditors. If you can't resolve payment of

Action to take if your builder goes bust

■ Change the locks on the house.
■ Secure all fencing and gates to the site and if possible install CCTV.
■ Fix notices on fencing denying access to unauthorised parties.
■ Notify your solicitor, the police, your insurers, and your warranty provider.
■ If possible quote the liquidator's contact address, c/o your solicitor.
■ Camp out on site or appoint a security guard.

materials on your site with suppliers, an offer may need to be made to the liquidator. But because the materials are now effectively second-hand (and may have been unwrapped) this is likely to be at a lower price than what they cost new. If you owe money to the firm of builders (probably for the very good reason that they let you down) you'll still have to prepare a counter-claim detailing what it's cost you to rectify the situation, such as payments you've had to make to other builders and subbies. You can also include a sum for distress and wasted time.

Where you owe money to a bankrupt contractor they may still try to get you to pay them. However, legally any money due to them should now go to the liquidator. They may suggest a solution whereby you employ them under a different name. But the trouble with this is that you could then fall foul of the liquidator. It could also cause trouble with former employees and subbies who are owed money. If they hold a grudge against their employer they may take it out on you by doing a bad job on your house.

Subbies with claims for money owed to them for unpaid work on your site should submit them to the liquidator, with a copy to you. You are not liable for paying them for work they did whilst employed by the contractor who is now bankrupt. Once their initial feelings of rage have died down, subbies may be willing for you to pay them directly for future work. If you're a stage in hand with the builder, he would have used the money you owe him to pay his subbies. Your objective is to get the house built, so it would make sense to negotiate a figure with them to finish the job.

HELP FROM YOUR WARRANTY PROVIDER

Warranty providers such as NHBC can offer considerable assistance in the event of your contractor going bust, up to the maximum liability of 10% of the value of the finished house. There are three help options:

- They pay you the money needed to complete the house to NHBC standards, plus the cost of making good any damage, or
- They reimburse any money you've paid to your builder, but can't recover from him, or
- They directly arrange for the outstanding works to be carried out in accordance with the original contract (excluding any extras), for which you pay.

Disputes

If you find yourself bogged down in a disagreement with your main contractor, and it simply can't be resolved, you may have a legal dispute on your hands. This means an independent third party capable of

The top hold-ups on site

■ **Failure to think ahead** – Thinking ahead prevents problems. You need to constantly 'walk through the house in your head'. Picture every part of the design, and consider where everything is going to be, and how it should look. Then ask yourself whether you've made this clear to your builders. Take tours of your finished house in your mind, one for each trade. Consider how deliveries of materials will coordinate with the availability of labour. Allow ample time for ordering any specialist fittings with long lead-in times.

■ **Money problems** – Cash flow is king. If possible arrange a mortgage where payments are advanced *prior to* completion of each stage.

■ **Delays with services** – No matter how urgent your requirement, utility companies are a law unto themselves and work to their own timescales. An early application is essential.

■ **The wrong builders** – Always check out references and former clients.

providing an objective professional opinion will need to be appointed to adjudicate, normally an experienced chartered surveyor or architect – but not the person who designed your house, in case it turns out that some problems are due to bad design.

Disputes should always be dealt with promptly, as they only get worse if ignored. If the problem is due to building work that's below par, and the builder has refused to rectify it, then a good first course of action is to ask the Building Control Surveyor's opinion. A site meeting will need to be held to discuss matters.

The final option of terminating the builder's contract is Big Potatoes, requiring legal advice. Once the dust has settled you'll need to appoint another builder to finish the job, for which they may well charge a premium. See website for further advice.

11 START ON SITE

Groundwork

Day One on site is usually a fairly harmonious occasion. Everyone is full of the best intentions and wants to crack on and make it happen. The first phase, the groundwork, comprises excavation for the foundations and drainage. This is followed by construction of the below-ground footings for the main walls up to damp-proof-course (DPC) level, and building the ground floor structure.

Strictly speaking, there's no obligation for a contractor to physically start work on a specific day. But it obviously isn't a brilliant start if you can't agree when he should turn up. But once the main contractor has taken possession of the site, it is effectively sub-let to them for the duration of the build.

After months of preparation, this is always one of the most exciting parts of the build. But it's also the phase where there's potential for major cost overruns to occur. Standard costings are normally based on conventional trench foundations (trench-fill or strip – see below). But if deeper foundations are deemed necessary, the costs for excavation and concrete will obviously increase. Worse, unstable ground could mean you'll need expensive special piled or raft foundations, causing groundwork costs to spiral by £10,000 or more.

One thing that can easily get overlooked at this stage, amidst all the excitement, is the common planning condition that your landscaping proposals and the main exterior materials (bricks, tiles etc) must be approved by the planners 'before work commences'.

Crowded town centre site can make excavating trench foundations difficult.

Preparing the site

The first job is for all shrubs, roots, and topsoil to be removed from the whole area that's going to be built on. This means stripping away the vegetation over the entire footprint of the house, extending at least a metre beyond the external walls (the area within the main walls is known as the 'oversite').

This must be carried out to a depth that will prevent later growth, usually at least 150mm. Fortunately, a digger can make short work of even the densest undergrowth. But before enthusiastically ploughing into the ground with your JCB it's important to consider what arrangements are in place for getting rid of the mounds of muck that will inevitably be generated. Sending soil away is expensive, but buying it back is even dearer. So if you can recycle some spoil, setting it aside for later landscaping, so much the better.

To estimate the cost of clearing and removing the oversite you need to know the volume of earth to be removed. For a detached house on a flat site with a 120m² footprint plus garage and driveway, this could amount to around 30m³ of excavated material, which is likely to cost at least £600 to remove. Where a sloping site has to be levelled, costs can escalate to ten times this amount.

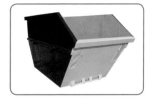

Skips

The time-honoured method of waste disposal is, of course, our old friend the skip. Skips can be hired in a range of sizes up to a cavernous 15m³, but  capacities are still widely quoted in 'yards' (1m³ = 1.308 cubic yards). Calculating how many you'll need to cart away all the excavated subsoil can be tricky, because it expands dramatically in volume once dug, as it's no longer compacted, and 1m³ can weigh as much as two tonnes. There always seems to be massively more of the stuff than you'd reasonably expect. To calculate the approximate amount of excavated waste, work out the volume of the trenches being excavated and then add 50%.

Alternatively it may be worth hiring a tipper lorry with a 'grab', which can be used to take soil and rubble to the local tip. These trucks can handle a large quantity of waste and can also be used to deliver bulk supplies such as ballast and sand. But unless carefully planned, hire costs can soon escalate.

Access

In order that heavy delivery lorries can get on to the site a firm driveway surface will be required. If none already exists you may be able to use a layer of clean hardcore from any demolished

The top ten trees that cause damage

(Source: BRE)

	Tree	Maximum mature height	Typical minimum recommended distance from house
1	Oak	16–23m	13m
2	Poplar	24m	15m
3	Lime	16–24m	8m
4	Ash	23m	10m
5	Plane	25–30m	7.5m
6	Willow	15m	11m
7	Elm	20–25m	12m
8	Hawthorn	10m	7m
9	Sycamore	17–24m	9m
10	Cherry and plum	8m	6m

outbuildings etc. Alternatively, order some one-tonne jumbo bags of granular sub-base, such as 'MOT Type 1', which can be rolled in place to form the base for the access area. If you need larger quantities, buy it loose in bulk, perhaps 10- or even 20-tonne loads. Any surplus hardcore can be retained for later use as a sub-base for patios or paved areas.

Trees

Trees that are very close to the construction area can cause problems. The minimum distance between a new house and a tree depends, for one thing, on the type of tree and how tall it can grow. It also depends on the nature of the soil. There are strict rules for this, set out in the NHBC handbook.

Tree roots can affect foundations in three ways:

- Roots can penetrate masonry joints at foundation level, ultimately causing cracking.
- Old roots can rot away, leaving gaps under the foundations – *eg* where trees have been felled.
- In clay soils, after a tree is felled and no longer sucks up thousands of litres of water, the sodden ground may 'heave' – *ie* swell and rise up.

Because of their large size, oaks, willows, and poplars are notorious for causing damage to buildings. These thirsty broadleaf species are the bad guys of the tree world.

Foundation issues are still the largest cause of claims on NHBC warranties. Consequently, warranty providers have become more and more wary of clay ground, and especially where there are tree roots. But guidelines vary quite considerably on how close you can build to trees without needing to construct special foundations. British Standard 5837 suggests the minimum distance should be the equivalent of the tree's anticipated height at maturity. Other guidelines suggest one-third to half the tree's mature height. But this clearly depends on the species and size of tree, and the type of subsoil. On some better-behaved soils, such as chalk, trees can be considerably closer to buildings than on clay.

Removing mature trees including stumps and all roots below ground, and filling the resulting cavities is an enormous task. So a good alternative (given the planners' fondness for retaining trees) may be to build beefier foundations that can cope with such threats – either by excavating to a greater depth or by piling.

Foundation types

The foundation method best suited to your site will have already been determined at the design stage. But in reality, ground conditions can be extremely localised so you can never be entirely certain what you'll find until you start digging. Conventional trench foundations deeper than about 2m are not normally acceptable. At this depth it generally becomes cheaper and easier to use piles. On difficult terrain, such as filled ground or where there's a serious threat from large trees, you may have to employ special foundations such as rafts, which spread the load of the building. Being hit with the shock news that you need a special foundation system may, at the time, seem like the end of the world, but in most cases it should easily be covered by the contingency fund. Even in extreme cases, the

additional cost should fall within 10% of the total construction budget.

Trench-fill foundations for a detached house can easily consume more than 50m^3 of ready-mix concrete. This means that, together with the excavation and spoil disposal costs, you probably won't have much change from £10,000, and the price tag can double where special foundation systems are needed. In terms of time, preparing foundations and pouring ready-mix takes three or four days on most sites. But again, for more elaborate foundations it can take two weeks or more to 'get out of the ground'.

Your chosen type of foundation can also be influenced by the method of construction selected for the main walls. For example, post and beam structures may suit piled or pad foundations whereas load-bearing masonry walls are traditionally built on trench or raft foundations, as these can readily provide a consistently level base. Foundation accuracy is even more important for modern panel timber frame construction– see Chapter 6.

The principal types of foundation are:

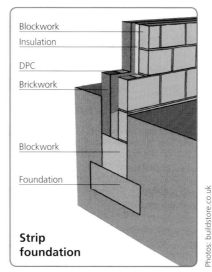

Blockwork
Insulation
DPC
Brickwork
Blockwork
Foundation

Strip foundation

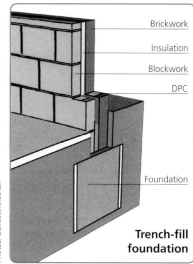

Brickwork
Insulation
Blockwork
DPC
Foundation

Trench-fill foundation

Strip foundations

'Deep strip' foundations are the least expensive and traditionally the most common type. A trench is dug, normally 600mm wide (sometimes 850mm wide on sand, silt, or soft clay) and typically at least 1m deep. Concrete is then poured in to a depth (thickness) of about 300mm (minimum depth 225mm), creating a concrete strip, sometimes reinforced with steel mesh. Upon this the foundation walls are then built up in suitable brick or blockwork. Strip foundations are often necessary in softer soils, such as sand, since they spread the load of the building out over a greater area.

Trench-fill foundations

Trench-fill foundations are excavated similarly to strip foundations but are filled almost to the top with concrete (normally to about 75–150mm below ground level). Although more expensive in terms of concrete, labour costs are significantly reduced because there's no need to build awkward below-ground walls in the trenches. This method of construction can get you 'out of the ground' in a day.

Trench-fill can be taken deeper than strip foundations and is often used in areas with heavy clay soil with trees nearby, or where there's a high water table. Whereas for strip foundations the minimum practical width is at least 600mm (because of the difficulty of laying bricks or blocks in a narrow trench), with this method trench widths can be as narrow as 450mm in normal ground conditions.

With trench-fill, the sides of the trench, as well as the

bottom, play a major role in supporting the load, so the trench sides must be capable of bearing loads. This typically makes clay and chalk soils suitable for this type of foundation.

Where soil is less firm, the foundations can be protected from potential ground movement by lining one or both trench sides with compressible polystyrene boards and a slip membrane (or a composite such as *Clayshield Prestige*). For additional strength, the concrete can be reinforced with steel mesh.

Trenches deeper than about 2.0m may need special shoring, but at this depth the combined costs of excavation, large volumes of concrete, and paying for the extra spoil to be removed starts to become prohibitive.

Where drainage pipes or service ducts have to pass through foundations of pure concrete, a suitable opening can be formed by temporarily positioning a 'dummy' pipe before the concrete is poured. The pipe should be covered with protective wrapping to create space around it to accommodate any future settlement without pipes fracturing. Above the opening, a lintel should be provided, or steel mesh to reinforce the concrete.

Pile and ring-beam foundations

Where foundations need to be more than about 2m deep, pile foundations are sometimes a better alternative. This is the case where ground is unstable and good support can only be found at very deep levels.

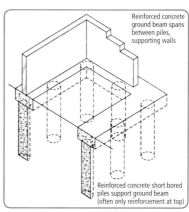

Reinforced concrete ground beam spans between piles, supporting walls

Reinforced concrete short bored piles support ground beam (often only reinforcement at top)

Piles can also provide a solution in crowded town-centre sites where physical restrictions mean conventional foundations can be difficult to excavate. Around

15% of new houses are now built this way, making them the most common type after strip and trench.

Piled foundations are usually designed and installed by specialist contractors using pre-cast concrete columns. These are dug, bored, or driven down into the ground using a crane-like pile-driver. The piles are placed about every 2.5m or so under the main walls, each one forced down until firm *bearing strata* is reached. Alternatively, a piling rig can drill holes about 200mm diameter and a steel cage of reinforcement is placed in the hole and filled with concrete.

Piles can be tied together at ground level with reinforced concrete ground beams (ringbeams), spanning like lintels from cap to cap to support the main walls. These beams can be delivered prefabricated or cast in situ.

Photo: Helical systems

Piling has become more common in house-building since the advent of short-bored piling systems that can be drilled using a hired mini-piling machine. Another recent development is screw-piles. Looking like a giant steel corkscrew, these are rotated into the ground by machine, with the advantage that they do not produce such large amounts of spoil.

Raft foundations

Nearly one in ten new houses are built with raft foundations, and in some areas with a history of mining this is the normal foundation method. A raft is basically a large reinforced concrete slab, rather like a super-strong solid concrete floor. Because they spread the foundation load over a larger area they're used where there's any weakness in the ground, such as on backfilled land or where geological conditions make the ground unstable or where there's groundwater on site. True to their name, rafts can 'bridge across' weak areas, so they're often appropriate on clay that's prone to excessive shrinking and swelling. If subsidence occurs, the raft should absorb the movement.

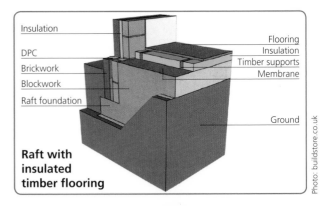

Raft with insulated timber flooring

Labels: Insulation, DPC, Brickwork, Blockwork, Raft foundation, Flooring, Insulation, Timber supports, Membrane, Ground

Photo: buildstore.co.uk

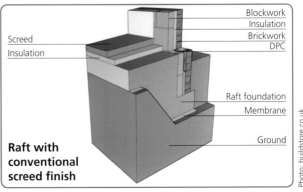

Raft with conventional screed finish

Labels: Screed, Insulation, Blockwork, Insulation, Brickwork, DPC, Raft foundation, Membrane, Ground

Photo: buildstore.co.uk

Construction entails excavating to a depth of about 650mm across the whole floor area and laying a bed of consolidated hardcore upon which concrete is poured. The structure is reinforced with steel mesh laid in sheets and the process is repeated with a final top layer of concrete. The edges of the raft, upon which both leaves of the main walls are built, are constructed thicker and deeper and are reinforced with metal 'cages'. These are stepped down and are known as 'edge beams'. Any areas that have to support the internal walls of the building are also stiffened in this way. Of course, this all consumes enormous amounts of concrete of a high-strength mix, making rafts relatively expensive.

Pad foundations

Where a house is built using a frame structure, such as traditional post and beam, it may be possible to support each post on a series of small individual pads rather than a continuous trench foundation, reducing the amount of concrete by as much as 75%. Pads can also suit modern lightweight timber-frame structures where the walls are built off reinforced concrete ground-beams spanning between the concrete pads.

They also avoid some of the problems

Photo: HSS Hire

Foundations

1st   Concrete piles or pad foundations with recycled
      aggregate
2nd   Concrete strip
3rd   Concrete piles with ground beams
      Concrete trench-fill
      Concrete raft

associated with pile-driving, such as the risk of damage from ground vibration.

The pads themselves can be circular, square, or rectangular. First a number of large holes are dug, about 900mm deep and 600mm square, and filled with concrete. Your engineer's calculations will determine the precise number required for the type of building, but some need to be spaced quite close together at 600mm centres. Although it's possible to excavate smaller pads by hand, larger circular holes are best drilled by machine. A hired mechanical post-hole borer such as a 'Skidster' can drill down to a maximum of about 1.4m depth x 600mm diameter.

Setting out

Conventional strip or trench-fill foundations are used in around 75% of new housing. Both require the same sort of trenches to be excavated. Setting-out is a task that self-builders often undertake themselves. The challenge is to draw your house as shown on the approved plans at full scale on the ground. Accuracy is crucial to ensure the walls are built in the right place and are square, which in turn depends on precise measuring, and, of course, accurate drawings. Get this wrong and it could have dire repercussions throughout the build, playing havoc with the roof design and floor calculations, even disfiguring carefully planned kitchen layouts. Your approved Building Regulations drawings should show all the dimensions you need. But for more complex designs, additional foundation drawings may need to be provided by the architect. These are the tools you'll need for the job:

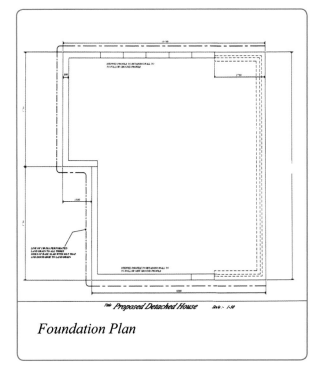

Foundation Plan

- A 30m-plus surveyor's tape measure
- Timber stakes approximately 2in square (50mm x 50mm)
- Several lengths of floorboard about 1m long
- A sledge hammer
- A large spirit level
- A spade
- A hired digger
- A builder's square (90° 'triangle')
- Ground marking paint
- Some string line

If you can also arrange some dry weather, plenty of strong tea and a glamorous assistant, so much the better.

Marking out trench positions

It's an exciting moment when you first see the positions of the walls of your new house marked out on the ground. Start by studying your plans for suitable landmarks – fixed points such as plot boundary walls. Then carefully measure the distance on the plan between a fixed point and the nearest main wall of your house. Working from scale plans, you should be able to transpose this measurement on to the ground and identify the position of one of the corners of the main walls. Hammer a wooden stake into the ground to mark the first outer corner of the house. Then bang a nail into the top of this stake, so it sticks up straight by about an inch (25mm).

Next, measure from this nail the exact length of the first wall, as shown on the plan. Tie a string line from this first stake and run it to a new stake marking the next corner, with a nail in the top as before. Go back and check your measurements and reposition the stakes as necessary so that they're perfectly vertical.

Using a builder's square, you can then mark a 90° right angle at the first corner, continuing to the next one until all the external corners are pinpointed with wooden stakes. A string can now be stretched around all the nails to mark the line of the eventual outer wall, which will be the shape of the finished building.

However, unlike developer-built 'boxes', self-build houses rarely have simple rectangular footprints. The trick with relatively complex shapes is to break them down into bite-sized rectangles.

Now is the time to double-check all the dimensions on site with those shown on your approved plans – moving things about now is going to be considerably less traumatic and expensive than trying to change the position of foundations once the concrete has been poured:

- Check diagonal measurements from the plans, and that diagonals are of equal length.
- Check distances from other boundaries.
- Check that opposite sides are parallel, *ie* consistently the same distance apart, as well as the same length (or as shown on the drawings).

Profile boards

The precise position of the trenches and walls can now be marked out. This is done using 'profile boards', which are horizontal strips of wood, about a metre wide, with four nails or screws lined up along the top. The space between the two outermost nails (about 600mm) will represent the total width of the trench to be dug, and the narrower space between the inner pair of nails will show the planned wall width (typically about 300mm). Each board has two vertical stakes for legs (about 450mm apart) so that it can be hammered into the ground.

The first board should be placed about 3m beyond one end of the wall, and another one similarly positioned beyond the other end, so that you have a pair of boards facing each other with their tops level. To mark the position of the trenches, simply run some string line between the boards at either end and pull tight (outer nail to corresponding outer nail etc). This creates 'tramlines', with the two inner string lines showing the finished wall positions and the two outer string lines the foundations. The line for the outer face of the wall should match your existing line between the corner pegs. More boards and lines can then be positioned, one pair for each wall.

Check with a builder's square that adjoining walls meet at a perfect 90° right angle. If any additional foundations are required for

internal, load-bearing 'sleeper walls', these can be marked by measuring from the string lines.

The positions where the digger will excavate can now be marked on the ground, using the outer lines strung between the profile boards as guides. This was traditionally done by sprinkling a line of sand or lime directly over the strings, although a can of spray marker paint may prove more resilient. If necessary use an upright spirit level and a straight edge, down from the lines to the ground.

Once a clear outline is marked on the ground the string lines can be removed. But the boards should remain in place, so that the inner wall lines can be put back once again after the concrete foundations have been laid, to help guide the bricklayers up to DPC level. The profile boards can then finally be removed.

NOTIFY BUILDING CONTROL 1
Start of excavation – two days' notice

Excavation

Once the trench positions marked on the ground accurately reflect those on the plan, it's almost time to let the diggers loose. But, at least two days before work begins, you must formally notify Building Control. They may choose not to visit at this stage, and instead wait until the foundations have been excavated, but that's their call, not yours.

If you're doing the groundwork yourself, hire a JCB or mini-digger. For conventional trench foundations, select the 600mm width bucket (rather than the narrower 450mm one, used for internal wall foundations) and excavate to the depth shown on the plans, taking care not to damage the

|—— trench width ——|

wall width

profile board 1   profile board 2
                              peg   trench line
                                         wall line
position of new wall   wall line
                                         trench line

Check before you dig (electric cable damaged by digger).

profile boards. The bottom of the trenches should be left as level as possible (a sloping trench base can cause all the wet concrete to slide down to one end). Trench sides should be cut square with the base.

Trench work is usually fairly safe down to about waist height (1m). Once you dig any deeper there's an increasing risk of trench collapse. If any trench sides are weak they may need shoring up with timber supports before the concrete is poured. In heavy clay ground or where there are trees nearby, special flexible polystyrene sheets made of compressible material are fitted against the sides of the trench, to help resist ground movement. Such 'slip membranes' allow the ground some degree of movement without it affecting the concrete foundations. On sloping sites small steps can be cut at intervals in the base of trenches in order to maintain some consistency of depth.

Several small steps are better than a few big ones, otherwise the walls may later crack due to markedly different depths. The height of each step should be no deeper than about 200mm (check on site with Building Control) and overlapped horizontally by the concrete above by about 1m.

The basic rule is 'the sooner out of the ground the better'. On easy sites, where trenches need be no more than a metre deep, the foundations for a whole house can be dug in a couple of days. Trenches should be left unfilled for the shortest possible time, to minimise the risk of accidents and problems like trench collapse and filling with water. If left unattended they must be covered or cordoned off – the second most common cause of injury on building sites (after falls from height) is from unguarded holes and trenches.

Once you've finished excavating, it's essential that Building Control are invited to inspect the trenches before the concrete is laid. Apart from checking whether they've been dug to the appropriate depth, they'll also want to be sure that the trench base is firm and secure and that the sides aren't about to collapse. You may be asked to provide temporary support to the trench sides or they may want you to dig a little deeper. Building Control have the right to request such modifications even if it means going beyond the foundation depths shown on your plans that they've already approved. This is to make sure the foundations are compatible with the actual ground conditions on site.

Unfortunately, this only works one way – you're allowed to increase foundation depths, but not to reduce them. This, of course, could throw out your careful calculations for ordering the correct volume of concrete, so it's best not to finalise your ready-mix order until after the inspection.

NOTIFY BUILDING CONTROL 2
Foundations excavated – one day's notice

Where naturally occurring chemical sulphates are present in the ground, they can attack cement-based materials including concrete foundations, blockwork, and mortar. If there's any risk of these chemicals being present the Building Control Surveyor will usually know, and a simple test can be carried out. The solution is to mix a suitable additive to concrete and mortar and to only use sulphate-resistant concrete blocks below ground. Where excavations expose any existing old foundations, concrete, or other buried materials, Building Control will normally require this to be broken up and excavated until firm ground is reached.

Before filling the trenches with concrete, it pays to plan ahead and calculate where you want the level of the concrete surface to stop. The DPC should be at least 150mm above the finished external ground level. Foundation blocks are normally laid in courses of 225mm (215mm + 10mm mortar joint) or laid flat in courses of 110mm (100mm + 10mm). With the trench-fill method, most builders pour the concrete right up to 75mm below the finished external ground level, equivalent to one 225mm course of blocks below DPC level.

To help calculate the number of courses of bricks or blocks you can make a simple tool called a 'gauge rod'. This is basically a giant ruler comprising a length of wood about 2m long marked with the heights of brick or block courses, including 10mm thick mortar joints.

Concrete mix

Once the Building Control Surveyor has approved the trenches, concrete should be poured with minimal delay, to reduce the risk of damage to vacant trenches. The ideal situation is to excavate the trenches and pour the concrete the same day. But before pouring, ensure the trench bases are clean, level, dry, and free of any loose material.

Concrete is delivered ready-mixed and poured directly into your trenches from a mixer truck.

Ready-mix is ordered by the cubic metre (m^3) and a full truckload comprises about $6m^3$. To calculate the volume needed, multiply the depth x length x width of the trench or slab. As a rough guide, a strip foundation in a trench 10m long x 600mm wide would need about $2m^3$ of concrete (assuming a concrete strip about 300mm wide sitting at the bottom of a 1.2m deep trench). The same trench-fill foundation would need more than $6m^3$ (or $5m^3$ for a 450mm wide trench). When calculating the amount needed always allow a little extra – better too much than not enough.

A typical foundation mix would be 1:3:6 Portland cement/sand/gravel (sand premixed with gravel is known as ballast). But when ordering it's often best to just explain what it's for and let the supplier work out the optimum mix. Standard foundation mixes are known as 'GEN 1', whereas a stronger 1:2:4 floor-slab mix is 'GEN 3'. Other specialised mixes are available for reinforced concrete and driveways etc. As a rule, the lower the GEN number the less it should cost.

Pouring concrete

Speed is of the essence when pouring concrete. A full truckload can take about 20 minutes to pump but the concrete starts to cure (set) fairly rapidly after placement, so avoid pouring one load then waiting several days for the next one. If there's a long delay between batches the first lot will start to harden and the new load won't bleed in with it, leaving a join that'll become a weak-point in the foundation.

Normally the mix should be of a 'thick lumpy custard'

consistency, not too wet, requiring just a little encouragement from a shovel to move it along the trench. Special 'self-placing' foundation concrete mixes are a recent innovation that flow easily around the trenches. Or you can use a concrete pump to achieve an even spread. Note that ready-mix should not be watered on site, as this will interfere with the mix, and watered concrete is a major cause of foundation failure.

Smooth the surface of the concrete so that it's level throughout the foundations – small discrepancies can be resolved with a bed of mortar so that the first course of masonry is laid level.

NOTIFY BUILDING CONTROL 3
After concreting foundations – one day's notice

SAFETY

Working with cement – wise precautions

- Wear protective gloves and boots – concrete burns the skin.
- Check the weather forecast before concreting – overnight frost can ruin concrete. Foundations poured in winter conditions should be covered over.
- In the event of trenches flooding you may need to hire a water pump.

Building the below-ground walls

After allowing two or three days for the concrete foundations to cure, the below-ground walls or 'footings'

can be laid in brick or blockwork up to DPC level. In the case of traditional strip foundations the walls must be built up centrally, so, for example, a 300mm wide cavity wall built on top of a 600mm wide concrete strip should have 150mm of concrete strip projecting out on

each side. If at this point the builders suddenly realise the trench location doesn't actually correspond to that shown on the drawings, they may be tempted to build off-centre, which can later result in structural instability.

However, with trench-fill foundations the concrete is substantially thicker, so strictly speaking it can be possible to build off-centre (useful where awkward neighbours won't permit temporary access). But any such deviation must first be checked with Building Control.

The architect should have specified bricks or blocks approved for use in damp underground conditions below DPC level, such as 'trenchblocks' or dense engineering bricks, and sulphate resistant cement. If the wrong materials are used they can erode very swiftly, with serious structural consequences.

Using full-width foundation blocks saves time. Otherwise the cavity between the two leaves will need to be backfilled up to ground level with compacted lean-mix concrete (the surface sloping slightly outwards). Where drainage pipes or service ducts need to pass through the masonry, concrete lintels will normally be needed to bridge over them. A space of at least 50mm must be left around pipes, and a flexible

pipe surround fitted to deter uninvited subterranean guests and creepy crawlies.

For below-ground work, start by constructing the longest walls, building the corners first. Use a string line to join one corner to the next and complete each course in between. The walls should be checked by measuring them diagonally whilst the mortar is still pliable so that the brickwork can be straightened in both directions if necessary. Continuously check vertical accuracy with a long spirit-level and only build two or three courses of brickwork at a time, to prevent the weight causing the courses below to slide out of true.

Airbricks

Your ground floor will most likely be of modern 'beam and block' suspended concrete construction – see below. Or it may be of traditional suspended timber. In both cases you'll need to install airbricks below DPC level, to maintain continuous air circulation to keep the underfloor void dry. These should be spaced roughly 2m apart, and sleeved right through both leaves and the cavity.

The damp-proof course (DPC)

The DPC prevents damp from the ground rising up through the walls. As noted earlier, this should be at least 150mm above finished ground level. But right now the excavated ground will be much lower than this, so you need to decide exactly where the final ground levels will be, once backfilled, and perhaps paved. Once the walls have been built up to the correct level, the DPC itself can be laid. This normally involves placing a long strip of black plastic along both the outer and inner leafs, bedded onto an even bed of fresh mortar. Laying the DPC is a key stage and Building Control will need to inspect it before it's covered.

**NOTIFY BUILDING CONTROL 4
DPC level – one day's notice**

The DPC should cover the full width of the masonry and project out about 5mm beyond any external face (but not into the cavity). Where the end of one

strip joins another, they should overlap by at least 100mm. Sufficient 'spare' should be left projecting on the inside to later overlap with the DPM flooring membrane by a minimum of 50mm, thereby forming a continuous barrier. Avoid having earth or flowerbeds banked up against the outer walls, unless additional protection is provided with vertical plastic sheets to prevent damp penetrating.

Service trenches

When preparing to bring in mains services there are dozens of ways to get it wrong, but just one right way. You need to consider the groundwork required and the materials to use, which will be dictated by each service provider. As noted earlier, where service ducts penetrate the below-ground masonry a lintel is necessary for protection.

Where service trenches are likely to conflict with other excavations, notably for drainage, it's usually better to delay installation of services rather than risk them getting damaged.

Once all the mains pipes and cables are in place, the trenches can be backfilled up to ground level. But remember to leave the boundary end open so that final connections can be made. It's also important to protect the surface of the trench from heavy vehicle traffic, which could damage underground ducting.

Electricity

Your plans should show where the foundation entry point for the mains electricity supply should be. This is normally where the distribution board will be located. A service trench will need to be dug from the plot boundary leading into the house. Suitable protective underground ducting can then be positioned in the trench. This needs to be fed through the foundation entry point into the house and raised above the ground floor in a gentle bend. You can then book a date for the electricity company to come out and lay their incoming armour-clad cable from the boundary. The cable will be

run through the ducting in the open trench and into the building, connecting to an electricity meter installed by the service provider.

Water

Cold water is supplied from the mains in the street, via a stopcock in the pavement or front garden. A new water meter will be installed just within the boundary of your plot by the water company, who normally also make the connections on both sides of the meter.

The usual point of entry for mains water into the house is at the kitchen sink. Blue alkathene pipework can normally be laid in the same service trench as the other utility ducting (assuming all service providers agree). Depending on the water company's requirements, pipes should be buried at least 750mm deep to

protect them from frost. To minimise the risk of leakage, it's best to fit the longest possible section of piping with as few compression joints as possible.

Where the supply enters the house, a protective duct with a gradual bend in it should be installed, ready to receive the incoming water

Photo: H+H Celcon

pipe curving upwards into the building. This blue water supply pipe can then be left temporarily capped off about half a metre above the finished floor level.

At this stage you also need to consider supplies for any garden water features you require, which may additionally need electrical power and lighting. And If it hasn't been arranged already, a builders' supply for mixing mortar can be run from a T-junction from the incoming blue water pipe and connected to a tap fixed to a timber post in the ground.

Gas

Mains gas supplies are run in yellow polyethylene pipework buried at least 375mm below ground. There's no need to prepare a special entry point into the dwelling for mains gas, because the incoming pipe terminates at a stop valve by the external gas meter. But gas supply pipes aren't normally laid in the same trench with the other services. At this stage all you need is a suitable open trench and an external gas meter box in position, either set within one of the main walls or partially recessed into the ground at the foot of the wall.

Phone and cable TV

While you've got an open trench, most phone and cable TV companies will provide you with duct piping to place in it, run from the boundary of your plot. A rat-proof pull-cord needs inserting through each section of the duct, so that the companies can later pull their wires and optic cables through once the property is complete.

Ground floors

When it comes to constructing the ground floor, there are two main options: a suspended floor made either from concrete beams or timber joists spanning between the walls. These days, solid concrete slabs laid on the ground are rarely used for new housing.

Which floor type?

Suspended floors have always been preferable on sloping sites or difficult terrain. Today ground floors constructed from pre-stressed concrete beams as floor joists, with the spaces in between filled with concrete blocks, are almost

universally used in new housing, and may even be required by warranty providers. Such 'beam and block' (B&B) floors offer several advantages. They can be laid very quickly, even in adverse weather, and can span greater distances than timber joists without support. Compared to traditional solid concrete floors, B&B is easier to construct and less prone to defects, plus the floor void is a handy place to run the service pipes. B&B flooring is actually quite efficient in its use of materials and is considered relatively eco-friendly.

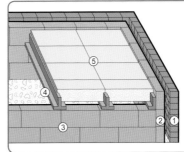

Beam and block floor
1 Outside wall
2 Concrete block inner wall
3 Sleeper wall
4 Concrete beam
5 Concrete block

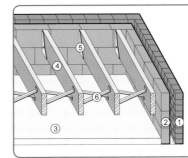

Wooden joist floor
1 Outside wall
2 Concrete block inner wall
3 Concrete oversite
4 Wooden joist
5 Metal joist-hanger
6 Herringbone strut

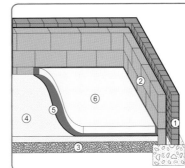

Solid concrete oversite floor
1 Outside wall
2 Concrete block inner wall
3 Hardcore infill
4 Sand binding
5 Damp proof membrane
6 Min 100mm concrete oversite

Diagrams: buildstore.co.uk

However, if you're a diehard traditionalist there's nothing to stop you building a Victorian-style ground floor of suspended timber joists and boarding. Although this is probably the least efficient and most expensive option, it's also regarded as the greenest, timber being a renewable resource.

One reason solid concrete slab floors have fallen out of favour is because they can suffer from serious defects unless built with considerable care. If the preparation isn't right – especially the compaction of the hardcore base – slabs can easily settle and crack up. Indeed, floor-slab settlement was until recently one of the most common sources of warranty claims with new construction. However, concrete slabs are still used for garage floors.

One consolation if you've already been put to all the expense and trouble of building special raft foundations is that your floor structure will already be in place. The raft will double as your new floor, just awaiting a layer of insulation and floor screed to finish it off.

Beam and block 'suspended concrete' floors

As with all suspended floors, the loadings are taken by the wall foundations, rather than direct to the ground as with a concrete slab. The main limitation is the span of the beams, which can normally manage up to 6m or so, although some deeper 225mm thick beams can stretch to around 8m. Even so, a load-bearing internal 'sleeper walls' with proper foundations will normally be needed to support the beams on their journey between the main walls.

Pre-cast concrete floor-beams have a profile rather like an inverted 'T'. They're typically 150mm deep, but are also commonly available in longer-spanning 175mm and 225mm sizes, and for infill can mostly accept the same standard 100mm deep blocks used in walls.

To ensure the correct beams are used for the required

span the manufacturers can custom-design them. All you have to do is supply a set of approved plans. Your drawings should show the location of internal supporting walls, and the details of any service pipes coming up through the floor, such as soil pipes for WCs. But you'll need to place your order up to eight weeks in advance.

Once the oversite area under the floor is clear of topsoil and vegetation, some thick polythene sheeting can be laid over the ground and weighed down with sand. The void beneath the floor should be a minimum of about 150mm, and should be ventilated with air ducts. Ventilation is particularly important to the underfloor space if the local area is at risk from radon or methane. These air ducts are of an odd-looking 'periscope' shape (known as 'cranked ventilators') that zigzag from the outside wall down into the void. This allows a free passage of air under the floor without letting light in, creating a microclimate as inhospitable as the moon, thus prohibiting plant growth.

FLOOR CONSTRUCTION

Constructing B&B floors on site could hardly be simpler. Although concrete beams are fairly heavy, it should be possible for two people to lift them into place. First, the beams are laid in rows with their ends resting on the inner leaf of the main walls, over the DPC (the DPC on the outer leaf may be a little higher). It's important to leave the correct intervals between the rows so that the infill blocks can simply be slotted into the gaps between the beams. This can be done quickly and cheaply with standard 100mm thick concrete infill blocks (440 x 215mm, typically 1350kg/m^2 medium density) laid in either length or width direction. For larger spans the floor can be made stronger by laying the beams closer together with the blocks laid sideways to the beams.

However there's no rule that says your floor has to be made with plain concrete building blocks. Larger purpose-

Photos: Millbank

GREEN CHOICE

Ground floors

1st	Suspended timber
2nd	Beam and block with hollow clay or concrete blocks
3rd	Beam and block with polystyrene insulation blocks
Not recommended	Solid concrete

Comments

Traditional suspended timber floors use renewable resources and can be insulated using natural materials. But a timber floor must be airtight (using PTG boards). Composite timber I-beams can be used instead of solid softwood joists to reduce cold bridging. Services can be easily incorporated through the void.

made floor blocks (560 x 440mm) of the same thickness can be laid twice as fast as the standard-sized variety, and special blocks containing integral thermal insulation can significantly reduce heat loss through the floor. At the floor perimeters, blocks can be cut to fit and holes for services can be made. Once all the blocks have been fitted, to prevent movement the floor surface should be grouted with a 1:4 cement/ coarse sand mix brushed into the joints.

INSULATION

Insulation boards are normally placed over the surface of the blocks. These are later covered either with chipboard flooring or a conventional cement screed (approximately 65mm thick). Kingspan and Celotex are well-known manufacturers of rigid insulation boards made from polyurethane (PUR) or polyisocyanurate (PIR). The target 'TFEE' U-value for floors is 0.13 W/m^{2}K which should be achievable using a minimum 130mm depth of PIR insulation boards.

However, a number of systems are now available which incorporate insulation within the floors themselves. High-performance 'jet floors', which use special expanded polystyrene or polyurethane infill blocks between the beams, are about 10% more expensive than standard blocks, but you may not then need to provide an added layer of surface insulation, just a screed.

To dispense with the need for a later wet screed, special highly insulated 'floating floor' panels can be laid on top of the blocks. These are made of an insulating material bonded to a base of either plywood or chipboard. Before laying the floor panels, a 500-gauge polythene sheet is laid as a vapour control to protect floor finishes from any damp in the structure below. The floor panels are then

Photo: Tarmac

laid, with their tongues and grooves glued with a PVA adhesive. Alternatively, conventional floorboards can be set on battens over insulation boards, instead of a screed.

Note that pre-stressed concrete beams are designed with a slight upward camber (around 13mm on a 4m long beam) designed to create a level floor when fully loaded with heavy blocks. When lightweight polystyrene blocks are used instead, the camber doesn't flatten out, so you rely on a screed to achieve a perfectly level finish. So if you want a screed-free dry finish, a levelling compound may need to be applied.

NB: The insulation around incoming water pipes and soil stacks must extend below the ground under the slab to a depth of at least 750mm, to protect against frost. Air must be prevented from entering the building via gaps by injecting expanding foam around pipes passing through insulation boards and where ducts pass through foundations.

Suspended timber floors

Although generally considered to be eco-friendly, there's one thing that's not terribly green about suspended timber floors –

GREEN CHOICE

Floor insulation

1st	Insulation of organic origin – from natural sources: eg loose cellulose fibre, sheep's wool, wood-fibre, hemp, cork.
2nd	Insulation of inorganic origin from naturally occurring minerals: mineral wool, fibreglass, vermiculite, foamed glass.
3rd	Insulation from a fossil-fuel origin: expanded polystyrene (EPS), polyurethane.
Not recommended	Extruded polystyrene, foam polyisocyanurate.

Comments

Cellulose fibre (recycled newspaper) or sheep's wool are preferred, as they're renewable. But neither will support loads and are therefore more suited to traditional timber floors (where they can be placed on supporting sheeting draped between floor joists). Cellulose fibre can be installed using a special blowing machine. Sheep's wool is pleasant to handle and doesn't release irritating fibres unlike glass fibre mineral wool, but is more expensive.

Hemp fibres can be used laid loose in lofts and flat roofs, or for walls and ground floors in combination with lime.

Woodfibre insulation boards are rigid and can be used in floors.

Expanded polystyrene (EPS) is the least damaging of the fossil-fuel-based insulation materials and can support floor loads.

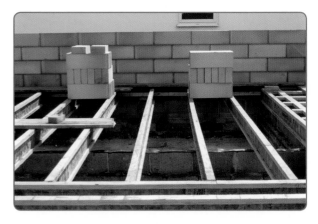

Netting slung between joists to hold mineral wool insulation.

the oversite ground under the floor has to be covered with a concrete slab. Today manufactured I-beams or eco-joists are commonly used in preference to traditional softwood joists, which can only manage relatively short spans. Otherwise, the ground floor construction is pretty much as per the upper floors – see Chapter 12 for construction details.

Briefly, these are the key requirements for new suspended timber ground floors:

■ Joist ends must be supported in metal joist hangers, and should not rest directly within the walls.
■ Your designer must specify the correct size and grade of timber floor joists for their span (normally spaced at 400mm centres).
■ The oversite ground must be cleared and covered with a concrete slab, minimum 50mm thick, laid over a 1,200-gauge polythene DPM and a hardcore base.

■ A space of at least 150mm is required between the underside of the joists and the surface of the oversite concrete.
■ A good cross-flow of air is required to the void beneath the joists, from airbricks built into the walls below DPC level.

Floor insulation is especially important to reduce heat-loss, given the need for a flow of (cold) air below the boards. The easiest solution is to wedge 50mm thick rigid polyurethane boards between joists, supported where necessary with battens. But there are alternative methods, such as laying 150mm deep cellulose fibre or mineral quilt 'loft insulation' over plastic netting strung hammock-like between all the joists.

Drainage

Drainage falls into two categories – foul waste and rainwater. Foul waste water from bathrooms, toilets, and kitchens is normally dispersed to sewage treatment plants via public sewers. However, in most districts

rainwater has to be dispersed separately unless there's a 'combined system'. This is to prevent a sudden tsunami of rainwater in severe storm conditions overpowering the sewer system, causing a deluge of liquid excrement downtown. Mindful of such horrors, Building Control will take a close interest in your new drainage system.

NOTIFY BUILDING CONTROL 5
Drainage commencement – one days' notice

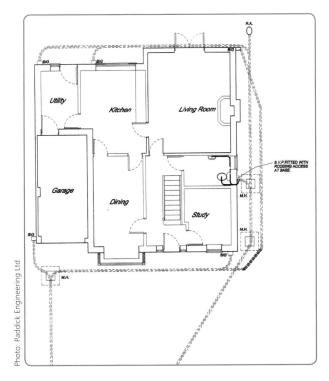

Photo: Paddick Engineering Ltd

Sloping sites

Steeply sloping sites give drainage designers the willies. To work properly, conventional underground pipes rely on gravity and have to be laid to precise falls. If you've got a site that slopes downward from street level to the house, the main sewer in the road may not be deep enough for conventional gravity-powered underground pipes to reach from the new house when set to the correct falls. In some hilly regions the waste has to be pumped uphill using a pumped sewage system, which can add up an extra £4,000 or more to the cost (plus maintenance). These systems often incorporate electrically powered macerators that function rather like kitchen blenders, mashing up the 'solids' to reduce the risk of blockages.

However, where a slope isn't too steep and the main sewer in the road is fairly deep, the relative levels may still be workable using a conventional drainage system.

Where a site slopes upwards from the road towards the new house, gravity should automatically propel the discharging waste on its way to the main sewer. The main worry here is that underground drainage pipes may be laid at too steep a pitch. This can cause solids and fluids to separate, leading to blockages. To prevent the waste discharging too rapidly, on some steeper slopes Building Control may agree to the installation of a 'tumbling bay junction'. This is a special inspection chamber incorporating a vertical 'drop down', which is constructed around the existing sewer pipe. The attraction of this arrangement is that it reduces the necessary gradient of the pipes and slows down the rate of discharge, allowing the effluent to enter the sewer at a controlled rate.

Excavation

Trenches are normally excavated about 450mm wide (a

digger's narrow bucket). This leaves about 150mm space either side of the pipe run. There are no set depths for pipework in the Building Regs but going much deeper than a metre means trenches can become tricky to work in. Drainage trenches need to be dug about 100mm deeper than the required pipe level, to allow for the granular bedding that the pipes will rest on.

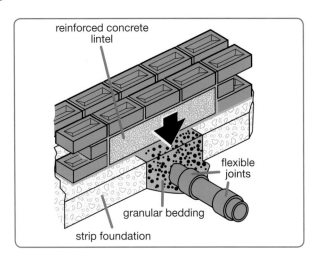

reinforced concrete lintel

flexible joints

granular bedding

strip foundation

Pipes

Modern plastic drainage pipes are lightweight and easy to handle. Although they're impact-resistant, they're vulnerable to damage when awaiting installation, and unless carefully stored can get blown into the path of vehicles. To distinguish them from other buried services, plastic drainage pipes are a brownish-orange colour similar to traditional terracotta clay pipes. Never use black soil pipes underground, or Building Control will insist they're replaced.

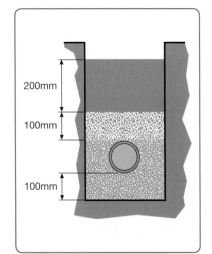

200mm

100mm

100mm

The diameter of pipes must be sufficient to cope with volumes of anticipated waste materials, something that should be shown on the drawings, although in most cases standard 110mm diameter will suffice. Pipes are available in generous lengths of 3m, 6m, and even a super-long 9m to minimise the number of joints. They're connected to each other with integral flexible synthetic rubber seals that allow for a small amount of movement whilst remaining watertight. To ease the connection at these push-in 'snap joints' use silicone lubricant. Don't use washing-up liquid, oil, or grease, as this will damage the seal, causing it to leak.

Pipe laying

Normally new pipes are laid starting at the highest point, *ie* nearest the new house. The approved drawings should clearly show the levels of the new drain runs. However, this should be checked on site, so the first task is to ascertain the height difference between the point where the waste water will leave the property and the point where the new pipe will join the public sewer in the road. The objective is to achieve the correct fall (gradient) so that the waste will be sent speedily on its way. As we saw earlier, where pipes are laid too shallow, the waste will hang around and the pipes will be prone to blockage. On the other hand, if pipes are laid too steep the water can be evacuated so swiftly that the 'solids' get left behind, again risking blockage. A gradient of about 1:40 is normally desirable, as consistent as possible along the full length of the pipes. In practice, pipes are generally laid to an actual fall on site of anything between 25mm and 90mm per metre run.

Once the trenches have been prepared, but before laying the pipes, check the gradient (shown on your plans) using a spirit level adjustable to preset levels. Then a bed of fine gravel or pea shingle can be laid to this gradient, to a minimum 100mm depth, compacted along the bottom of the trench. The pipes should be carefully laid on top. Sharp inclines should be avoided, although a few gentle bends may be permitted provided they don't prevent rodding. As noted earlier, pipe connections are a tight fit, so the joints at

the ends of pipes first need to be smeared with special collar lubricant. It's obviously essential to keep gravel out of the collar joints or they won't stay watertight for long.

Where new drainage trenches pass within a metre of the foundations of your main walls (or those of the neighbours), if they're deeper than the foundations there can obviously be some risk of structural damage. So the Building Regs require that such pipes must be encased in concrete, at least up to the level of the base of the nearby foundations.

Trees
Sometimes, there's no option but to run new pipes fairly close to trees, for example where they're growing in the garden next door. Where pipes are most at risk from roots, they can be protected by encasing them in concrete. Alternatively a 'root barrier' can be constructed. This is a deep, narrow trench filled with concrete between the pipes and the trees, which acts as a shield.

SVPs – soil stacks
The main purpose of soil and vent pipes (SVPs) is to connect to upstairs WCs for dispersal of foul waste (euphemistically referred to as 'soil' or 'solids'). But 'grey' waste water from pipes serving baths, basins, and sinks also usually connects to the SVP – see Chapter 13.

The SVP is usually 110mm diameter grey or black plastic, and is the only part of your drainage system that needs to be vented, being open at the top, at roof level. This helps prevent the dreaded problem of 'siphonage', which can occur when large deluges of waste water surge down the pipes pulling along all the air in their slipstream. Without ventilation, this can literally suck the water out of traps and gullies behind, thereby allowing the sweet aroma of drains to waft up into your bathroom. To prevent such horrors, the top of SVPs should project at least 900mm higher than any nearby window – including roof windows. In most new houses SVPs are run internally, venting via special roof tiles. Happily, this also saves your beautifully designed home from being scarred by lots of ugly external pipework.

At its underground base, the SVP connects to the main drainage system via a gentle bend in order to minimise

foaming from detergents and prevent the accumulation of solid waste. Where the SVP runs internally it needs to pass out of the house through the below-ground wall or foundations (see below).

Pipes through foundations
Pipes passing through walls or foundations must be specially protected from the potential risk of damage from settlement in the wall. This can be achieved by leaving a small opening for the pipe with lintels supporting both leaves of the walls above, as you would for a window or door opening.

Alternatively, as we saw earlier, pipes running through concrete foundations can be protected by being placed within a circular plastic duct. A gap of at least 50mm 'wriggle room' is left around the pipe to allow for settlement and expansion. Such openings not only need to exclude burrowing creatures, but must also be sealed against any risk of radon or methane gas in the locality, preventing it seeping in and accumulating in confined spaces. The choice of material will depend on local conditions, so ask Building Control what they recommend.

Another measure used to prevent pipes fracturing is for the section of pipework running up to the building, or to an inspection chamber, to be kept fairly short, about 600mm, with a joint at each end. Because both the joints are flexible, they act like a pair of hinges, allowing the pipe and the building freedom to move independently of each other. Similarly, where pipes are encased in concrete a gap of about

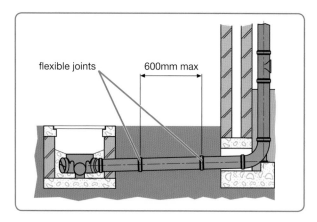

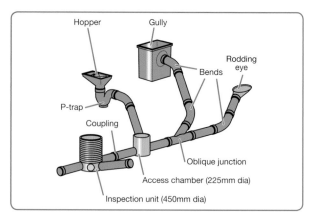

12mm must be formed in the concrete around pipe joints, to allow some flexibility.

Inspection chambers and rodding eyes

Inspection chambers serve two main purposes: first, as a junction where several underground pipes can join up into one main outgoing drainage pipe; and second, to provide access for clearing blocked pipes.

Your approved drawings should clearly show where new inspection chambers are to be located, and should also note their depth or 'invert levels' (the distance below ground of the channel at the bottom of a chamber). Part H of the Building Regs recommends using inspection chambers at changes of direction in the foul water drainage runs, so that all parts of the drainage system are accessible.

Ready-made circular polypropylene units are suitable for drains up to 1m deep. These mini-inspection units are lightweight and easy to install. Their diameter is about 450–550mm and they typically have five inlets and one outlet. Anything deeper will require a full-size rectangular 'manhole' type inspection chamber. For depths up to 2.7m, the minimum internal dimensions for a rectangular manhole are 1,200 x 750mm. Traditionally these were built of brick, but today they've been largely superseded by preformed concrete sections or in-situ concrete poured around a plastic liner. Anything deeper than 2.7m is a major project best left to professional drainage contractors.

These heavy 'manholes' can be prone to settlement, so pipes entering the chamber should be fairly short, say 300–600mm in length. These 'rocker pipes' allow slight movement of the chamber or drainage system without imposing any stresses on the joints.

Where space is very limited, Building Control may accept small plastic 'rodding eyes'. These are short branch pipes that come up to the surface with an airtight cover through which drain rods can be inserted for clearing blockages (by pushing the blockage along the pipe and into the nearest full-size inspection chamber).

NOTIFY BUILDING CONTROL 6
Before drains are covered up – one day's notice

Testing and backfilling

Building Control will need to inspect new drainage work before it's covered up, to ensure that all joints and seals are secure with no leakage. Unless you request a test now, they may wait until completion. But leaving trenches exposed for weeks on end is to invite accidents and damage. Once the new drainage work has been officially approved the trenches can be backfilled.

To protect flexible pipes from harm, they must be fully surrounded by about 100mm depth of pea shingle/gravel. No stones larger than 40mm should be used, particularly immediately above them. This is followed by 300mm of soil, which must be free of rubble. The earth excavated from the trench can be used for backfilling, but heavy compaction should not take place until there's at least 600mm depth of soil

over the pipe. In areas where there could be damage from loadings above, such as parked cars, the pipes should be protected by concrete.

Private drainage systems

When it comes to disposing of sewage, connecting to a mains drainage system is normally the easiest and most cost-effective option, as well as the least energy-intensive. Of the 4% of the UK's population not served by mains drainage, the majority use septic tanks – see below.

Most private drainage systems rely on a primary treatment stage where physical settlement removes 'gross matter', followed by a secondary biological treatment using bacterial micro-organisms, that process the effluent until it can be safely disposed of into the soil. Sometimes a third stage of treatment further purifies the effluent.

Whichever sewage disposal system you choose Building Control will have to give their consent, and the planners will want to be assured that it won't cause a nuisance. Should you want to discharge effluent into a watercourse you will additionally need consent from the Environment Agency (or Scottish Environment Protection Agency). N.B. It's an offence to make a sewage effluent discharge to surface waters without an environmental permit.

If there is no mains drainage to the site, or within striking distance, you've got three main options:

Septic tanks

A septic tank is a large, bulb-shaped, plastic chamber buried in the ground, into which foul waste is collected. They're surprisingly compact for what is effectively a private sewage works, typically measuring only about 3m

Photo: blackwatersolutions.co.uk

Photo: blackwatersolutions.co.uk

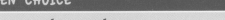

Grey-water systems

Foul water is divided into 'black water' from toilets and 'grey water' from baths, showers, and basins.

It can make environmental sense to recycle the less polluted grey waste so it can be used for flushing WCs, cutting water use in a new home by about a fifth. Recycling systems require a settlement tank and filtration treatment so that every time you flush your loo it isn't submerged in bath bubbles. Waste water is usually collected in an external underground settlement tank where it's disinfected and pumped into a small header tank for storage.

At present, however, up-front installation costs are fairly high, with a long payback period. In other words it's likely to cost more to install than you'll save in water bills over many years. Also, current systems don't score well from a lifecycle perspective because they consume chemicals to disinfect the recycled water and use energy for pumps.

On the plus side, a well-designed grey-water system can provide sustainability and independence, and may even impress your friends. Which is not necessarily the case with some more extreme green measures, such as composting toilets that use sawdust, earth, or ashes instead of water to neutralise excrement.

Photo: www.aqua-lily-co-uk

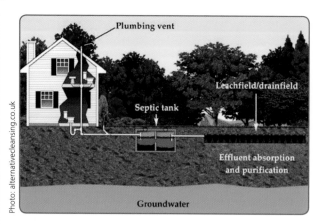

Photo: alternativecleansing.co.uk

deep x 2.5m long. Tanks range in capacity from 2,800 to 6,000 litres or longer, but as a rough guide an average family would need a 4,500-litre tank.

Above ground, all you see of them is a manhole for access. But the really smart thing about septic tanks is that they require no power supply or pumps. Instead nature goes to work in the form of millions of friendly 'anaerobic' bacteria (ones that live without oxygen) to break down the sewage into relatively harmless run-off. The solids are retained as a crust or sludge whilst the effluent can be dispersed into the ground via a network of pipes or 'land drains'.

Septic tanks should be located at least 12m away from the house, and are reasonably easy to install and relatively inexpensive, costing from around £700 plus a similar sum for excavation and installation. The need for a land drainage system, however, could bring the total cost up to around £5,000. The only running costs are pumping out from a sludge tanker lorry once or twice a year at around £100 a shot. Some more expensive versions offer increased aeration that speeds up the bacteriological process.

But no matter how efficient a piece of kit you install, before you can discharge liquid waste into the subsoil Building Control will need proof that the ground is capable of absorbing it. This can be checked with a percolation test – digging a trial hole, filling it with water, and timing how long it takes to drain away. If the subsoil is too impermeable, such as dense clay, or the water table is very high, the effluent won't be able to discharge and would eventually block the tank. Where ground conditions are not suitable, or there's a watercourse nearby, the authorities may require extra refinement of the effluent, for example by discharging the effluent through a filter bed before it's passed into weeper drains. In some cases, specially-designed reed-beds can process the outfall from septic tanks – see below.

Mini treatment plants (MTPs)

As with septic tanks, MTPs process raw sewage and churn out sterile effluent that can be safely passed into the ground. Another similarity is the requirement for sludge to be pumped out from time to time. Where they differ is in the MTP's requirement for electrical power and the need for periodic maintenance (usually carried out under contract by

the installer). But as a result, the quality of the processed final effluent is claimed to be sufficiently sterile to be discharged into a ditch or pond. However, it isn't exactly pure drinking water that comes dribbling out the other end, so the discharge may need to be via a filter bed or weeper drains.

The secret of mini treatment plants is that, in addition to using the same bacteria as septic tanks, they speed up the treatment by employing *aerobic* bacteria (that live in oxygen) to help break down and neutralise the sewage.

There are two main MTP systems, and both work by stirring up the mixture to expose the effluent to the air and water. Rotary biological compactors (RBCs) use large motorised plastic paddles that slowly revolve half-submerged in the effluent, whereas sequence batch reactors (SBRs) aerate incoming sewage in a single chamber.

The cost of installing an MTP is relatively high at around £5,000, plus running and maintenance costs. But machines with more expensive price tags generally produce higher-quality treated effluent.

Cesspools

A cesspool (or cesspit) is nothing more than a big tank where raw sewage is deposited until a tanker comes along to pump it out and cart it all away. Although this is the solution of last choice, there may be little alternative where the ground isn't suitable for discharging processed effluent, such as waterlogged or rocky ground. However, excavation and installation costs in such ground can be relatively expensive. On very rare occasions a cesspool may temporarily be required in a built-up area where new connections are banned until the public sewers or the local sewage works have been upgraded to cope with higher capacities. A new 18,000-litre fibreglass cesspool costs about £3,000 plus a similar amount for excavation and installation, and also has to be emptied fairly frequently, sometimes as often as once a month.

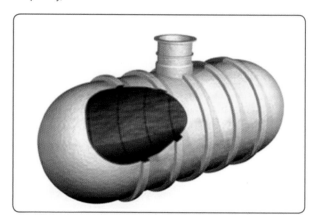

Reed beds, ponds, and leachfields

Problems of effluent quality from septic tanks can be solved by discharging the treated liquid via a series of reed beds. Reed beds, being rich in microbes, can soak up and neutralise a surprising amount of effluent, naturally converting it into harmless plant material and gases without using any power. They're normally lined to prevent seepage and are then filled with gravel and soil as a base for the reeds and plants. The more effective systems use floating beds of reeds growing on special mats, returning good quality water to the environment whilst providing a wildlife haven. The main drawback is the relatively high cost, plus the fact that many self-build plots don't have the necessary space – you need to allow at least 20m² to serve an average family home.

Photo: alternativecleansing.co.uk

Photo: reedbeds.com

Perhaps an ideal green waste system, space permitting, would be to process effluent via an on-site treatment plant, then pipe it to a specially constructed reed bed or willow wetlands, until finally progressing to an aerated pond. Seeded with aquatic plants, treatment ponds can look beautiful, but are expensive and require a very large surface area of at least 50m². It's claimed that specially designed ponds can be used for all three stages of treatment if they're aerated to remove ammonia.

However, the preferred option for final dispersal of effluent, where the soil is free-draining, is to allow it to soak

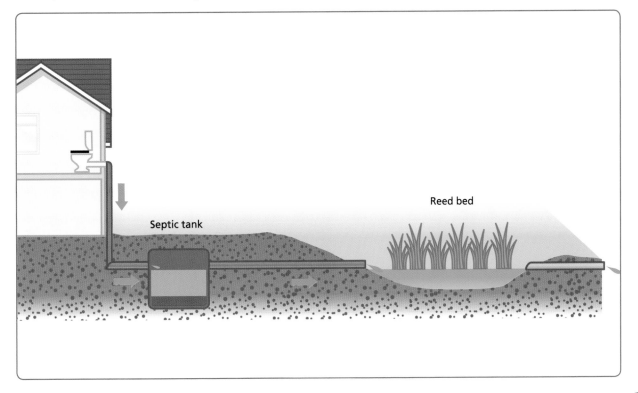

Septic tank

Reed bed

Rainwater harvesting

Rainwater harvesting is a way of collecting rain from roof surfaces to use it for flushing toilets and supplying washing machines (rainwater is soft and ideal for washing clothes). It's even possible to harvest rainwater for drinking. There are three main levels:

■ On a basic level rainwater can be diverted to inexpensive water butts (up to 220 litres' capacity, and made from recycled plastic). Used for watering gardens, this can cut household water consumption by about 5%. To prevent water butts overflowing when full, an overflow leading back to the downpipe is required.

■ Rainwater for use in WCs and appliances can be stored in a tank of about 2,000 litres ($2m^3$) located either below or above ground (the latter requires protection from freezing). The advantage of systems that collect water from roof slopes via discreet grilles inset between courses of tiles is that they store water in loft tanks in the eaves and don't require electric pumps. Although rainwater is less prone to pollution than ground water it will still need filtration and treatment to remove leaves and moss from gutters. Stored rainwater is then either pumped directly to the points of use or pumped up to a header tank serving the bathroom. Recycling surface water from patios and gardens is probably not cost-effective. Being of a lower quality than rain collected from roofs, it needs to be passed through special sand filters to remove particles.

■ Recycling rainwater from roofs for drinking or cooking requires additional filtering and purification. You can buy small purpose-made on-site treatment and filtration systems so that water can be stored, filtered, mineralised, and pumped to drinking-water taps. But this is not generally cost-effective.

Rainwater harvesting can help solve dispersal problems and cut water charges. In some areas water companies even make a small annual payment in recognition of the reduced demands on their systems (although after use it gets disposed of via the foul drains).

Rainwater harvesting is obviously better suited for properties located in areas of heavy rainfall and with generously sized

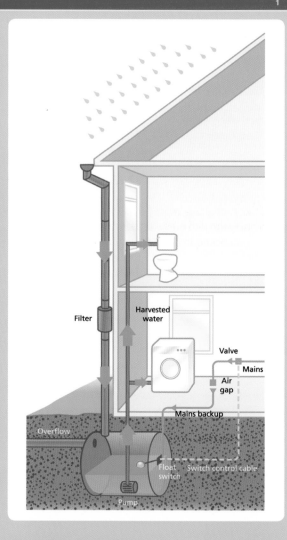

roofs, although you're still likely to need mains backup for dry spells. The downside is that payback periods on new homes can be as long as 20 years, because efficient new WCs and appliances use less water, making your water saving projections less impressive! It also takes energy to construct, install, run, and maintain such a system. But as water shortages become more common in future it should make increasing sense to make use of all that relatively clean rainwater that would otherwise vanish straight into the ground.

Photo: Rainharvesting Systems

Photo: Rainharvesting Systems

into the soil via leachfields. These comprise a series of perforated pipes in trenches surrounded by gravel, which can remain completely unseen below the grass in the garden. Simple and efficient, this provides good quality tertiary treatment at low cost – and a stimulating conversation point whilst taking tea on the lawn.

Surface water

Rainwater that collects on hard surfaces like roofs and paving, known as surface water, will require a suitable dispersal system. Although some districts have combined systems that can take both foul waste and rainwater through the same pipes, most Councils won't allow surface water to discharge into public sewers.

Where connecting to the main system is prohibited, the best solution is to divert rainwater from the house into a handy ditch or nearby stream. If there isn't one, then the usual arrangement is to construct a soakaway, at least 5m away from the house. But where ground is poorly drained, for example sites with heavy clay subsoils, a larger area will be needed to disperse the water through filter beds. These may be in the guise of lawns made from gravel that soak up the water, although not all sites are large enough to accommodate such features.

Where gardens slope downward towards the house there's likely to be a risk of storm water cascading down and ponding around door openings. To prevent this, drainage channels may need to be installed near the main walls. Where back inlet gullies are fitted flush with the paving or patio surface at the base of rainwater downpipes they can help channel surface water away. Conversely, on sites that slope down to the street additional precautions may be required to restrict surface rainwater from gushing on to the road in storm conditions.

Soakaways

The traditional method of dispersing accumulated rainwater is via a soakaway. This comprises a large hole in the ground, filled with rubble. Run-off rainwater from roofs and hard surfaces is piped into the soakaway, where it collects and

then gradually seeps into the surrounding soil. In recent years they've become more widely used in urban areas, where space permits, to help prevent overloading public sewers with water that doesn't require treatment. They can either be constructed traditionally in the form of pits or trenches filled with rubble (approximately $3m^3$), or you can buy ready-made concrete chambers with holes in the walls.

Alternatively, a network of pipes with holes in them can do the same job. Some form of inspection access should be provided (in case they need jetting to clear them).

Soakaways are most effective in non-clay areas with a low water table (in areas with a high water table there's a risk they can start working in reverse, with ground water filling the soakaway).

You need to arrange for Building Control to check the soil quality and advise on the required depth and distance from the house. This may involve digging a trial pit and filling it with water three times in succession to monitor the rate of seepage. In most cases where the soil drains well, and the roof area is less than $100m^2$, you should be able to construct a traditional type of soakaway.

Photo: jastimber.co.uk

Main trades needed on site

- **Groundworkers:** Excavation of trenches for foundations and drain runs, excavation for soakaways and septic tanks, etc.
- **Bricklayers:** Building the walls up to DPC level (although groundworkers sometimes lay foundation blocks).
- **Labourers:** Laying beam and block ground floors, backfilling drainage trenches.
- **Plumbers:** Laying and connecting drainage pipes.
- **Carpenters:** Installing suspended timber ground floors.

Who does what? See full list at www.Selfbuild-Homes.com.

12

THE BUILD – EXTERIOR
The walls, windows, and roof

Photo: Skaala Windows

The walls

Once Building Control have inspected and approved the walls built up to DPC level, you're free to forge ahead and start constructing the shell of the building. With a bed of mortar laid over the DPC and a few courses of bricks or blocks, things will now rapidly begin to take shape.

Where you're building in modern timber frame or non-standard construction (see Chapter 6) the outer leaf of the main walls will most likely be built in brick, stone, or blockwork. So although this chapter focuses predominantly on masonry walls, much should still be relevant. But regardless of your chosen method of construction, a strange thing can happen at this stage. You notice that the floor space appears incredibly small. Thankfully this uneasy feeling that the rooms have somehow shrunk should pass once the walls are all nicely plastered and decorated, at which point everything should magically resume its rightful proportions.

Any load-bearing internal walls can be built at the same time as the main outer walls and bonded into them.

Inner SIPs leaf clad with blockwork.

Photo: ClaysLLP.co.uk

Many standard materials come in matching sizes so that they all fit neatly together. Designed to make life easier for volume house-builders, standard window heights relate to standard brick courses, and internal doorway heights relate to the dimensions of concrete blocks. Because modular-sized door and window frames are compatible with courses of bricks and mortar joints, it should make it easier to achieve neat brickwork above the openings. Even some staircases are manufactured to fit standard room heights. The trouble

Setting-out the walls

Cavity walls can be built very quickly. Some highly motivated brickies, if well supplied with the raw materials, can steam ahead at a staggering rate. A team of two brickies plus a labourer should normally be able to lay up to 1,000 bricks a day or 30–40m² of blocks, or more likely a mix of the two.

As the walls grow upwards, check that the external openings for windows and doors are properly set out and accurately recreated from the drawings.

Where the outer leaf of the wall is brick, every dimension should be based on a standard brick size (or other specified size) plus a standard width of mortar joint, usually 10mm. This way, the brickwork should finish neatly at window and door openings as the scale drawings come alive in 3D.

is, self-builders are not modular people, so you may not want a standard-shaped house made from the cheapest materials and off-the-peg fittings. You may instead prefer to use reclaimed bricks or bespoke window units.

But fitting expensively glazed new windows and premium quality doors into the walls as construction progresses can only too easily result in them becoming damaged. So, for peace of mind's sake, simple 'dummy window' templates made from timber can temporarily be built into the walls. For the main entrance fit a cheap temporary door until it's safe to take the wraps off your hardwood masterpiece.

Bricks

Brickwork typically accounts for as much as 70% of the appearance of a building (although less than 5% of its cost). So picking the right bricks is a critical decision. In terms of size, the standard UK metric brick is 215mm long x 102.5mm wide x 65mm high. With a standard 10mm mortar joint, this gives a working size of 225mm x 75mm, which means you need 60 of them to build one square metre of wall. Most bricks are also available in greater depths of 73mm, and some as large as 80mm.

Some people are attracted to the 'instant character' that reclaimed bricks can add to a new property. But ancient clay bricks tend to be rather fragile and porous, and require the use of traditional lime mortar, rather than rock-hard modern cement, which can damage them. A good compromise might be to buy modern ones manufactured with a traditional appearance. These ancient-looking bricks, some as short as 2in (50mm), are perfect for recreating an 'Olde Worlde' period house.

Decorative timber beams with blockwork inner leaf.

Modern timber frame clad with blockwork.

Brickwork accounts for up to 70% of a building's appearance.

But appearances aren't everything. New bricks are technically rated for both frost-resistance and salt content. Frost can attack damp bricks, causing the faces to crumble. Salts within bricks can chemically react and expand, causing cracking and spalling. In most parts of the country standard 'M-rated' bricks should be sufficiently frost resistant, but if you're building in an area that's prone to severe frost, or in mountainous regions, go for the harder F (or FR) variety. Salt ratings are either N (normal) or L (low). In very exposed locations, such as some coastal districts, L-rated bricks may need to be specified, along with sulphate-resisting cement.

Different batches of facing bricks can vary slightly in colour, so wise brickies pick from mixed batches. This avoids the problem of 'shading', where half a wall has obviously been built using a batch of a different shade that the supplier has had kicking around for a year or two.

If you decide to use a more expensive type of handmade brick, it shouldn't make much difference to the amount of labour required. But if you decided to instead use a natural stone outer leaf, the cost of both materials and labour would increase; this is due to the requirement for a backing wall of blockwork plus the extra cost of stone. Building in stone is an increasingly rare and expensive skill, perhaps adding as much as £10,000 to the price-tag of an average house compared to walls of standard brick and blockwork.

Concrete blocks

The cheapest type of outer leaf construction is blockwork – except, unlike brick or stone, it can't be left 'fair faced' and needs to be clad or rendered (see Chapter 14). Rendering

Styles of finishing

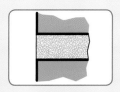

FLUSH

Flush finishing is achieved by drawing a strip of wood about 12mm wide, 6mm thick and 100mm long along the joints.

BUCKET HANDLE

Formed by pulling a suitable shaped piece of metal (or a bucket handle!) along the joints. Also known as the 'hollow key' style.

WEATHERED

A weathered finish is good for throwing off rainwater from a wall, and is considered to be fairly durable. It is sometimes used on stacks to improve weather resistance. The vertical joints can be struck to the left or the right but must be kept consistent, otherwise the wall will look peculiar!

STRUCK

Struck jointing is not ideal for most exterior facings as it leaves the upper edge of the lower brick exposed to the weather.

RECESSED KEY

Recessed key jointing suffers from the same drawback as 'struck'. Here the mortar is raked out and is then pressed back evenly using a special 'chariot' jointing tool.

---

involves two more trades, the plasterer and decorator, and means the scaffolding will have to remain in place longer. This can add at least £5 per square metre to the total cost. So in the majority of properties blockwork construction tends to be restricted to the load-bearing inner leaf of main walls and some internal walls.

The standard size of a concrete block is 440mm long x 215mm wide x 100mm deep, the equivalent of six bricks, and you need ten of them to build one square metre of wall. So constructing walls using relatively large blocks gives a rapid sense of progress, and you start to see the fruits of your labour within a few short hours.

As we saw earlier, you can't just use any old blocks left over from the last job. The drawings will specify blocks of the correct strength (eg 7 N), density (eg 2,000kg/m³) and thermal efficiency (eg 0.15W/mK) designed to support loadings and meet energy targets. Well-known brands are H+H Celcon, Hanson Thermalite, and Tarmac Hemelite. Some lightweight blocks have a rippled or 'striated' face, ready for rendering. Others have a warm, buff, stone-like colour. But with ever-increasing thermal insulation standards to meet, aerated blocks used on the inner leaf of main walls can be very prone to shrinkage cracking when drying out after plastering, one reason why blockwork inner faces of main walls are today almost universally dry-lined with plasterboard.

Mortar

A typical mortar mix for new brickwork would be 1:4 Portland cement/sand (a stronger 1:3 mix would be used in masonry that gets very wet, such as below-ground work). This produces a strong mix with good frost-resistance when set. However, it can also be fairly brittle. Adding a small amount of lime produces a less rigid mortar that is very slightly 'plastic',

Checking brickwork

In Britain we tend to take neat brickwork for granted. But it's important to nip any problems in the bud before the wall grows too big. Here are some useful checks to help sort out the good guys from the bad:

- Check that the DPC has been correctly installed.
- Check for mortar smudges on brickwork.
- Pay particular attention to the pointing, looking for uneven horizontal mortar joints ('beds') and for vertical mortar joints ('perps') that line up neatly.
- Stand at the end of a wall and take a squint along it to see if there are any bulges. Continually monitor the accuracy of each vertical and horizontal plane.
- Blobs of excess mortar, known as 'snots', should be scraped off the inner cavity walls as work progresses, as such obstructions can later risk rainwater bridging across the cavity on to the inner wall. Excess mortar and debris built up at the bottom of the cavity can compromise the effectiveness of the damp-proof course.
- Insulation should be added as the wall is built.

allowing any small settlement at joints to 'self-heal', so that walls are less likely to develop cracks. The lime also gives mortar a traditional quality, lightening its appearance and improving workability. So a mix of 1:1:6 Portland cement/hydrated lime/sand is a good general-purpose blend (non-hydraulic or semi-hydraulic lime can also be used), especially when trying to match the appearance of older buildings.

Although 'anti-freezing' agents can be added to the mortar, bricklaying should really only be carried out when temperatures are above 2°C, and fresh work must be protected against frost, especially overnight.

The type of sand used for brickwork is 'builder's sand' as opposed to 'sharp sand', which is coarser and more suitable for making concrete. Where consistency of appearance is important, the colour of the sand used in your mortar mix should be taken into account, as colours vary considerably depending on its region of origin.

Mortar makes up around 20% of the area of a brick wall, so getting the finished joints looking right is vital to the appearance of the entire building. So the rough mortar edges on fresh brickwork should be carefully finished before the mortar gets a chance to harden. It's often best to go for a simple, traditional style, such as a plain 'flush' finish.

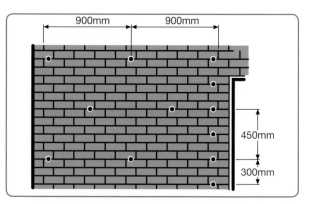

900mm | 900mm | 450mm | 300mm

| Butterfly Tie | Double Triangle Tie | Vertical Twist Tie |

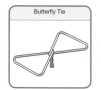

Wall ties

For cavity walls to be stable and strong, the two leaves are joined together with wall ties that bridge across the void. To achieve structural stability, the ties must be placed at regular intervals and fully bedded a minimum of 50mm into mortar joints, sloping at a slight gradient downwards and outwards so that any moisture heads away from the rooms. The first row is inserted at ground level.

As a general guide, ties are fixed in the wall about 900mm

apart horizontally and about 450mm vertically, staggered in a 'domino five' pattern. Around door and window openings the density is normally increased and the ties are spaced vertically about every 300mm, positioned no more than 225mm in from the edge. You need about four wall ties per square metre.

Today 'double triangle' wall ties are the most widely used type. These comprise a

Good brickies protect cavities from mortar 'snots'

Included	Not included*
Setting out the walls on the foundations.	Erecting scaffolding or towers.
Fitting air bricks.	Supplying materials.
Laying the DPC.	Constructing timber templates for windows and doors.
Building in templates or frame ties to door and window openings.	Plant hire.
Installing cavity insulation.	Cutting indents into existing walls.
Fitting lintels.	
Building in restraint straps.	
Bedding on the wallplate at roof level.	*Unless agreed as an 'all-in-rates' package.

225mm long strip of stainless steel wire about 2mm thick with a small fold at each end. The squiggly bits in the middle of ties are known as 'drips' (because any water travelling along it should drip off before it gets a chance to soak into the inner leaf). Of course, big blobs of bricklaying mortar landing on them will mess things up, so it's important that ties are kept clean. Wall ties are designed to suit particular cavity widths, so where wider cavities exceed 100mm, longer 250 or 275mm ties can be used, while for walls in severely exposed locations stronger flat 'vertical twist' types are recommended.

Keeping warm

The Building Regs are very hot on the subject of insulation. In order to comply with SAP standards (see Chapter 6) your approved drawings should clearly specify the required levels of insulation. The amount of heat that's allowed to escape from each part of the building's 'thermal envelope' – the walls, roof, floors, and windows – is explained in terms of 'U' values. The bigger the number, the more heat can leak out. So if, for example, the walls have a stated 'U' value of 0.30 it means 0.30W is the maximum amount of heat permitted to pass through each square metre of wall (technically that's watts per square metre x degrees Kelvin). A modern cavity masonry wall should be able to achieve a nice low 'U' figure of, say, 0.25 without too much trouble (compared to a Victorian solid brick wall that may be hard pushed to go much below 1.25). However to meet the target 'TFEE' U-value of 0.18 W/m^2K you may need wider 125mm cavities, insulated with, for example, 75mm PIR board leaving 50mm air space.

Cavity insulation

It's normally the bricklayer's job to install the insulation in the cavities as the walls are built. However, it's important to note that with modern timber frame construction, the wall cavities must not be filled – the insulation is already fitted within the timber panels that form the inner leaf. With conventional masonry walls, there are two principal ways of installing cavity insulation: 'full fill' and 'partial fill'.

Traditionally, full-fill cavity insulation uses mineral or glass fibre wool batts (usually yellow) stuffed into the cavity as the

Photo: Kingspan

wall is built. But modern partial-fill insulation is generally considered superior. Here, rigid foam boards are fixed against the inner leaf with special clips on the wall ties.

Some newbuild warranty providers require 'full fill' to be avoided altogether. This is because insulation works by trapping air within its body, and to work well it needs to remain dry. But if damp penetrates the outer leaf, the insulation stuffed into the cavity could become limp and soggy, so won't be of much use. With the full-fill method you also sometimes find hidden gaps, where some parts of the cavity are 'better stuffed' than others.

The attraction of partial fill is that it leaves a clear 'defensive space' between the insulation and the outer leaf. To accommodate a decent thickness of insulation as well as an effective air void, cavities are now designed to be at least 100mm which means an overall wall thickness of 300mm to 350mm. The insulation boards are typically 50 to 100mm thick rigid foil-backed polyurethane sheets (Celotex and Kingspan are well-known brands). To be effective the boards must be firmly attached and not left hanging limply where the sun don't shine. Polyurethane is one of the most efficient insulators, being roughly twice as efficient as EPS or wools (but about three times as expensive as EPS).

But these aren't the only methods. On some new housing estates cavities are filled with tiny lightweight water-repellent polystyrene granules blown in by machine once the walls have been built. Despite fully filling the void, major housebuilders are confident these will not allow damp to pass, even in the most dire weather. Or if you prefer an organic house, sheep's wool or natural cellulose fibre made from recycled newspaper can be blown into the cavities once the walls are completed.

Photo: excelfibre.com

Thermal bridging

The more insulation you stuff into the fabric of your house, the greater the relative importance of any remaining cold areas. The weakest links are found anywhere that cold can bridge across the cavity from the cold outer walls to the inside walls of rooms. Metal wall ties and timber studwork are two of the usual suspects that can act as 'thermal bridges' resulting in 'cold spots' on the walls in rooms. When the warm, moist air in a typical home hits one of these cold spots, it condenses into water, causing damp patches, which can eventually grow black mould.

The obvious solution is to dry-line the inner face of the walls with foil-backed plasterboard, thereby forming a barrier between the warm humid air in rooms and the wall structure, adding an extra layer of internal insulation. Insulation can also be applied externally, either in the form of purpose-made boards finished with render, or in the traditional form of weatherproof timber boarding or tile hanging.

However, there are other parts of the structure where thermal bridging can be a problem, notably to the reveals of door and window openings. The solution here is to install plastic 'cavity closers' that fit snugly into the ends of cavities, sealing them off. These contain special strips of polystyrene insulation and have the added benefit of acting as a DPC, thereby preventing damp from crossing the cavity and cutting heat loss at the same time. Another weak point is found at the top of window and door openings where steel lintels can allow cold to penetrate, so these now come ready-packed with insulation.

Other high-risk areas are where the walls meet the roof, due to insulation being skimped in awkward corners (which can lead to damp mould staining to some upper bedroom walls), and to the ground floor where the screed floor finish meets the walls, which can be solved by a strip of insulation around the edge, between the screed and the cold wall. N.B. thermal bridging is now a factor in achieving 'TFEE' energy efficiency targets – see Chapter 6.

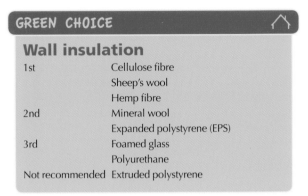

Condensation and ventilation

People are sometimes surprised to learn that an average home produces well over ten litres of water vapour a day. Most of this humidity is generated by ordinary human activities like breathing, sleeping, and sweating, supplemented by moisture emitted from tumble driers, washing machines, showers, gas heaters, and houseplants. Dogs and children are especially rich sources of airborne water vapour.

And it all has to go somewhere. The worry is, if you don't expel all this warm, steamy air it will seek out cold surfaces and condense back to water, causing damp, as described above.

So a big part of the solution to condensation is to do exactly what you do in the car when it steams up: wind down the windows and turn up the blower. In other words, improve the ventilation. Ventilation allows condensation to harmlessly evaporate before the dampness can cause problems, and is essential in most lofts and below suspended floors. This is why the Building Regs have strict requirements for ventilation.

The other main part of the solution is to make sure the house is well insulated, because if the inside surfaces are relatively warm there should be no surface condensation.

Preventing condensation

Condensation can be prevented by:

- Good ventilation, to prevent build-up of excessive moisture.
- Effective insulation, to help raise internal surface temperatures.
- Room heating, to make indoor surfaces warmer.
- Making sure the structure is airtight, with high vapour-resistance on the inside so that no moisture can escape into the structure. The inside of wall surfaces can be made vapour resistant using foil-backed plasterboard or a polythene vapour barrier, as in timber frame construction.
- Ensuring that walls have a low vapour-resistance on the outside, so that any trapped moisture can easily pass out of the building and evaporate outdoors. Traditional 'breathable' materials such as permeable soft bricks and lime mortar allow this; but modern hard, non-absorbent materials, such as hard bricks, cement, and gypsum plaster, do not.

However, if there are gaps in the construction, moisture can be driven through them, propelled by the higher internal vapour pressure towards cold outer layers, such as within a cavity wall. Water vapour can even diffuse through plasterboard (unless it's foil backed). It can then condense behind the scenes, reducing the effectiveness of the insulation. This is a particularly serious concern in timber frame structures, where any such damp can cause decay to a building's structure. Which is why a polythene vapour control layer is placed on the warm inner side of the insulation, effectively wrapping timber frame buildings in a plastic sheet. But even then careless fitting of services such as electrical sockets can cause leaks, which is why pipes and cables should be installed in a service void, inside the vapour control layer.

Beams and lintels

Regardless of all the complex desk-calculations, some builders on site always know better and prefer to stick with the same familiar types of lintels they've used for years. Funnily enough, these usually happen to be the ones that are cheapest and most easily available. So Building Control will want to check that the beams and lintels used on site match those shown in the approved plans.

Steel lintels are preferred for openings in outer walls. These are pre-insulated and can be neatly hidden within the brickwork. Alternatively, for the visible outer leaf, you may prefer the look of natural stone lintels (and matching sills), or those made from granite mixed with cement to resemble stone at a fraction of the price. Or perhaps chunky 'railway sleeper' timbers, which are extremely robust and can look the *cat's pyjamas* when displayed above windows and doors on the outer wall.

When specifying beams and lintels, their suitability for purpose will, of course, depend on how long the span is, and how much load they need to support. Popular steel lintels by major suppliers such as 'IG' and 'Catnic' are available off-the-shelf in lengths from 600mm to 4,800mm, with a range of sizes in between in jumps of 150mm. The optimum size of concrete lintel to fit single-leaf internal walls is 70mm deep x 100mm wide, and these are available in a range of lengths from 600m to 1,800mm.

Where wide openings need to be spanned, the load transmitted at each end of a beam can be so great that it can

potentially crush ordinary bricks or blocks, particularly if it's a steel beam. To prevent such calamities, the ends need to sit on top of 'padstones' that spread the load on the wall. These should be specified and shown on your drawings, normally in the form of precast reinforced concrete padstones, dense concrete blocks, or super-hard engineering bricks.

Another way to help spread the load is by increasing the amount of the beam that projects onto the walls at each end beyond the minimum 100mm to 150mm or more. The recommended bearing should be clearly specified in your Building Regs drawings. However, for a relatively small opening like a typical door or window, ordinary brick or blockwork can normally take the weight of lintels unassisted.

Cavity trays

A cavity wall is designed to allow any moisture that gets into the cavity to pass harmlessly down to the ground and out again. But where an opening such as a window or door interrupts the cavity, there's a potential risk that moisture could instead find its way into the building.

Modern practice is to fit purpose-made plastic trays over window and door lintels, so any water that runs down the cavity will be caught and dispersed out of the wall through 'weep holes'. These are small vertical gaps left between a few bricks (at vertical 'perpend' joints). They act as tiny drainage outlets, usually with small plastic vents pushed into the holes. Weep holes should be installed no closer together than 900mm intervals, with at least two provided to drain each cavity tray above openings.

Trays should actually be provided over anything that interrupts the cavity – lintels, ducts, even recessed meter boxes – since these can all potentially direct dampness on to the inside wall.

Illustration: Cavity Trays Limited, Somerset (www.carytrays.co.uk)

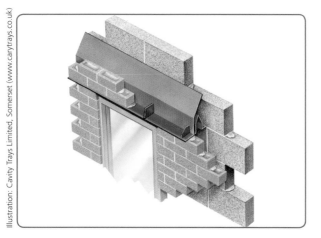

Lateral restraint straps – the rules

To make a building structurally strong, the walls are tied into other parts of the structure. Horizontal L-shaped steel restraint straps are built into the inner leaf blockwork and then fixed across floor joists, ceiling joists, and roof rafters.

■ Straps should be built into the blockwork of the wall, with the short 'L' turned down inside the cavity against the inner leaf, rather than being screwed in later.

■ Where joists run parallel to the wall, straps should be spaced every 2m or closer.

■ The long part of the strap should extend across no less than three rafters or joists (either above or below them) with timber noggins fixed under the strap between the joists or rafters.

Internal walls should also be tied with vertical straps to ceiling joists on each storey.

Some 'intelligent lintels' are now designed with a dual function, acting as their own cavity trays. Alternatively, plastic sheets (damp-proof membranes) are sometimes laid over the lintel and turned up at the sides, in effect acting as trays, collecting any moisture and directing it to the outside.

Scaffolding

Scaffolding needs to be ordered well in advance so that the 'first lift' can be erected in good time for the bricklayers setting-out the first-floor walls. Any platform over 2m high must have metal edge-guarding fixed to the sides, and protective netting to catch falling debris where people are

likely to be passing underneath. Erecting scaffolding is definitely not a DIY job, and must be carried out by a licensed specialist firm with plenty of insurance. Specify that it is erected in accordance with BS EN12811-1:2003.

If you need to build close to the boundary with next door and can't come to an amicable arrangement with the neighbours about entering their garden to help build your house, some of the scaffolding may have to be erected internally. This is because there is no automatic right of access to next door's land for the purpose of building (although there is for carrying out maintenance – The Access to Neighbouring land Act 1992).

Wall plates

When wall construction finally reaches roof level, a strip of 100 x 50mm softwood is normally bedded in mortar along the tops of the walls. This is the wall plate, and its job is to provide a secure base for the roof timbers to rest on. In order to hold the wall plates in place, they're clamped vertically to the walls below using 30 x 5mm L-shaped galvanised steel restraint straps. These straps should be placed no more than 2m apart, and no further than 450mm in from the ends of the walls, extending at least 300mm down the wall. It's normally the brickies' job to bed wallplates and strap them down, the carpenter having cut them to the correct size.

Internal masonry walls

Interior dividing walls built of blockwork are generally preferred to hollow timber stud partition walls. Non-structural walls built from lightweight concrete blocks are relatively quick and cheap to build, as well as providing reasonable sound insulation between rooms and a general feeling of solidity. But because they impose greater loads than timber studwork, they're often confined to the ground floor. In mainstream housing it's generally considered better to build timber stud walls upstairs, as they weigh less than half that of equivalent solid masonry walls. See 'Timber studwork walls' in the next chapter.

However, where upstairs walls are built in blockwork, to keep the weight down thinner 75mm blocks are preferred to the standard 100mm blocks used at ground level. This works best where they're built directly on top of the ground-floor walls. Otherwise additional support will need to be provided. For example, upstairs walls are sometimes built directly above doubled-up floor joists, bolted together. Alternatively if an upstairs masonry wall runs at right angles to the

The most common classifications are C16 and C24, the latter costing slightly more as it's stronger and can be used over larger spans. The strength class depends on both the species and the grade of the actual piece of wood. In all there are 16 strength classes ranging from C14, the lowest softwood strength, through to D70, the strongest hardwood strength class.

Floor joists

The approved drawings should clearly show the required joist sizes, and explain how they're to be spaced and fixed. Floor joists traditionally comprise 200 x 50mm '8 x 2' softwood, spaced at either 400, 450, or 600mm centres ('centre' measurements are taken from the centre of one joist to the centre of the next). To achieve longer unsupported spans (6m plus), manufactured timber I-joists can be used. Despite their thin appearance they're actually incredibly strong and light. Although more expensive, they take half the time to install, don't suffer from twisting and warping, and services are easy

floor joists, it can be built off a minimum 75 x 75mm timber sole plate or a steel beam. Alternatively, where your upper floors are of concrete 'beam and block' construction, it may be possible to build solid masonry bedroom walls off them without the need for support from walls directly below.

Where internal walls are supporting loads from ceiling and floor joists (and sometimes also from the roof), they're usually built of dense concrete blocks. Or you could perhaps opt for good old-fashioned 'naked brickwork' left exposed as a design feature. But with solid masonry walls of all types, door openings need to incorporate lintels, usually of a simple reinforced concrete type.

to run through preformed 'knock-outs', rather than cutting or drilling notches. Alternatively lightweight, 'open web' Posi-Joists or Eco-joists with a central W-shaped metal web also allow long spans, and can accommodate wide bore pipes and ducting.

Once a room becomes wider than the maximum span for a floor joist

Intermediate floors, windows, and doors

Floors above ground level are often referred to as 'intermediate' floors because they perform a dual function as floors and ceilings. The main considerations are their structural strength, sound insulation, and incorporation of service routes. Traditionally they're constructed from softwood joists spanning between the structural walls of the building. Although comparatively rare, upper floors are sometimes built from concrete beam and block in the same way as ground floors. For obvious reasons, fixing the timber floorboards or floor panels in place is normally left until the building is dry and weathertight.

Grading timber

Most construction-grade timber is spruce, known as carcassing, deal, or whitewood. Joinery-grade wood is usually pine (redwood), which is more durable than spruce, and is used for windows, skirtings, floorboards etc, often sold with a planed finish. For structural purposes new timber is supplied ready strength-graded.

Timber strength class	C14	C16	C18	C22	C24	C27
	< Weaker				Stronger >	

(typically about 4.5m for softwood) extra support will be required, either from a structural beam or a load-bearing internal wall. On external walls, the Building Regs require joists to be supported from galvanised steel joist hangers (small purpose-made cradles), which are fixed to the walls.

A good chippie will lay traditional softwood floor joists so that the top edge is level across all of them, to reduce the risk of squeaky floors. This can be done by running a string line over all the tops and adjusting the joists with packing underneath. To minimise twisting and distortion in timber joists, small struts are fitted between the joists in an 'X' pattern known as herringbone strutting. For spans up to 3.5m, one line of strutting is adequate, but for every additional 1.5m further struts are needed. Modern construction makes use of ready-made lightweight steel herringbone straps designed to fit standard joist spacings. Where floor joists rest on internal walls additional strengthening is recommended to prevent 'rotation', usually by jamming small timber off-cuts called 'noggins' between the joists.

Windows

As a rule of thumb, your windows will cost at least twice as much per square metre as the walls they sit within, although many self-builders will spend considerably more on high-performance bespoke units.

Mass-produced timber windows are commonly made from Scandinavian redwood, factory-vacuum treated with preservatives. Locks are now fitted as standard, and opening casements are draught-proofed. Windows and other joinery items are supplied with a base coat of wood stain that you can later choose to finish with a suitable stain or paint, or a white-primed base that's

Window fixing – points to check

- The sills should have a thin 'drip groove' underneath, set back a few millimetres from their outer underside front edges.

- The sills should project well clear of the wall, so that rain can drip off freely without soaking into the wall.
- Timber window and door frames should be protected by DPCs or plastic cavity closers.

- Where the outer wall below the window is to be tiled or timber clad, a lead flashing 'apron' should be fixed under the sill (to be dressed down over the cladding).

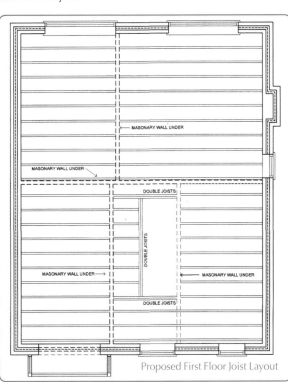

Proposed First Floor Joist Layout

MASONARY WALL UNDER
MASONARY WALL UNDER
DOUBLE JOISTS
DOUBLE JOISTS
MASONARY WALL UNDER
MASONARY WALL UNDER
DOUBLE JOISTS

specifically designed for painting. Window units should be delivered factory pre-glazed with a protective coating to the timber.

The standard height options for off-the-peg window units are 450, 600, 750, 900, 1,050, 1,200, 1,350, and 1,500mm (sizes rising in 150mm jumps, equivalent to two brick courses). The standard width options are less logical, at 488, 630, 915, 1,200 and 1,770mm.

Actually fitting the windows is often left until later, in order to protect them from damage. Both UPVC and timber windows are available that simply clip into the cavity closers installed earlier as the main walls were built. When window and door frames are fitted, the cavity closer should overlap them by at least 25mm.

Glazing

To reduce heat-loss most sealed double-glazed units have vacuum-sealed cavities with generous air gaps of 16mm,

20mm, or 24mm. Special 'low-E' glass has a microscopically thin coating applied to the inside of the inner pane that allows the sun's rays to pass through but reflects infra-red radiation back into the room. 'Low-E' coatings are available either in a 'hard' form or as the more efficient

'soft' variety. Specifying cavities filled with argon or krypton gas further slows the transfer of heat across the cavity because these inert gases are more 'syrupy' than air.

To prevent condensation forming on cold winter mornings around the perimeter of windows, where the glass is cold at the edges, special 'thermal break' spacer bars are fitted. These separate the cold outer frame from the warm inner parts, with a layer of resin or plastic sandwiched between the inner and outer sections.

Perhaps surprisingly, when it comes to the subject of noise reduction, secondary glazing is actually superior to double glazing. With an optimum space of 150mm between the secondary glazing and the glass in the main window, this can provide effective insulation against noise at high frequencies, as well as deadening low frequency sound, such as road traffic.

External doors

Standard sizes

The most common off-the-peg door sizes are 1,981 x 838mm (78 x 33in) and 1,981 x 762mm (78 x 30in). For taller, wider folk, 2,032 x 813mm (80 x 32in) doors are available. Standard external door thickness is 44mm. Including the frame, most UPVC exterior doors are available sized 2,085 x 920mm (82 x 36in), or 2,085 x 1190 mm (82 x 47in) for French doors.

Security

The NHBC newbuild standard is to fit five-lever locks to all external doors (plus a cylinder rim/night latch to the main

---

Windows and Building Regs

Photo: velfac.co.uk

From a technical viewpoint, new windows need to comply with Part L1A of the Building Regulations for thermal insulation (Part J if your house is Scottish). Unless an approved mechanical ventilation system is fitted, habitable rooms must normally have opening windows, pre-fitted with small trickle vents to provide background ventilation.

There are minimum size requirements for window openings. For all habitable rooms, window-opening areas should be equivalent to at least 5% of the room's floor area. Scottish standards stipulate a minimum glazed area in each room equivalent to 15% of its floor area. Any glazing in 'critical' areas (within 800mm of floor level, or within doors or side panels etc) must be toughened or laminated.

Windows in each habitable room normally require at least one opening pane to allow escape from fire (for the actual sizes and locations see Part B1). Openings within 1m of a boundary are normally limited in size to 5.6m².

As we saw in Chapter 6, when it comes to energy efficiency it's the performance of the overall building that determines compliance with the Building Regs. However, the total energy performance of a building will be strongly influenced by the thermal efficiency of the glazing. So the SAP calculation for your property will take into account all the key performance characteristics of the windows – ie U-values, solar heat gain ('g value'), and daylight transmission. The target 'TFEE' U-value for windows is 1.4 W/m²K, (or 1.0 for doors). This should be easily achievable, e.g. with argon filled glazed units, and pre-insulated doors.

One of the problems with U-values is the difficulty of comparing different windows – because frames and glazing have separate U-value ratings. The BFRC rating system (British Fenestration Rating Council) uses Window Energy Ratings that take account of the performance of the whole window, including heat gain through the glass. Ideally aim for a high A or B rating, but not lower than C. See www.bfrc.org.

Temporary boarding until safe to fit main door.

entrance door). Locks should be specified to comply with BS 3621. The easiest locks to use are mortise types with lever handles that automatically operate a latchbolt and deadbolt.

The roof

Once the new building has finally got its 'hat' on, you'll have passed a major psychological milestone. At last you'll be able to glimpse the light at the end of the tunnel. But first there's the small matter of actually constructing your roof.

There are two conventional types of roof structure: traditional 'cut timber' roofs that are hand-built on site; or modern, lightweight, prefabricated 'fink truss' roofs, which are by far the most widely used type in new mainstream housing. However, there's a recently developed alternative method of roof construction using Structurally Insulated

Panels (SIPs), which comprise a sandwich of two sheets of timber boarding bonded to a filling of rigid foam insulation – see Chapter 6.

Trussed rafter roofs

The attraction of using modern ready-manufactured trussed rafters is that they can span a very respectable 8m or so without needing extra support from internal structural walls. Although manufactured trusses are relatively expensive to buy, they're quick and easy to install, with consequent savings in labour costs. But when building your own home it's well worth specifying 'room-in-roof' (RiR) trusses. These provide you with a ready-made shell for a loft conversion, making it a relatively simple matter to complete the job and occupy rooms in the roof at a later date. For an extra couple of thousand pounds or so, this can provide a very cost-effective solution, but their extra weight means you'll also need to budget-in a decent sized crane.

When trusses are delivered to the site, they need careful handling to prevent damage. They must be stacked well clear of the ground on bearers and sheltered from the weather.

Constructing a trussed rafter roof is essentially an assembly operation, with individual trusses typically spaced at 600mm centres. But because some larger ones can weigh 30–40kg, lifting them into position at a great height is no mean feat, requiring cranes to be hired. Once the trusses are all neatly lined up, they can all connected together with binders and straps and each one skew-nailed to the timber wall plates. But to prevent the trusses collapsing like a row of dominoes at the first gust of wind, additional bracing is needed.

It's the bricklayers' job to fix horizontal 'lateral restraint

straps' into the cavities of the main walls at rafter, ceiling, and first-floor levels. These help tie in the roof and floor structure to the gable or party walls.

As noted earlier, these straps should run across at least two sets of rafters and joists. The brickies will leave them ready for the chippie to screw them in, and pack timber noggins under each strap in the gaps between the rafters etc.

Traditional 'cut timber' roofs

For a one-off house, the hand-built approach offers considerable design flexibility. Here, the carpenter cuts the various timber components to length on site. Traditionally the rafters are spaced about 400mm apart. At their tops, they're nailed to a timber ridge board. At their base, the lower rafters are joined to the ceiling joists, with the resulting triangular structure resting on the timber wall plates. But left like this rafters have a disturbing tendency to slide down and

push out walls, so in larger roofs additional support is required about halfway up the rafters in the form of large horizontal 'purlins'. These in turn may need support from struts resting on load-bearing internal walls. But today, steel beams are often employed as purlins, so that no extra support is necessary.

Given that you're paying for a skilled carpenter to custom-build your roof at considerable expense, it seems a shame to hide it all away behind giant sheets of plasterboard. A hand-built roof structure can be a work of art, so why not leave some of it on display for all to admire? Your chippie may even be able to conjure up a medieval oak 'kingpost' roof structure, the perfect backdrop to that suit of armour and stag's-head trophy you've always promised yourself!

DIY or sub-contract?

If your roof is a fairly simple design and you're a bit of a hotshot at joinery, there's no reason why you can't make the timber framework components of a cut roof yourself. However, if the thrill of working at height holds little appeal, you might be better off leaving the actual installation as well as the tiling to an experienced roofer. Roofs can sometimes be deceptively complex, especially where there are hips or

valleys to contend with. Plus there's often considerable time pressure to get the building's 'hat' on before the weather turns, so it's normally best to leave this to the professionals.

Underlay

Roofs should have a secondary barrier beneath the tiles to help keep out severe weather. Today, modern high-performance lightweight 'breathable' roofing felts such as Dupont Tyvek, Klober Permo, or Monarflex are normally specified. Rather like high-tech mountaineering clothing, these 'breathable membranes' cleverly prevent rainwater from getting in, yet allow water vapour to escape outwards from the loft by permeating through the material.

Traditionally, lofts needed to be ventilated to prevent

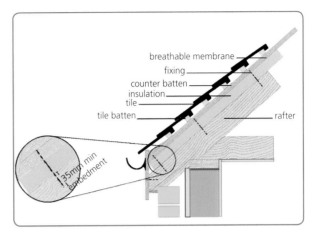

build-up of damp from condensation. However, ventilation may be omitted if a 'breathable' underfelt is used beneath the tiles or slates in a 'warm roof'.

Warm roofs

Conventionally, lofts are cold, draughty places, with rolls of mineral wool insulation strewn loosely above bedroom ceilings in a bid to keep the rooms below warm. Today, 'cold roofs' are being superseded in new house construction by 'warm roofs'. Here, the insulation is moved higher up, being laid on the outside of the roof structure, so that the whole house, including the loft space, is kept warm. This is ideal if you want to occupy the roof space for living

New roof with multifoil insulation.

accommodation. Warm roof construction also reduces heat loss by increasing airtightness. Unlike conventional cold roofs, you don't have to go to all the trouble of ventilating the space below the insulation as it's at a similar temperature to the rooms, with minimal risk of condensation.

Warm roofs are constructed with the insulation laid on top of the rafters, e.g. comprising rigid polyurethane foam boards. Although any water vapour that gets into the insulation from the rooms below should be able to escape through the breather membrane (see below), as a precaution a plastic sheet vapour barrier can be laid under the insulation (on the warm, inner side). On the outer side of the insulation boards, timber 'counter-battens' (minimum 38mm x 38mm) are nailed down the line of each rafter to create an air space above the insulation. These are fixed through the insulation into the rafters below using special helical fixings (except where nailing into rigid structural boards under the insulation). Finally, sheets of breather-membrane underlay are spread over the top of the counter-battens, held in place by conventional horizontal battens on to which the tiles or slates are hung.

There are, however, other materials that can be used to insulate a new roof. Thin sheets of 'multifoil' that look a bit like a BacoFoil and Kleenex sandwich can instead be laid over the rafters. Despite being only about 30mm thick, these shiny 'radiant heat barriers' can provide the equivalent of more than 200mm of loft quilt insulation. The target 'TFEE' U-value for roofs is 0.13 W/m2K. This should be achievable with about 250mm total thickness of PIR insulation boards or about 350mm depth of mineral wool – see Chapter 6.

Battening

The roof tiles or slates are hung from rough-sawn timber battens running horizontally. These are fixed to the tops of the rafters with galvanised steel nails through a layer of breather membrane underlay (or as described above for warm roofs). Battening helps improve the lateral stability for the roof structure and as a rule, it's best not to use anything less than 50 x 25mm tanalised softwood.

However, before setting out the battens you first need to know the required spacing between each row. This is known as the 'gauge', and the precise figures will vary according to the specific type of slates or tiles you're using. It depends on a tile's recommended lap (how far the tile above overlaps the one below) in relation to angle of roof pitch.

Gauge, pitch, and lap figures can found on the manufacturers' websites. The gauge will be equal between all the battens except usually for the bottom one, where shorter eaves tiles are fixed.

Once all the rows of battens are nailed in place they can double as a sort of giant roof ladder that tilers make good use of when scampering up and down roof slopes.

Valleys and flashings

Traditionally plumbers were the trade that specialised in leadwork, which included valleys and flashings on roofs. But today roofers or specialist leadworkers are more likely to

perform this role. Because the detailing at junctions such as chimneystacks, wall abutments, and valleys can be complex, working drawings may be needed to help get it right. Valleys are found where one pitched roof joins another at an internal angle. To form a traditional 'open valley' a lead lining is laid over a

Photo: Helical systems

Photo: Helical systems

timber 'valley board' in strips no longer than 1.5m and a minimum of 100mm wide. The lead used should be Code 4 or thicker and strips higher up should be lapped over the ones below by at least 150mm (225mm for shallow roofs less than 30°).

Modern fibreglass (GRP) valley linings and flashings are a cheaper alternative to traditional lead and easier to fix, although good old-fashioned lead is superior and will last a lifetime.

TYPES OF VALLEY

Open

Mitred

Purpose-made valley tiles

Tiles and slates

Your specific choice of tiles or slates will determine the correct technique for fixing them in place. The easiest type to lay are large interlocking tiles that simply hook over battens, with a single lap over the course below and linking into their neighbours on either side. Smaller traditional plain tiles or slates are generally considered more attractive although are more expensive, requiring greater numbers of battens. However, it's reckoned that using dearer handmade plain clay tiles or natural slate, rather than the ponderous-looking, larger concrete type, can add as much as £10,000 to the finished value of a large detached house – thanks to the enhanced kerb appeal.

For a traditional-looking roof at a budget price, artificial reconstituted slate or stone tiles may be worth considering. These are made from crushed slate/stone with added colour, resins, and glass fibre reinforcement for strength and acrylic coating on the surface to protect them from UV light and restrict water absorption.

Laying roof tiles

Start with the first row of tiles at the lower edge, above the gutter. Mark the centre of the lowest batten then loosely place a row of tiles onto it.

To ensure the roof looks properly balanced, start at the mid-point of the batten either with the first tile centred, or with a joint between two tiles. Then spread the tiles symmetrically out from the centre in each direction, and reduce the gaps to create a perfect fit. Aim to get them finishing equally on either end of the roof slope. At the end verges special 'one and a half' width tiles are normally placed every other course to achieve a neat finish at the edges, and at the eaves shorter eaves tiles are used.

Should tiles need to be cut in order for a row to fit neatly, the trimmed tile is usually placed in the next-to-last position along the row rather than at the end, so the outer tile doesn't have a cut edge exposed at the side. Subsequent rows above are fixed with staggered joints lapped according to the manufacturers' guidance.

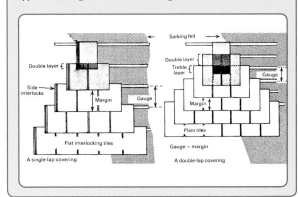

213

Ridges and verges

The most exposed part of the roof is at the very top – at the ridge. Traditionally ridge tiles were kept in place by being bedded in mortar. But modern 'dry-ridge' tiles use special screws and fixing wires that tie them to the ridge timbers below.

The verges are the side edges of the roof above an end gable wall (or a lean-to). They normally overhang the wall below, typically projecting about 50mm. The end roof tiles that form the top of the verges are tilted up slightly to discourage rainwater from pouring over the side of the roof. Traditionally, verges are pointed up with mortar, which over time can develop cracks and erode. So modern a 'dry-fix' alternative has been developed using plastic cover strips secured with clips, or special one-piece tiles called 'cloaked verges' that wrap over the edges. However, modern preformed verge-closers can look rather clumsy, making them unsuitable for traditional architecture. Beneath the verges, you've normally got a choice of bargeboards or decorative brickwork – see 'The roofline' below.

Photo: SwishBP.co.uk

Photo: SwishBP.co.uk

The roofline – eaves, fascias, soffits, and bargeboards

The term 'roofline' refers to the detailing at the bottom and side edges of roofs. The style you pick can have a major impact

on the way your house looks, so it's important to get it right.

Fascia boards are horizontal strips that run along the eaves at the feet of the rafters. They're commonly made from 25mm thick timber or UPVC and come in handy for supporting the guttering. The most widely adopted standard eaves style in housing since the 1930s has been 'box eaves'. Here, a traditional fascia board is fixed along the ends of the projecting rafters. This leaves a gap underneath which is filled with a strip of plywood soffit to form a box shape, hence the name. The soffit closes off this entrance to the roof void, yet provides ventilation, with small vents fitted at intervals.

However, it's now very common for designers to dispense with fascias altogether, instead adopting a simpler, traditional method where the projecting rafter feet are left exposed. But the bricklayer will need to fill any gaps between the rafter feet on top of the wall, to prevent a mass influx of birds, bats, and bees into your loft. Any necessary loft ventilation can be provided with insect-proof wire mesh.

Bargeboards are similar to fascias but instead adorn the A-shaped edges of gable roofs, nestling under the verges. The heyday of the bargeboard was the early Edwardian era when large, elaborately moulded bargeboards would project assertively over gable end walls, to make a bold design statement. But once again, many house designs now dispense with bargeboards, perhaps substituting some fancy 'dog's tooth' brickwork under the pointed up verges. At the end of the day, it's a matter of taste.

Flat roofs

Flat roofs are, of course, not actually flat. By definition, anything up to about 12° pitch counts as flat. They need to be built to a slight slope or 'fall' so that rainwater can easily disperse, rather than collecting and 'ponding',

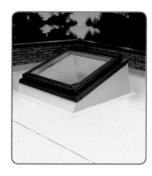

which over time damages the surface. The Building Regulations encourage a minimum fall of 1:40.

When building a new flat roof structure, the joists are traditionally 50mm thick and either 150mm, 175mm, or 200mm deep, depending on the span. They're normally set perfectly level, so that the ceiling below will also be level. To create the required fall for the decking, tapered strips of timber the same width as the joists, called 'firrings', are nailed along the tops of the joists (timber merchants can supply firrings cut to the correct fall).

Materials traditionally used for covering flat roofs include metals such as lead, zinc, and copper. Of these, lead provides a highly durable solution, and although relatively expensive, is well worth the investment if you can stretch to it. Copper is considered to be the Rolls Royce of flat roof coverings, but is even more expensive and famously develops a natural pale green patina over time, which may not suit all tastes.

The cheapest and most common covering materials for domestic flat roofs is roofing felt and fibreglass. These are normally laid in triple layers bonded to the deck, with hot bitumen tar. One major reason for felt roofs having a short lifespan results from the effect of the sun's heat, cooking the felt in summer and freezing it in winter. The continual expansion and contraction is very damaging, so felt roofs need protection, *eg* with a solar-reflective finishing layer of white mineral chippings.

Today, however, much of this work can be avoided by fitting two to four layers of high-performance glass-reinforced polyester, which doesn't need to be covered with chippings or solar-reflective paint. Artificial rubber 'EPDM' (see boxout) is a good compromise in terms of cost and quality, with a projected lifespan of over 40 years (but usually only guaranteed for 30).

If your flat roof has to double as a balcony or walkway, the surface layer will need to be formed from a harder-wearing material, such as asphalt or purpose-made paving slabs. These are placed above small raised supports to allow rainwater to disperse invisibly away under the paving, along the waterproof sub-surface.

FLAT ROOF INSULATION

Modern flat roofs have slabs of insulation laid above the deck. They are therefore 'warm roofs' with no requirement for ventilation beneath. As with pitched warm roofs, a special plastic sheet is first put in place on the deck to form a vapour barrier. Then thick polyurethane foam insulation boards are laid on top followed by the surface layers of the roof covering material.

One of the major weak points on flat roofs tends to be at junctions where the roofs meet adjoining walls. Here, getting the detailing right is very important. An upstand formed from the roof covering material, typically felt, should be dressed up the wall by at least 150mm and fixed into a chased-out mortar joint, bedded in mortar. The sharpness of

the 'corner' where the felt is folded up the wall is reduced by fitting a small strip of timber or plastic 'angle fillet' under the felt. To complete the job, a lead flashing is cut into the wall above and dressed down over the joint, finishing no closer than 75mm above the roof surface.

Photos: Cellotex

Rainwater fittings

Selecting and fitting the gutters and downpipes is sometimes a bit of an afterthought. But badly installed, cheap fittings can cause leaks, resulting in ugly damp staining, spoiling the ship for a ha'p'orth of tar. Traditionally fitted by plumbers, the guttering and downpipes are today

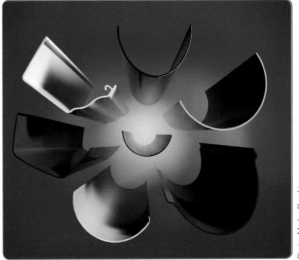

Photo: Marley Plumbing

Photo: Marley Plumbing

Photo: Marley Plumbing

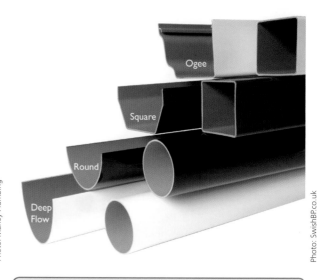

Ogee

Square

Round

Deep
Flow

Photo: SwishBP.co.uk

Photo: Marley Plumbing

installed by anyone competent enough and able to work at height (from scaffolding) – ideally a roofer.

When it comes to selecting the style that best suits the character of your home, there's a wide variety of guttering on the market. Most popular is good old black PVC, which is relatively cheap and adequate for most purposes. But you could equally opt for low-maintenance aluminium, galvanised steel, traditional cast iron, or even copper. You'll also need to decide between the various shapes, such as moulded or ogee, half-round or squareline, in colours such as black, white, or brown.

The Building Regulations stipulate that rainwater systems (*ie* your gutters, downpipes, and gullies) should be able to put away at least 75mm of rainfall per hour. So it's important to make sure that the guttering is fully supported by brackets, so that it doesn't sag and is set to the correct falls. It must be able to cope with potentially heavy loadings of snow and ice. Also, gutters must be served by sufficient

GREEN CHOICE

Rainwater fittings

1st	Galvanised steel
2nd	Cast iron
3rd	Coated aluminium
Not recommended	PVC, zinc, copper

Comments
  Galvanised steel rainwater goods are widely used in mainland Europe.

Installing a rainwater system

- Guttering needs to be laid to a fall of at least 10mm for each 3m run, so that rainwater will disperse easily.
- Gutter brackets can be screwed to the fascia. Where there's no fascia board, special brackets are available for fixing directly to rafter feet, or into the brickwork. Brackets should be set about 900mm apart – or closer where there are junctions to bays etc.
- Start at the highest point of the run, marking the position of the clip, and do the same at the lowest point by the downpipe. Run a string between the two marks to get the right fall.
- At its lowest point, the gutters should be no more than about 50mm below the edge of the tiles. The bottom edge of the roofing felt should lap down into the gutters.
- Assemble the sections of guttering and fit them to the brackets.
- Connect the gutters to downpipes. Because the eaves normally project out, overhanging the wall below, downpipes often need a 'swan neck' near the top to bring them back to the main wall.
- Plastic downpipes should simply slot together, requiring support brackets every 2m or closer.
- At their base, the downpipes normally discharge via a back inlet gulley into the underground surface water drainage system.

Photo: Wavin Hepworth

downpipes to encourage rainwater to glide effortlessly away. In exposed locations, high-capacity 'deepflow' type guttering should be specified.

It's worth noting that where dormer windows are built up from the wall below (as opposed to being set back further up a roof slope), they tend to cause complications with the rainwater system, often needing a separate gutter and pipes either side.

Flues and chimneys

Few self-builders can resist the appeal of a traditional fireplace. Large brick or stone inglenook cottage fireplaces are especially popular. The chimney breast can be built at the same time as the walls, incorporating ready-made concrete flue liners usually 225mm diameter, inserted section by section. These are fairly straightforward to construct, being pieced together as you go, and the joints sealed with special fire-cement, extending right up through the chimney. The resulting space between the liner units and the surrounding brickwork can then be back-filled with a weak mortar mix.

Good chimney design

- The chimney should terminate at least 900mm above the main roof ridge to avoid turbulence and 'smoke blowback'.
- Lead flashings should protect the vulnerable joint where the stack meets the roof, around the base. Flashings are fixed into a mortar joint in the chimney brickwork about 150mm above the level of the roof covering. They should be embedded at least 25mm into the masonry, fixed in place with metal wedges, and sealed with mortar.
- The top brick courses of a chimney stack should project about 30mm so that they overhang to throw rainwater clear of the main stack.
- The flaunching (the big lump of mortar at the base of the pots) should slope outwards.
- To prevent water soaking through the masonry and down into chimney breast in the rooms below there should be a damp-proof course (DPC) through the stack brickwork approximately 150mm above the roof, and another one below the brick head.

Main trades needed on site

- **Bricklayers:** Building the walls up to wall plate level, and building up gable end walls after the carpenter has fitted the rafters.
- **Carpenters:** Making templates for window and door frames. Fitting upper-floor joists, doors, and windows. Building timber roof structures (except for battens), cutting and fixing timbers to form the base for valleys. Fitting fascias and soffits.
- **Scaffolders:** Erecting and later dismantling scaffolding.
- **Roof tilers:** Laying underfelt membrane, laying warm roof insulation, fixing counter battens, tile battens and laying tiles, bedding ridge tiles and verges. Most will also do small amounts of leadwork, like fixing valley linings and lead flashings, but leadwork on flat roofs and large valleys is a specialist trade.
- **Groundworkers:** Constructing soakaways for rainwater drainage.
- **Labourers:** mixing mortar, fitting guttering, hod carrying, and general shifting of materials.

Who does what? See full list at www.Selfbuild-Homes.com.

13

THE BUILD – INTERIOR

Photo: Kingspan

Photo: oakwrights.co.uk

With the luxury of a roof over your head, now is a good time to sit down and take stock. The first thing to assess is whether your programme is still on target. The timetable leading up to the scheduled completion date may now need to be amended if, for example, exceptionally vile weather has held up progress. Thankfully, with the remaining work now largely indoors, bad weather can no longer be an excuse for delays. This is also the time to grit your teeth and revisit your budget, particularly if there have been any 'extras'. So get the calculator out and update all those projected expenditure figures that by now may be looking a tad optimistic.

Back at the design stage, you may have had to 'guesstimate' figures for things like kitchen units or some of the bathroom fittings if firm prices weren't known at the time. Although this is a useful way of buying time in the early stages, there inevitably comes a day of reckoning. So you now need to firm up any estimated costs.

But there's one more thing that it's essential to check right now – are there any potentially crippling delays in the pipeline for specialised fittings or materials that are on order? Delivery dates for big-ticket items like custom-designed windows and high-performance glazing need to be carefully coordinated. You obviously want key deliveries made in good time so as not to hold things up on site, and yet not too soon because of the risk of damage or theft.

The masterplan

From now on there will be a succession of trades buzzing around the site, hopefully not getting under each other's feet. As a result it can be fairly easy to lose track of precisely who's supposed to be doing what, and when they're meant to be doing it. Some tasks need to be carefully synchronised so that they're done in the correct order; others are less time-critical. A rough guide to the order of internal works would be as follows:

- External doors, windows, and glazing fitted and protected
- Upper floors boarded
- Ground floor screed
- Internal partition walls and staircase carcass
- First fix – plumbing, heating, electrics
- Plasterboard ceilings and walls
- Plastering

- Second fix – plumbing and bathroom fittings, heating and electrics
- Second fix internal joinery – doors, stairs, skirting, kitchen units, architraves
- Loft insulation (unless already fitted in a warm roof)
- Tiling, painting and decorating, external finishing

First fix

By now, the structure that's been slowly rising from the rubble of your site should be starting to bear some resemblance to the beautiful home you sketched out on paper so long ago. But of course, it won't feel like a home until the shell has been fitted out internally – starting with first fix.

Joinery

When it comes to joinery, the first-fix stage is a bit of a mixed bag. Unless your internal walls are of masonry construction (normally blockwork), your timber studwork partition walls will need to be in place before the electrician and plumber can start to work their magic.

The upstairs floors can be boarded and the staircase carcass installed at this stage, although sometimes these are left until the 'wet trades' are out of the way.

TIMBER STUDWORK WALLS

Many traditional builders are suspicious of stud walls, regarding them as 'cheap and nasty', and, given the choice, most homeowners prefer solid walls (normally built at the same time as the main walls). Nevertheless, stud partitions are usually cheaper to build, and the perceived drawbacks

---

GREEN CHOICE ⌂

Internal walls

1st	Timber frame
2nd	Sand-lime bricks
3rd	Hollow concrete blocks
	Concrete blocks with recycled aggregate

Comments
  Framed partitions of timber or steel have less
  environmental impact than solid masonry walls.
  For masonry, lime mortar is preferable to cement.

---

can be overcome. Lightweight timber studwork walls typically comprise a framework of 100 x 50mm softwood timber studs, clad on both sides with plasterboard sheets.

Resistance to sound can be improved by stuffing insulation between the studs and using double layers of special 'soundshield' plasterboard.

A quicker modern alternative used by mainstream housing developers is the 'space-saving' metal frame partition system wall. These tend to be thinner at around 50–75mm and are also clad with plasterboard. You can even buy 'off-the-shelf' ready-made plasterboard sandwich partitions with a cardboard cellular core, which just need cutting to size. But

since one of the routine complaints with new mass-produced housing relates to 'paper-thin hollow walls', where the whole wall shakes every time a door's slammed, self-builders sometimes prefer to take the opposite approach and specify thicker internal walls with luxuriously deep door linings that add to the feeling of quality.

DOOR LINERS

Once the internal wall structures are complete, but before being plastered or boarded, the door linings can be fixed in place. Timber door linings are sold in standard sizes designed

to accept internal doors manufactured to predetermined heights and widths (standard internal doors sizes are 1,981mm high by either 762 or 686mm wide and 35mm thick). But it's fairly simple for builders to construct them to pretty much any size you want from timber boards. So for self-builders this can be a good opportunity to make a bold architectural statement by fitting some custom made or perhaps super-wide 838mm doors (with the bonus of complying with the accessibility requirements in Part M of the Building Regs). See 'Second fix'.

TIMBER FLOORING

Sometimes temporary plywood sheeting is placed over the floor joists until all the messy plastering is out the way. Alternatively, new floors can be laid now and protected with tarpaulins. Up to you.

When it comes to selecting your timber flooring, you have two main choices: floorboards or chipboard panels. For a traditional floor with an attractive stain or varnish finish, softwood or more expensive oak floorboards are ideal. Although chipboard panels are cheaper and quicker to lay, they need to be carpeted or

covered with wood flooring, laminate, or floor tiles – which will add significantly to the cost. The main problem with chipboard is that no matter how carefully it's installed it always seems to eventually develop creaks and squeaks.

Softwood planed 'tongued and grooved' (PTG) floorboards are commonly available in 18 x 121mm or 20 x 144mm sizes sold in varying lengths. To prevent shrinkage problems and unsightly cracks, new packs of boards should be opened and allowed to acclimatise to room temperature before laying, by storing them for as long as possible in a centrally heated environment.

SOUNDPROOFING

Internal walls should have a minimum sound resistance of 38dB, and intermediate floors 40dB. Whereas this can be easy to achieve with most internal walls, floors can be a little more challenging. The main drawback with solid or laminate flooring is noise from walking, known as 'reverberated noise'.

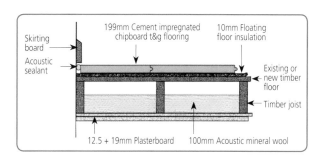

199mm Cement impregnated chipboard t&g flooring
10mm Floating floor insulation
Skirting board
Acoustic sealant
Existing or new timber floor
Timber joist
12.5 + 19mm Plasterboard
100mm Acoustic mineral wool

If noise transmission is an issue, special sound-deadening acoustic floor panels such as QuietBoard can be fitted instead of conventional chipboard panels, or laid as a 'floating floor' on a layer of resilient rubber matting. Made from high density cement-impregnated chipboard (as per Building Regs Part E), these look similar to standard tongued and grooved chipboard panels. A standard panel size is 1,200 x 600 x 19mm. Alternatively, sound-deadening acoustic insulation can be stuffed in voids between floors, or acoustic matting can be installed between the joists and floorboards to cushion vibration.

KITCHENS

The original kitchen layout drawn up at the design stage may only show basic stuff like how many units will fit within the available room space. But when it comes to actually installing the units there are a surprising number of factors that need to be carefully coordinated. So a detailed kitchen plan should be provided, to ensure that all the electrical sockets and plumbing connections will end up in the right place and at the right level.

But the fact is, details sometimes get missed on drawings. So it's well worth going to the trouble of physically 'marking out' the room that will become your kitchen prior to plastering. This involves transposing the positions of all the base and wall units shown on the plans by marking them on the walls. Take a walk around and note the window and door openings, marking the positions of the boiler, pipework, and wall sockets. Then check the positions of all the base units, wall units, and appliances to make sure they all fit without blocking doors, windows, and radiators, remembering to factor in the dimensions of floor coverings and plasterboarding yet to be fitted. Where walls are of timber studwork, sufficiently deep

Photo: oakwrights.co.uk

Photo: oakwrights.co.uk

Kitchen design

- **Internal room measurements** must be accurate and in metric (usually mm). Ensure your kitchen supplier visits the site to take their own measurements, so that they can't blame you later if things don't fit.
- **Note the internal heights** of windowsills, and the available ceiling space.
- **Check design restrictions** such as the position of boilers, doors, windows, supply pipes, and waste pipes.
- **Mark the position** of hot and cold supply pipes and waste pipes.
- **Electric sockets and switches** must be well clear of hobs and sinks.

Photo: Air Uno

- **Cookers** need a minimum 300mm of clear worktop space either side and should not be located next to a sink or beneath a window.
- **Wall units** must not be fixed directly above a hob/range or above a sink.
- **Fridges or freezers** shouldn't be next to a cooker.
- **Door swing openings** for all base and wall units should be marked on your plan.

Photo: oakwrights.co.uk

fixing points will be required. Another way to provide the necessary support for wall units or appliances is by lining stud partitions with thick plywood sheeting.

Kitchen base units are normally 700mm high and 600mm deep. Wall units are also 700mm high but only 300mm deep. If there's a standard width it's the 'appliance friendly' 600mm unit, although 300, 400, 500, 1,000mm units are also available.

Laminate worktops are usually sold in 38mm thicknesses, but dearer hardwood beech or oak ones can be 28mm or 38–42mm. Beech and birch are cheaper than oak and walnut. Iroko is a popular mid-priced alternative. A good quality worktop can outlive the units; for durability pick granite, marble, quartz, laminated hardwood, or solid resin. Granite tends to be the most expensive choice but is particularly hardwearing and better from an environmental viewpoint.

STAIRS

As we saw at the design stage, for many self-builders the staircase is one of the major focal points of the house, so only a handcrafted work of art will do. This may be a big ticket item, but that doesn't mean you have to blindly chuck money at it. For example, there's no point going to the expense of

Photo: Mak Kingsley Architects

Stairs and the Building Regulations

Briefly the requirements are:

- Landings should be at least the width of the stairs (both in length and width). No doors should open outwards on to landings, potentially swiping people off their feet.
- The pitch of the staircase must be no steeper than 42°.
- Headroom – there must be minimum 2m clearance above each step.
- The tread of each step must be a minimum of 220mm deep (strictly speaking this refers to the 'going', which is the tread without its 'nosing', the bevelled front bit that sticks out).
- The height of each 'riser' must be a maximum of 220mm (the riser is the vertical part of each step).

buying a beautiful hardwood staircase if the natural grain is going to be buried under several layers of paint, rather than being stained or lacquered.

If by now your budget is looking decidedly tight, there are some excellent 'off-the-shelf' stairs that can be modified and adapted to suit your design. But from a green perspective, note that some cheaper staircases have plywood risers, made of uncertified tropical hardwood – softwood is preferred.

Integral garages

As a rule, the Building Regs pay scant attention to garages. However, where you've got one incorporated within your home there's suddenly a potential risk of fire – which means your design will have to comply with a number of safety requirements.

First of all, the internal walls that separate the main house from an integral garage must provide a minimum 30 minutes' fire resistance. Fortunately, an ordinary blockwork wall should have no trouble resisting the spread of fire, but timber stud partitions must be boarded on both sides with special pink-coloured fireboard (or two layers of 12.5mm thick plasterboard, with joints staggered), followed by a

Timber floor to house but solid concrete garage floor.

layer of skim plaster. The same plasterboarding solution applies to garage ceilings, although additional fire-stopping insulation may need to be fitted at ceiling level.

It isn't essential to have a doorway between the house and the integral garage, but it does make life considerably easier. It also makes it easier for a vehicle fire to spread rapidly and engulf your home. So internal doors to garages must be self-closing fire doors with 30 minutes' resistance ('FD 30'). While you're at it, it's a good idea to specify a door frame with a combined intumescent strip and smoke seal that acts as a barrier to poisonous exhaust fumes.

Garage floor surfaces should be a minimum of 100mm lower than the house. Most garage floors are not much higher than external ground levels, having been specially built to a lower level than the rest of the ground floor at foundation stage, and traditionally comprise a simple concrete slab. But this makes them vulnerable to ingress of storm water. So a ramped entrance to, for example, a suitable beam and block floor can provide an alternative solution. Where it's not possible to meet the 100mm drop rule in this way, it should be possible to gain an extra 75mm or more by not screeding and insulating the garage floor. Another way to comply is by placing a raised concrete threshold to the doorway that effectively raises the adjoining floor height.

Of course, any rooms located above integral garages can get pretty nippy unless the floors are insulated to minimise the chill factor. This is normally achieved by wedging rigid polyurethane boards between the joists (or laying loft insulation quilt).

Services

The time is now right for the electrical cables and pipework to be routed through the building's exposed skeleton. Once this is done, these incomplete services will spend several weeks awaiting their 'second fix', which will only take

Photo: Cytech-europe.co.uk

place once the walls and ceilings have been plasterboarded and all the messy plastering and screeding is safely out of the way. Given that your new house will need to pass an airtightness test (see next chapter), it's important that electricians and plumbers seal any holes cut in the building's envelope.

INTERNAL CABLE AND PIPE RUNS

Electric cables and water supply pipes are routinely concealed within floor structures. Cabling through ground

floor voids should be run in small ducts such as plastic overflow piping, to keep it safe and dry, and pipework needs to be lagged. To avoid weakening timber joists it's important that rules are followed when cutting and drilling joists for pipe and cable runs. However, waste pipes are normally run above floor level and boxed in, with service hatches provided for future maintenance. Talking of which, one prediction you can make with some certainty is that technology will continue to advance, making present systems redundant for phones, computers, TV, and security. So it's a good idea to 'future-proof' your house by running all pipes and cabling in accessible service voids so that they can be replaced or 'rewired' in future with minimal disturbance.

ELECTRICS

DIY electrical work is restricted under Part P of the Building Regulations. Although there are some grey areas, this effectively excludes DIY self-builders, and most general builders, from undertaking electrical work, beyond replacing the odd switch or socket. So with the best will in the world, your role will be largely restricted to monitoring progress. As far as new houses are concerned, the electrical work is included within your main Building Regs application. But Building Control Surveyors have got enough to do without testing the electrics, so you'll normally only be allowed to employ registered contractors.

An electrical installation certificate must be provided to Building Control at the end of the job by a 'competent person' before you can obtain your completion certificate. This confirms that the works have been designed, installed, and tested to comply with BS7671 requirements. A 'competent person' is defined as an installer registered with a government approved self-certification scheme (*eg* ECA or NICEIC) or someone qualified to sign a BS7671 installation certificate.

Sound source – integral speakers.

Sound insulation to stud walls

Photo: OPUS.eu

So the best thing you can do is to provide clear layout plans marked with the correct coloured symbols showing the exact positions for all your new sockets, switches, and light fittings. Otherwise the electrician will most likely do it the easiest way, rather than how you actually want it.

The electrician's first job, after checking your plans, will be to route all the cables for the different circuits around the building and fix the various mounting boxes in place. Once the plasterboarding is done and the wet trades are finished, he can return to do the second fix, installing covers to switches, sockets, and ceiling roses. So if you suddenly realise that you need some extra power sockets or light fittings, now's the time to mention it. Don't wait until everything is all beautifully plastered.

One consistent complaint about new homes relates to the inconvenient positioning and insufficient numbers of electrical sockets. So it's best to be generous with the number of DSSOs (double-switched socket outlets) in your rooms. Depending on room size, a modern household requires about three or four DSSOs for each bedroom, five or six each for kitchens and living rooms, and a couple for halls and landings. However, there are limits to what you can request. The Building Regulations require power sockets to be positioned no lower than 450mm above the floor, and light switches no higher than 1,200mm from the floor, and no sockets are allowed in bathrooms, other than shaver sockets (run from the lighting circuit).

It's worth checking that your consumer unit has at least 20% spare capacity to allow for any additional circuits that may be required in future, should you one day want to build an extension. Note also that whilst it's common practice to set mounting boxes into plasterboard dry lining, special care must be taken in timber frame properties not to puncture vapour barriers. Where you've got bare blockwork walls, in order that that sockets and switches don't stick out beyond the finished face of the plasterwork they normally need to be recessed, so it's necessary to chase them into the wall surface.

Where cables are run along wall surfaces they should be covered in protective plastic sheathing (or conduit or trunking) prior to walls being plastered. Some electricians

prefer chasing out grooves in blockwork to accept cables, but care must be taken not to crack lightweight blocks. Cables to power sockets should rise up vertically from below, whereas cables to light switches are run down from the ceilings. Wise electricians know that taking time to clearly mark every cable to show where it's come from and where it's going can save a lot of hassle later at second fix. If possible keep a plan with circuits identified in different colours.

GREEN CHOICE

Electric cable

It's best to specify PVC-free cable, which is widely used on large public buildings. As well as dispensing with plastic, this is much safer in fires, because it doesn't give off large quantities of toxic smoke.

GAS

Copper gas supply pipes are normally run from the new external meter along the outer walls, fairly close to ground level. As a safety precaution, pipe runs inside the house should be kept to a minimum and at least 25mm away from electric cables. Gas plumbing must always be done in copper pipe, not plastic, and no gas pipes may be run in unventilated floor voids. Work on gas appliances and pipework must by law only be carried out by qualified Gas Safety registered engineers.

First-fix plumbing

The only restriction on carrying out DIY plumbing work is competence. Plumbing is a skilled job, so you have to know what you're doing; but modern push-fit plastic pipes have made some of the work quicker and easier, so this is a job that self-builders often feel confident about

tackling themselves. But it's essential to first tell the main contractor – you can't suddenly announce your newfound interest in plumbing when they've already scheduled a plumber to do the work.

GREEN CHOICE

Pipework

1st	Polyethylene
	Polybutylene
	Polypropylene
2nd	Stainless steel (for exposed internal pipework)
3rd	Copper

Comments

Polyethylene pipework is an alternative to PVC for above-ground drainage for cold water only. Blue polyethylene is now generally used for external mains supply.

Polybutylene (PB) is used in Hep2O push-fit plumbing systems for heating and for hot and cold water supply pipes. Its environmental impact is substantially less than copper. But note that the first metre run of pipework from boilers must be in copper.

There are three main areas of plumbing: hot and cold water supplies, waste pipes, and central heating (CH). At first-fix stage, each hot and cold water pipe supplying fittings should be left terminating above floor level with a small isolation valve (a stop-tap). This allows you to pressurise the system at an early stage, and also makes it easier to replace sanitary fittings in future. Where hot or cold pipes are run in cold spaces such as lofts or ground-floor voids, they need to be lagged with insulation.

COLD WATER SUPPLY

At this stage, your water supply will consist of nothing more than a forlorn-looking length of blue polyethylene supply pipe sticking out of the floor. A stopcock (isolation valve) will need to be fitted where the cold supply enters the building, typically in the kitchen or utility room. But in most new houses there's no need for a 'rising main' to supply a traditional cold-water storage tank in the loft, as mains water is supplied direct to taps, WCs, and boilers etc. When it comes to running pipework to baths and basins etc it's customary for the cold taps to be on the right and the hot supply on the left (even with mixers).

SHOWERS

Showers need a consistently powerful water supply for which modern mains pressure hot-water systems are well suited. Otherwise flow rates may need to be beefed up by fitting powerful pumps. The temperature of the water supply must be thermostatically controlled, to prevent unwitting shower-users getting scalded because someone else has upset the hot/cold mix by running a nearby tap.

A common weak point with showers is leakage from shower trays. Ceramic or stonecast trays are preferable to cheaper acrylic ones that can be prone to flexing, allowing leakage at seals. All-in-one moulded cubicles are an even safer option. Alternatively, you could dispense with

Photo: Kermi

the tray altogether and build a 'wet room' with a drain fitted inside the floor void, which must be able to cope with a flow of 30 litres per minute. The floor needs to be waterproof and should slope towards the drain at 1:40 for a metre around it. This is easier on a solid floor where you can form a 25mm depression into the screed, rather than having to build up timber floors.

WASTE PIPES

Waste water from bathrooms, kitchens, and WCs usually discharges via branch pipes into the nearest soil and vent

pipe (SVP). From there it's invisibly whisked away, transported through the underground drainage system to the sewers. With your underground drainage pipes already installed (in Chapter 11) the pipework for all the waste plumbing can now be pieced together. In modern properties the waste pipework is normally run internally, so there should be one or two orange-coloured 110mm plastic

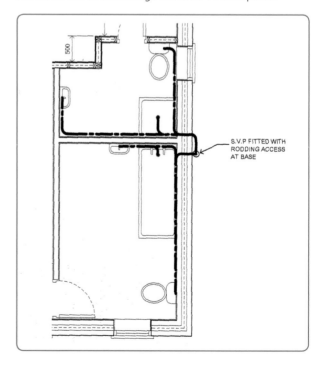

S.V.P FITTED WITH RODDING ACCESS AT BASE

connectors poking through the ground floor (with their openings temporarily protected) awaiting connection to new internally-run soil stacks. However, in some cases designers favour the traditional arrangement where the SVP and waste plumbing are run on the outside of the house.

But whether your system is indoors or not, the branch waste pipes serving baths and sinks etc are only effective up to certain lengths before there's a risk of siphonage occurring. If pipe runs are too long, the protective seal of water in traps to baths, sinks, basins, and toilets can literally be sucked out, allowing foul odours to enter the house. This happens when a deluge of outgoing waste water in a long pipe causes pressure to build up in its slipstream.

To avoid such horrors, there's a maximum distance that you can safely locate your new sink, loo, and bath from the waste stack. These distances are typically 3m for baths or showers, less than 2m for most basins, and 6m for WCs.

If this restriction cripples your otherwise excellent bathroom layout, it may be possible to work around the problem by fitting special 'anti-siphonage' traps or bigger bore pipes. Or you can cheat by using non-return valves that don't need a water-sealed trap at all, such as 'HepVO' valves.

If your design boasts lots of en-suite bathrooms and plentiful WCs, you might worry that each will require an ugly SVP poking up through the roof. Happily, as long the house has one conventionally vented SVP there's a useful short cut. For subsidiary bathrooms and WCs you can use shorter 'stub stacks' that are capped with an 'air admittance valve' (AAV), also known as 'Durgo valves'. Unlike SVPs these don't need to be permanently open, as they have a special one-way valve that opens automatically to relieve excess pressure and prevent siphonage. Air is admitted into the system before it seals closed again, without emitting unpleasant smells. AAVs can be neatly boxed in or placed unobtrusively in bathrooms and loft spaces, but must be located at least 150mm above the flood level of the highest sanitary fitting in the house. However, they're not suitable for outdoor use as frost can damage them.

Heating and hot water

A heating system has to perform two tasks – to supply domestic hot water (DHW) and to provide space heating. Many designs turn out to be very good at one of these, but not at the other, because each makes very different demands. Whereas space heating isn't needed for several months a year, domestic hot water is required all year around in short sharp bursts.

As we saw in Chapter 6, renewable energy can make a

Photo: Mark Kingsley Architects / JM Collingwood

significant contribution to both your space and water heating requirements. Solar panels alone can generate up to half your hot water. But even where you invest heavily in more than one renewable energy source, it's unlikely to make you fully self-sufficient. In any case, you'll still need to install a hot-water storage cylinder and some means of warming rooms, usually by radiators or underfloor heating. So no matter how green your home, a boiler is normally necessary to be certain of all-year-round warmth and hot water.

Photo: Kermi

DOMESTIC HOT WATER (DHW)

Today, all new boilers are of a condensing type. These can be up to 98% efficient as they recycle the heat from hot flue gases. Where mains gas is not available, oil or LPG/propane fired boilers are available, and biomass is an increasingly popular fuel.

There are two main types of system for new houses – combination boilers or unvented systems with a separate hot-water cylinder. Unlike traditional gravity-fed systems, neither require conventional cold-water storage tanks in the loft (although you may want to install one for emergency water supplies). Instead, both systems rely on

direct mains pressure water supply (the average UK mains pressure is 3 bar).

COMBINATION BOILERS

'Combi' boilers combine both central heating and DHW in a self-contained unit, eliminating the need for separate cold-water tanks and hot-water cylinders, along with all the associated pipework. Combis are very efficient at providing unlimited hot water directly a tap or shower is turned on, at a similar pressure to the incoming mains supply. When hot water is simultaneously demanded at taps and for room heating, DHW is prioritised.

Combis are bulkier than ordinary boilers since they contain a built-in hot-water vessel, and although dearer to buy they're cheaper and easier to install. Their main limitations have traditionally been their low flow rates, so they can struggle to cope with demands in larger houses, especially when several hot taps are run simultaneously. A family home of four people typically needs around 230 litres of hot water a day, heated to about 60°C. So for properties with multiple bathrooms a more powerful boiler and cylinder system is likely to provide a better solution.

UNVENTED SYSTEMS

Today the majority of new homes are equipped with a mains pressure hot-water system comprising a condensing boiler and separate cylinder capable of storing hot water at mains pressure. One cylinder should happily serve several bathrooms, and provide consistent, balanced, hot and cold flow rates, which suits modern showers and mixer taps.

Because there are no separate loft tanks for emergency expansion they're referred to as 'unvented' systems. Any expansion is instead taken by a vessel connected to the hot-water cylinder, although some unvented cylinders, such as Megaflow, can accommodate water expansion internally. In a typical system, the incoming cold mains supply can be

heated either indirectly from the central heating boiler or directly in the cylinder by an electric heater. When you open a tap, the stored hot water is forced out by incoming mains cold water, thereby providing hot water at mains pressure. Such systems need at least 2.5 bar of mains water supply pressure to work well.

The plumbing is more complicated than for a traditional cylinder as you need to install pressure release valves, an expansion vessel, and a pressure regulating valve. But the lack of loft tanks means that although these cylinders are relatively expensive to buy overall costs including installation should not be massively higher than a traditional system.

A neater solution for some properties can be found with 'all-in-one units' that combine the functions of a boiler and mains pressure hot-water cylinder in one large boiler box.

SPACE HEATING

The sign of a well-insulated, low-energy house is where you only require occasional top-up heat to keep room temperatures nice and cosy. More energy is needed to heat the hot water than for space heating.

As we saw in Chapter 8, the amount of space heating required can be minimised by fitting whole-house mechanical ventilation systems with heat recovery (MVHR). But no matter how energy-efficient your home is, there's still the question of how best to deliver heat to the rooms when it's required.

RADIATOR CENTRAL HEATING SYSTEMS

Central heating systems serving radiators are still the most popular method of space heating in new homes, sometimes combined with UFH (see below). Radiators needn't, of course, be white and oblong. Instead this can be an

Photos: Kermi

opportunity to add some seriously cool style or retro chic to your new home.

But unless you provide the plumber with a clear set of drawings showing exactly where you want radiators positioned and pipes routed, it'll probably get done the way that suits them – probably with the all the rads bunched back-to-back on internal walls, to save running much pipework.

To reduce heat loss and prevent freezing, pipework serving radiators should be lagged, particularly where it's run within cold floor voids. For timber floors, lagged pipes can be installed within the floor space. For concrete ground floors the pipework needs to be run in special ducts in the screed, or can be surface-run along walls and boxed in.

UNDERFLOOR HEATING (UFH)

After a slow start (it was invented by the Romans), underfloor heating is today increasingly popular in new properties, and is widely installed in kitchens and conservatories. Although a UFH system is likely to add up to £2,000 to the cost of installing a standard radiator system, it adds a touch of luxury, plus there's the added benefit of freeing up wall

Photos: Wavin Hepworth

space with no bulky radiators to get in the way of your furniture.

For new homes, 'wet' UFH systems are a more popular choice than cheaper 'dry systems' that use thin mats containing electric elements (but are less energy efficient and far more expensive to run). Wet systems employ a series of narrow plastic water pipes run in loops embedded within the floor, so the heat is released slowly. They're normally laid within the screed or below chipboard floor panels, but placed above the insulation layer. Most floor finishes are compatible, but UFH is especially effective when used with stone or tiled surfaces that would otherwise be cold. Warm water is pumped at low pressure through concealed pipework, and where buried in the floor screed it acts as a heat reservoir, rather like a storage heater.

Typical floor surface temperatures are a comfortable 26–29°, and are temperature-controlled by room thermostats. Underfloor systems are relatively energy efficient, since to heat an entire room they need only be set to a relatively low room temperature (10–15°) compared to radiator systems (18–30°). This suits modern condensing boilers, which are at their most efficient at lower temperatures, and also makes UFH compatible with solar heating systems.

Photo: Wavin Hepworth

The main downside is the relatively slow response time, so this may not be ideal for people who pop briefly in and out of their homes, only occupying them for a few hours a day. That said, intelligent controls can be programmed to anticipate temperature changes, boosting warmth when needed. Many users find it best to leave them on continuously, just set at a lower temperature at night. For many self-builders, the optimum arrangement is to have UFH downstairs and radiators upstairs.

OPEN FIREPLACES

As a focal point you can't beat a good old-fashioned open fire. However, with around 75% of the heat disappearing straight up the chimney, the average fireplace is spectacularly energy inefficient.

You can increase the efficiency of a real fire either by installing a back boiler, or by containing it within a stove. Wood-burning stoves are popular and can enjoy over 80% efficiency. Part J of the Building Regulations requires that a CO detector is fitted in rooms containing solid fuel appliances.

CONTROLS

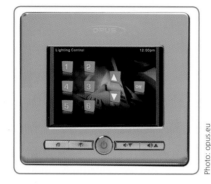

Photo: opus.eu

A good control system that effectively manages both hot water and room heating is essential for energy efficiency. You need to be able to control time and temperature for both functions separately. For room heating, the best arrangement is to have a central thermostat in addition to adjustable thermostatic radiator valves (TRVs) fitted to each radiator, allowing a custom temperature to be set for each room. Most boilers have built-in clocks and programmers as well as a thermostat control that shuts off when the water gets to a certain temperature. Thermostats should also be fitted to hot-water cylinders.

Essentially, programmers are sophisticated on/off switches for the whole heating system, overriding other controls. But many programmable controls are overly complex and not easy to use, which can defeat the point of installing them.

When it comes to the task of wiring heating controls to boilers, programmers, and pumps, the question of which trade is responsible can be a bit of a grey area. In practice, the job often gets done by plumbers, but the electrician will need to leave a fused spur socket nearby.

ROOM VENTILATION AND HEAT RECOVERY SYSTEMS

The Building Regulations lay down requirements for room ventilation. These are discussed in Chapter 8 along with mechanical ventilation and heat recovery systems.

Plastering, screeding, and internal finishes

A few days of the 'wet trades' and you'll notice a profound transformation. As the resulting pinky-brown veneer on your walls dries out, it will create a feeling of home, albeit one in need of decoration and fitting out. This miraculous achievement is, of course, all down to the

plasterers, who as a trade often work in pairs or small gangs, supplying materials as well as labour.

Plasterboarding

It's now standard practice in mainstream housebuilding for the blockwork inner leaf of main walls to be dry-lined. This at once boosts the thermal insulation of the walls whilst providing a nice flat smooth surface. It also prevents the risk of serious shrinkage cracking when plastering direct to lightweight blockwork. It also cuts out long delays prior to decorating while you wait for traditional two-coat plasterwork to dry out. A skim plaster finish is normally applied, although there are some advantages to decorating directly to the prepared plasterboard surface. Normally the

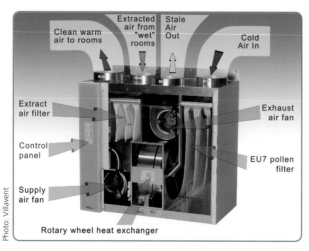

Photo: Villavent

Clean warm air to rooms | Extracted air from "wet" rooms | Stale Air Out | Cold Air In

Extract air filter

Control panel

Supply air fan

Exhaust air fan

EU7 pollen filter

Rotary wheel heat exchanger

walls can be boarded ('tacked out') along with the ceilings.

Plasterboard sheets can be fixed to a framework of timber battens with dry wall screws, or can simply be squashed onto blobs of plaster known as 'dabs'. It's good practice to leave a tiny gap of about 2mm between adjoining sides of boards. Working with enormous sheets of plasterboard can be very hard graft, but should be within the capabilities of most self-builders. To save time, it's best to use larger 2,400mm x 1,200mm sheets designed to suit standard 600mm stud spacings, but you'll need four hands rather than two. A useful trick of the trade is to use a simple foot-lifter or 'seesaw' made from a spare offcut of wood which can be wedged under the wall boards to lever them against a prepared vertical framework.

After fixing the sheets in place plasterboard jointing tape should be applied to all the joints, especially at the corners where walls abut ceilings or other walls. This binds the whole surface together to prevent subsequent shrinking and cracking to the finished skim-plastered surface. Where you prefer to decorate direct to board surfaces, tapered edge boards should be used so that the taped joints and screw heads are safely recessed and won't stick up. The joints are then taped and filled with jointing compound, and later filled again before being sanded to a smooth finish, ready for sealing and decorating.

Ceilings

Where you're boarding a ceiling and there's a cold space on the other side (eg a bedroom ceiling below a cold loft or the roof rafters) it's best to use special foil-backed plasterboard. The foil acts as a vapour barrier, helping prevent warm moist air getting through from the house into the cold zone and

Photo: Envirograph

condensing into damp. Alternatively, a large polythene sheet can be fixed to the joists or rafters before boarding to act as a 'vapour check'.

Once fully 'tacked out' and the plasterboard joints taped, ceilings are normally finished with a thin coat of smooth plaster. This is applied by hand-held trowel, the plasterer standing on planks laid across small scaffold towers. Coving and decorative mouldings can be fitted afterwards.

Plasterboard is naturally fire-resistant and can normally protect the timber joists above for at least half an hour. But despite being reassuringly non-combustible, to be fully fireproof the surface needs to be skim plastered to protect the paper lining. In locations especially at risk from fire-spread, such as integral garages, you need to specify special pink-coloured fire-shield boards (or else a double layer of standard 12.5mm plasterboard). Also, any exposed steel beams must be encased in fire-resistant plasterboard with a skim plaster finish. This is because steel reacts badly to fire, buckling, warping, and bowing with potentially lethal consequences.

Hard plastering

A large percentage of the snagging issues that arise after completion typically relate to plastering defects. Professional plasterers reckon that as much as a quarter of their work comes from being called out to rectify DIY plastering

disasters. So this certainly isn't a skill to learn on the job, and is normally best left to the trade with the magic touch.

Although the main walls in new homes are today mostly dry-lined, many homeowners prefer the appeal of 'good solid' walls. Of course, with timber frame construction you have no choice. But where your internal walls are built of blockwork, or your main masonry walls are not dry-lined, they'll require the traditional hard plaster treatment. Here, two coats will be needed, a base coat to cover the rough blocks and a lightweight gypsum plaster coat for a smooth finish. This costs roughly the same as dry-lining and skimming but the drying-out process with traditional two-coat plasterwork can mean having to wait anything between two and six weeks before decorating. A 'breathable' water-based paint should be used rather than vinyl emulsion, which can trap in any residual

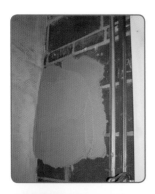

Photo: Wavin Hepworth

moisture. If your base coat is of sand/cement render, it may take even longer to fully dry out. Contrast this with a 2mm skim coat that can normally be decorated within 24–48 hours.

Whether you opt for hard plaster or skim, to achieve a neat finish you need to fit corner beading of galvanised metal or plastic to any outer corners of walls, door openings, and window reveals. These are set in dabs of plaster and carefully aligned using a spirit level (or can be anchored with galvanised masonry nails).

Floor screeding

With concrete beam and block floor construction, the most popular method of finishing is to apply a fine sand and cement screed, which provides a smooth, flat finish. This is a job best carried out by the plasterers. However, there are alternatives to screeding, such as placing floorboarding over battens and a layer of polythene sheeting.

Insulated floor ready for screeding, right with radon barrier.

Screed is typically laid to a thickness of about 75mm over insulation panels (the minimum depth required is 65mm). If you plan to run pipes and cables through the screed, purpose-made plastic ducts should be embedded within it, with access covers for maintenance (unprotected copper pipes in concrete floors will corrode).

The screeding mix should be quite dry, without too much water, and can be mixed on site from cement and sharp sand. A mix of 1:3 cement/sand is better for flexible floor coverings like vinyl, and 1:4 for a rigid tile floor. To avoid sandy patches and hollow areas, the mortar must be well mixed. To be on the safe side, it is best delivered ready-mixed in truckloads. The final surface can be smoothed using a power float, but extra time must be allowed for the newly laid screed to thoroughly dry out before covering with tiles or timber flooring etc. This normally means waiting at least two or three weeks, which is why it makes sense to coincide the job with the drying-out period for plastered walls and ceilings.

To assist the process, it's best to keep windows and doors open as much as possible. Whilst waiting for trapped moisture to fully evaporate, you can get on with external finishing work, landscaping or outbuildings.

In addition to 'plain vanilla' floor screeds, a range of more sophisticated specialist self-smoothing screeds is now available which cure within a few hours. These can be mixed and poured and then spread using a steel float. They're available in more exotic colours than the drab grey/green of the traditional sand/cement mix. These self-levelling screeds are easier to use but are relatively expensive.

Freshly screeded floor – no shoes allowed!

Photo: potton.co.uk

Second fix

Once all the plastering and screeding is safely out of the way, the electrician, plumber, and chippie will all need to return to complete their work. But you need to

Energy-efficient lighting

With an eye on the future it makes sense to fit low energy lighting throughout. The minimum at the time of writing in Part L1A of the Building Regs is for at least 3 out of 4 of all new internal light fittings must be low energy i.e. lamps with more than 45 lamp lumens per circuit watt (and a total output greater than 400 lamp lumens). Light fittings may either be dedicated ones (that only take low energy lamps) e.g. pin base CFL (compact fluorescent), or standard fittings with low energy lamps, e.g. bayonet or screw base CFL or LEDs.

With external lighting all lamps should have sensors so they automatically cut off when not required (in daylight or when there's no PIR stimulus), but also need to be controllable manually by switch if lamps are more powerful than 45 lumens per circuit-watt.

keep your eye on the big picture, because second fix is notorious as the big money-guzzling phase when costs can easily run out of control. This is the stage when the temptation is greatest to spend wildly on extravagant fittings. Seasoned self-builders therefore try to postpone some of this expenditure until more funding becomes available once all the VAT has been reclaimed. It's particularly easy to get carried away with things such as designer appliances, exotic lighting, and multimedia systems. Nice as such luxuries are, however, they rarely add much to a property's value.

Although kitchens and bathrooms are one of the last areas to be completed, they depend on some fairly elaborate plumbing and electrical connections all being in the right place. To avoid the need for expensive alterations later, it's essential to carefully plan all these connections well in advance.

Bath TV – bring it on!

Comfort Intelligence Home System - Cytech-europe.co.uk

Some weeks before all the pipework, cabling, and fittings are fully installed, and everything s connected up, you need to contact the utility companies to arrange dates for supplies to go live. Needless to say, this won't happen overnight. To prevent any unnecessary delays, double-check you have all the necessary meter boxes and ducting in place and in accordance with the service provider's specification.

Electrics

The second-fix electrics are normally a fairly straightforward process – assuming there are no last-minute changes of plan. This is the stage when all those protruding bare cables and empty mounting boxes that have lain dormant since first fix can finally be connected up. As lots of gleaming new switches, socket covers, and light fittings appear, you start to get a glimpse of your new home lurking behind all the dust sheets.

But as always, you need to be ahead of the game. So before work starts mark up the plans and then mark the walls and ceilings to show where any special fittings need to go. The electrician should normally fit your lights. But before installing expensive, fragile ceiling lights it's a good idea to temporarily install some cheap plastic pendant bulb-holders. The only permanent feature lights to fit at this stage are those requiring recessing that would be difficult to do later.

Photo: opus.eu

Photo: Kermi

At the end of the job, once the circuits have all been checked and tested, the electrician must provide an electrical installation certificate, which will need to be passed to Building Control.

Smoke alarms

Smoke alarms must be wired to the mains. To work effectively, they must be positioned correctly, sited away from the corners of rooms (which smoke tends to avoid) but not too close to light fittings (which can obstruct them). There should be at least one smoke detector fitted on each storey, no more than 3m away from each bedroom. In circulation areas (lobbies, halls, and landings) they must be fitted within 7.5m of the door to each habitable room (Building Regs Part B1). Locate them away from cooking areas, heaters, and bathrooms, since steam and fumes can set them off, and don't fit them in garages or kitchens.

Plumbing

Second fix heating and plumbing is a little more arduous than for the electrics. All the pipework tails will need

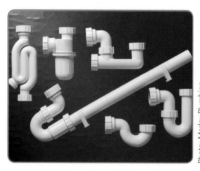

Photo: Marley Plumbing

connecting up to the various new kitchen and sanitary fittings. The heating and hot-water systems should be fired up and the new boiler tested by the plumber – sometimes with a little help from the electrician. With oil or LPG, the fuel storage tank will obviously need to be filled before you can test the system. Dishwashers and washing machines should be connected and checked to confirm that they work properly – a task that's often left to the client.

In an ideal world new plumbing systems would be leak-free from the word go. But of course, in reality a loose connection here or an unsoldered joint there isn't unusual. Small leaks usually manifest themselves within the first couple of days, so make sure the plumber hasn't in the meantime disappeared without trace.

Bathrooms

If space permits, it's normally a good idea to have a separate shower cubicle within a bathroom rather than just a mixer and shower screen over the bath. Or you may have opted for a Continental-style walk-in wet room, with fully-tiled

waterproof floors and walls. Either way, it's easy to spend a king's ransom on the latest sanitary ware. But bathroom fashions change depressingly swiftly so it's rarely cost-effective to splash out wildly on designer baths, exotic basins and super-slick toilets. Savvy developers know that much of the wow-factor in bathrooms is actually created by clever use of the right accessories.

When the job is done pay careful attention to the detailing at the edges of baths and shower trays. Probably the most common cause of leaks in bathrooms comes from the joints where they abut walls, particularly if using plastic fittings. Never rely on grout, but rather use purpose-made sealing trim strips, or a suitable silicone mastic sealant.

Joinery

Second-fix joinery is mainly about finishing off a whole range of smaller jobs such as fixing skirting boards, architraves, picture rails, and boards to window sills, plus hanging internal doors, boxing in pipes, and fitting staircase newel posts and balustrades. But the biggest outstanding task is normally to install the kitchen. If your builder is contracted to both supply and fit the kitchen units, there should be no doubt about who's responsible for sorting out any gaps and loose fittings.

Before the base unit carcasses can be installed, your final floor levels must be known and the floor tiles laid. Even where base units' legs are to be boxed in with plinths, the floor surface must be level. Check that the flooring extends fully under units, as this often gets skimped.

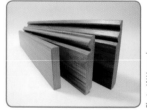

Photo: W.Howard

Photo: Mark Kingsley Architects / JM Collingwood

Floor finishes

Natural timber flooring improves with age and is easily maintained. Hardwood planed tongued and grooved planks (PTG) are excellent for durability, as well as adding warmth and beauty. But budget constraints may require a degree of compromise, in which case softwood PTG with a lacquered finish can

be a good alternative. Cheaper versions use a veneer with a 4–5mm layer of hardwood glued to a composite backing. Shrinkage is a factor on all wood floors although it's less of a problem with some types of hardwood. This can cause unsightly gaps to appear, unless flooring materials are allowed to acclimatise to centrally-heated room temperatures for at least a week prior to laying.

In kitchens you need a strong base such as limestone, terracotta, ceramic or terrazzo tiles, marble, or slate. Allow for the finished level of floor tiles to match that of any carpeting in adjacent rooms.

To lay vinyl coverings on a concrete floor, a latex levelling screed can provide a suitable surface, with a recommended overall thickness of about 8mm. With timber floors, vinyl should be laid on plywood sheeting.

Internal doors

The majority of factory-made internal doors are of the hollow, lightweight variety. These comprise a simple timber frame with a honeycomb cardboard core, clad on both sides with moulded panelled surface covers. The main alternative is solid wood doors, which have a more expensive feel and make a pleasingly chunky sound when they shut. These are commonly made from pine (clear or knotty) or from better quality softwoods such as Douglas fir, sometimes with oak veneers. Solid wood doors are available either for staining or as paint-only grade. Traditional four- or six-panel designs are popular, as are farmhouse-style 'ledged and braced' doors. From a green perspective the best options are hardboard-faced hollow core or solid softwood.

Fire doors are only likely to be required where you have more than two habitable floors (or an integral garage).

Where, for example, the roof space provides a third storey of accommodation, a fire-proof internal escape route is required. Fortunately this can normally be achieved with a conventional staircase and landing layout, which doubles as an 'escape corridor' leading down to a ground-floor entrance hall. However, all the habitable rooms along this route must be fitted with fire doors, incorporating intumescent strips (self-closers are not required).

If money is tight, it might be worth fitting cheap temporary doors with a view to replacement at a later date when funds permit. Otherwise there's nothing to stop you making your home super-stylish with some high-quality hardwood or specially glazed doors. Or how about some curiously dimensioned reclaimed jobbies from the local salvage yard? As long as it's all planned in good time, so that door openings in internal walls are built to the right dimensions, fitting them shouldn't cost significantly more than the cheapest option.

The task of hanging doors is said to be a good test of a joiner's skill. This is a job that may look easy, but there's an art to getting it right. So once hung, take a look at the cut of the door and see whether it's fitted with a consistent gap all round. Note also whether the surrounding architraves are neatly mitred at the top corners and are uniform and level. And check that the way each door is hung corresponds with light switches. Above all, check that the door opens on the correct side, swinging open to left or right as specified.

Photo: William Ball Kitchens

Main trades needed on site

■ **Carpenters/joiners:** Building studwork internal walls, fitting door linings, loft hatches, and doors and windows. Fitting all internal joinery, hanging doors, and installing kitchen units.

■ **Electricians:** Running first-fix cables and fixing mounting boxes. Second-fix wiring, switch and socket covers and lighting. Testing completed system.

■ **Plumbers/heating engineers:** Running first-fix hot and cold supply pipes, heating pipes, and waste pipes. Fitting water tanks. Second-fix plumbing and connecting up bathroom and kitchen fittings. Installing UFH and radiators. Testing completed plumbing, heating, and hot water systems.

■ **Plasterers:** Plastering walls and skimming plasterboard, screeding concrete floors.

■ **Labourers:** Fixing plasterboard sheets to ceilings and walls, laying insulation. Carting around bathroom and kitchen fittings.

Who does what? See full list at www.Selfbuild-Homes.com.

14 COMPLETION

Photo: oakwrights.co.uk

Photo: William Ball Kitchens

It's towards the end of the project at the decoration stage that many self-builders decide to roll up their sleeves and tackle some of the finishing work. If you're confident in your own ability to do a good quality job, there's nothing to prevent you from taking responsibility for the final lap – as long as you provide your builders with sufficient notice. With their sights firmly set on the next job, they might be only too keen to have everything wrapped up as swiftly as possible. So bringing forward the completion date could be a tempting proposition, allowing them to get their hands on the final payment sooner than anticipated. And not having to fork out for the cost of all the decorating could breathe some life back into your depleted bank account.

But first there's some unfinished business to attend to. Back at the design stage (Chapter 6) you solemnly promised that your design would be eco-friendly and meet all necessary green targets, even to the extent of being zero-carbon. Now you have to prove it. This means demonstrating that your new home is sufficiently airtight and doesn't leak heat. Only then can your 'as-built SAP' finally be calculated.

Air leakage testing

Air leakage is the uncontrolled flow of air through gaps and cracks in a building, causing cold draughts to enter and allowing heat to leak out. This is a different kettle of fish from ventilation, which is the controlled flow of air into and out of the building to maintain comfort and a healthy atmosphere.

To meet air leakage targets and comply with the Building Regs (Part L1A), with very few exceptions new homes now have to be air tested – even one-off self-build houses. To measure the amount of air leaking to the outside, the building is pressurised by a large fan installed in the front-door opening. The test is based on an artificially created pressure difference between inside and outside of 50 Pascals. This is more

Photo: Lowenergyhouse.com

Common sources of air leakage

- Joist hangers cut into the inner blockwork leaf of the main walls allowing draughts into floor voids.
- Window and door frames poorly sealed to surrounding walls.
- Loft hatches without draught-stripping.
- Suspended timber ground floors.
- Cable and pipe service entries to ground floors.
- Cables, pipes, and recessed lights run through ceilings to cold lofts.

Common air leakage paths

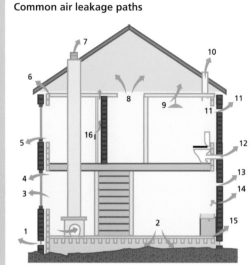

KEY
1. Underground ventilator grilles
2. Gaps in and around suspended timber floors
3. Leaky windows or doors
4. Pathways through floor/ceiling void into cavity walls and then to the outside
5. Gaps around windows
6. Gaps at the ceiling-to-wall joints at the eaves
7. Open chimneys
8. Gaps around loft hatches
9. Service penetrations through ceilings
10. Vents penetrating the ceiling/roof
11. Bathroom wall vent or extract fan
12. Gaps around bathroom was pipes
13. Kitchen wall vent or extractor fan
14. Gaps around kitchen waste pipes
15. Gaps around floor-to-wall joints (particularly with timber frames)
16. Gaps in and around electrical fittings in hollow walls

than most houses experience in reality, so the actual rate of air leakage should be substantially less than the figure you achieve. At the time of writing, the official 'notional' target to beat is $5m^3$ per hour (that's the number of cubic metres of air passing through $1m^2$ of the building's exposed external surface per hour). A low-energy house should achieve $3–4m^3$ with an MVHR system fitted (mechanical ventilation with heat recovery).

To meet the target, you have to focus on the task of sealing the building from the design stage. On site, contractors may need to be reminded about the importance of this, especially with penetrations cut for services, which typically account for half of all air leakage. Plumbers and electricians are used to banging holes in the structure for pipes and cable runs, without paying much attention to sealing them up afterwards. And if the external envelope is leaky, voids behind dry-lined walls can transmit draughts throughout the building, blowing out at skirtings and electrical sockets. With timber frame construction, it's a good idea to provide a service void for pipes and cables on the inside of the main walls, so the airflow-retardant membrane doesn't need to be punctured by services.

Insulation

Insulation is not an optional extra. Building Control will not pass the project without the work having been done in accordance with the approved drawings and design SAP calculation. Most insulation will have been fitted as the walls, roof, and floors were built. But where you have a conventional 'cold loft', rolls of glass fibre or mineral wool insulation will

need to be laid to a depth of 300mm or more. The stage at which this is done isn't critical, but builders can sometimes be prone to skimping this task, so always take a quick peak up in the loft prior to completion. This isn't the most pleasant job in the world, but it's one that, if need be, you can do yourself.

Interior finishing

This is the final phase of your project being a building site. Once decorated, it will at last become a home. But the style of your new residence will depend to a large extent on your chosen decorative theme. This was probably one of the first things you decided on, long before the first sod was turned on site. But interior design can be a dangerous subject. What's seriously in vogue one year may turn out to be nothing more than a passing fad – think textured ceilings, woodchip wallpaper, repro dado rails, and avocado-coloured bidets! And the more radical a style or theme, the more rapidly it tends to go out of fashion.

Decorating

As already observed, many self-builders take on the decorator's role towards the end when there's less time pressure. But achieving good results requires a lot of hard graft as well as skill. In fact, to achieve a good quality result you need to spend about two-thirds of your time on preparation compared to only one-third on the finish. In other words plenty of laborious rubbing down and filling. Decorators are responsible for snagging problems left by other trades, but there's a limit to how much rough plasterwork and unprepared timber you can cover up with bucket loads of filler and paint.

Interior design themes

MINIMALIST
Perfect for those who love clean, simple design, but hate clutter, pets, and children. It helps if your favourite colour is white, and you lean a bit to the 'A.R.' side. Art gallery minimalist decor means a simple *objet d'art* or classic item of furniture can become a focal point. Suits modernist and Art Deco architecture.

Photo: Mark Kingsley Architects / JM Collingwood

Photo: PGmodel.com

DESIGNER-INDUSTRIAL
This theme, best suited to kitchens and bathrooms, features plenty of matt or polished heavy metal fittings and bare reclaimed expanses of brick. Exposed steel beams can blend well with Victorian-warehouse style.

GOTHIC
Darkly atmospheric, pointed arches, high ceilings, rich colours, shimmering candlelight, and heavy furnishings and fabrics. Suits people called 'Herman'.

CONTINENTAL
A blend of world cultures. Perhaps a Mediterranean cobalt blue and white theme, with patterned tapestries and Far Eastern mosaics. The world is your cultural oyster.

PERIOD HOUSE STYLES
These can be difficult to recreate, because all your furniture and fittings need to be in keeping. Incongruous modern elements such as TVs and PlayStations risk compromising the period integrity. Done to perfection, you could end up in a living museum.

Photo: Crown Guild carvers

This trade is one of the cheapest in terms of labour costs. Decorators normally supply their own brushes, filler, and sandpaper, but paint is normally provided by the client or main contractor, unless you request otherwise.

Time was when all the visually important internal joinery was universally finished in good old gloss white. But to add some charm to the interior design you might want to try something a little more interesting, such as a matt white or soft sheen cream finish. To really do justice to all that expensive virgin joinery you may prefer a transparent varnish, stain, or waxed finish that enhances the timber rather than hides it. But this means any faults and blemishes in the wood will stand out, so extra careful preparation is needed to achieve a good, natural finish.

Varnish is popular for joinery and timber floors as it protects the wood and highlights the natural beauty of the grain. It's often applied before the walls are decorated, since careless emulsioning could leave indelible marks on the bare timber that show through no matter how many coats you apply. Varnish combines polyurethane with wood stain, so that the stain effectively sits on the surface of the timber rather than being absorbed into it. Coloured varnishes can be used to modify the colour of the joinery so it blends in with other woods, for example giving softwood the appearance of oak. Wood stains are easy to use and resistant to fading. They consist of dyes that are spirit- or water-based.

When it comes to emulsioning the walls, you can't just slap paint on to freshly plastered surfaces. This is because the plaster is very porous and will suck out the moisture, causing the paint to blister, crack, and peel. Sometimes you can actually hear the plaster sucking up the liquid! Skimmed plasterboard can be decorated once it's dried to a lighter pinky colour, normally within a couple of days. But plastered masonry walls needs to be left for a few weeks. Once dry to the touch, plastered surfaces can be decorated with emulsion, but it's best to use microporous paint, not vinyl emulsion, which prevents walls from breathing and traps any residual moisture. New plaster can be painted with three coats of emulsion, the first primer coat diluted with water to the consistency of milk.

Tiling

A professional tiler takes up to about an hour to fix one square metre of tiles, plus another ten minutes or so to grout them. But tiling walls and floors should be well within most DIYers' capabilities, albeit at a slightly less impressive pace. Bathroom tiling gives walls and floors an essential water-resistant quality, whilst providing an opportunity to create a suitably aquatic theme. Wall tiling to showers and other high-moisture areas is best carried out to a backing of special *Moistureshield* type plasterboard, and tiling to timber floors or boxed-in pipes is best laid over a water-resistant plywood base.

Tiles should be centred horizontally on walls to achieve a finished result that looks balanced. It's especially important with kitchen walls for the tiles to fit neatly in relation to electric sockets and switches.

Tiles and mosaics can also look good laid diagonally, although this requires more cutting. Always use a

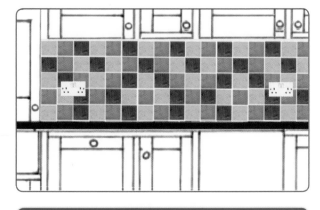

GREEN CHOICE ⌂

Floor finishes

1st	Linoleum
	Natural cork
	Natural stone
	Timber
2nd	Ceramic tiles
Not recommended	PVC vinyl
	PVC faced cork
Comments	Use natural adhesives

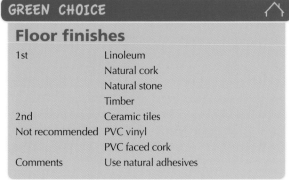

Photo: potton.co.uk

waterproof adhesive and grout (not just water-resistant) as anything else will absorb water and grow mould. Also, be sure to use the correct tile spacers to help even out any inconsistencies in tile shapes.

When it comes to floor tiling, ceramic, quarry, or stone tiles are usually laid on a screed. To successfully tile timber floating floors the surface must be as rigid as possible and a suitable flexible adhesive used.

External finishing

When it comes to painting the outside, you ideally want to order up some weather that's calm, dry, and reasonably warm. In other words, your work programme will need to be fairly flexible so you can spring into action when the right weather conditions do finally appear. But there may be other time pressures. Timber fascias and bargeboards need to be decorated after the roof tiling is complete but before the guttering is installed. Thanks to 'working at height' legislation the days of daredevil painters being sent on

dangerous missions up impossibly high ladders are long gone. So if the scaffolding has already been taken down, suitable scaffold towers can be used for all work at first-floor level and above.

Painting timber

The traditional way of finishing external timbers is with oil-based gloss paint, applied in three coats – primer, undercoat, and gloss. But today, new timber windows and doors are manufactured ready-primed with a honey-coloured stain basecoat, which can be finished on site with topcoats of stain or paint. However, any bare timbers, for example eaves or exposed rafter feet at roof level, will need the full treatment – which means knotting and priming (or a stain basecoat).

The main alternative to paint is wood stain. Stains are dearer to buy than paint, but quicker to apply, and should let you dispense with the traditional third coat. Because stain soaks into the timber, it's far less prone to peeling than paint or varnish. The thicker stains are classed as 'medium build' and have a light treacly consistency. The thinner 'low-build' variety is used on sawn timber, which sucks up a surprising amount of the stuff. Traditionally wood stains have been spirit-based but, like paint, they are now available as water-based acrylics, normally containing a fungicide to inhibit mould growth.

Tile hanging

Vertically hung wall tiles are a good way of visually breaking up a large expanse of brickwork. Tiling helps insulate external walls whilst allowing them to breathe, and requires very little maintenance, which may account for its enduring popularity.

Plain tiles are typically sized 265 x 165mm and each tile is hung on a batten and double-nailed in place. A layer of breather membrane underlay is first laid below the preservative-treated battens, which are commonly fixed to the blockwork outer leaf. Above the tiles there should be a horizontal strip of lead apron (Code 4 lead) placed under windowsills and dressed over the tile course below by about 100mm, in order to make the joint watertight.

Weatherboarding

There are different styles of timber cladding. Traditional overlapping 'barn style' cladding uses rough-sawn planks, fixed horizontally with a 30mm overlap. You can either use tapered 'feather-edge' boards or square-edged 'clapboarding'. Planed-timber rebated 'shiplap' cladding, on the other hand, uses interlocking boards that

are laid completely flush. Similarly, square-edge boarding or tongued and grooved planed timber boards can be close-butted, and are sometimes fixed diagonally or vertically.

Starting at the bottom, boards can be nailed horizontally to 50 x 50mm battens over a breather membrane, usually to a blockwork wall surface. At least 150mm should be left between the bottom edge of the weatherboard and the

ground. The corners can either be mitred or finished with a corner trim.

It's best to specify naturally durable 'fit and forget' timber such as western red cedar, which needs no painting. Cedar can cost five times the price of untreated softwood boarding, but it lasts for 60 years, although its original golden honey colour will, over time, become a silvery-grey. Oak and chestnut are less expensive 'maintenance-free' alternatives, as is iroko. Native larch is also popular and is significantly cheaper than cedar or other hardwoods.

Otherwise, softwood boarding normally needs to be treated with preservative to protect it against rot. Ideally it should arrive on site pressure-impregnated, prior to painting or staining. Redecoration will be required every five to seven years, but using low-maintenance wood stain or microporous paints can help extend this.

Alternatively, you may want to consider pre-painted fibre-cement boards that are manufactured to resemble timber but with none of the maintenance headaches. A recent development that arguably offers the best of both worlds is 'improved timber' boarding, such as Thermowood.

Render

A lovely smooth white or creamy render finish can make even the plainest property look smart and stylish. Render is normally applied directly to blockwork but can equally be spread on metal lathing nailed to battens or to insulation panels. It is built up in layers, starting with a base coat, about 7mm thick, using a 4:1 mix of sharp plastering sand to cement plus a waterproofer/plasticiser. The blockwork is first wetted and sometimes a PVA coat is applied to

Types of weatherboarding

Feather-edge boarding
(planed board)

Shiplap boarding
(flush)

Square edged boarding

Clipboarding
(square edge)

Shiplap boarding
(rebated)

Photo: Renderplas

Photo: oakwrights.co.uk

improve the bond. The surface of the base coat needs to be scratched to provide a key for the top coat, which can be applied once the 'scratch coat' has set and cured (normally after a day or two). The top coat uses a weaker 5:1:1 mix of plastering sand, cement, and lime, plus a waterproofer/plasticiser. The lime gives the finished render flexibility, to reduce cracking.

Although there is no set overall thickness, the two coats combined would typically total about 15mm. For a smooth finish the final coat is levelled with a straight edge before being rubbed down with a float or trowel to help close the surface, and then sponged over for a slightly sanded texture. At the base, rendered wall surfaces should project outwards just above DPC level with a 'bellmouth drip' to disperse rainwater.

But you don't have to use a conventional sand/cement mix. Modern hardwearing acrylic based polymer renders reinforced with alkali-resistant glass fibre are available in a wide variety of colours and finishes. There are also flexible render systems that are purpose designed for use with polystyrene backings, such as PIF construction. These can be supplied through-coloured and with a choice of various finishes, such as lumpy roughcast, but are relatively expensive.

Render is not normally maintenance free, requiring decoration every five years or so (although some masonry paints claim up to 15 years before needing recoating). Before painting newly rendered walls the underlying surface must be allowed to fully dry out, and then sealed with a stabilising solution to bind the surface and make it less absorbent. But if you've already taken the scaffolding down, painting the walls will take considerably longer.

Photo: Peter Kent Architects

Landscaping

For many self-builders, landscaping offers a creative opportunity to let rip and design a stunning garden backdrop that perfectly sets off their new dream home. A spot of planting may even provide welcome therapy after the stresses of the last 12 months.

To make it more enjoyable, this is often something that can be tackled at a leisurely pace once the new home is occupied. However, it may be financially beneficial to get the serious landscaping work done as part of overall build, so that your VAT claim can include the main garden materials. And the planners may have originally stipulated that certain works need to be carried out as a condition of your consent. This will be more of an issue in Conservation Areas, where they're likely to require that any garden walling is built in traditional materials, which can easily cost two or three times the price of standard work. Retaining walls are especially expensive.

GREEN CHOICE ⌂

Wall cladding

1st	Naturally durable timber (non-treated)
	Reclaimed bricks
2nd	Clay tile or natural slate
	Lime render
3rd	Cement render
Not recommended	UPVC, non-sustainable tropical
	hardwood, treated softwood

Comments

If possible specify materials which are 'self-finished' and don't require painting or staining and maintenance. Durable timber (timber that doesn't require treatment or painting) is the preferred cladding material as it can be used in its natural state without an applied finish. Durable timbers include oak, cedar, sweet chestnut, and larch, specified to avoid sapwood. Locally sourced or European timber is preferable.

Also recommended are tile hanging and/or self-coloured renders, preferably 'through colour' lime render. On timber frame walls this can be applied on a stainless steel mesh backing with a ventilated cavity behind.

There are two types of landscaping – soft and hard. Soft landscaping concerns shrubs, trees, topsoil, architectural gazebos and suchlike. Hard landscaping relates to paths, driveways, garden walls, patios, slabs for greenhouses etc.

A certain amount of preparation should have already been done laying water pipes and electric cables in trenches to supply water features, garages, and exterior lighting. But trying to mould a beautiful garden out of a building site means you'll probably still be faced with several weeks of hard work wielding spades and pushing wheelbarrows. A minimum of about 150mm of topsoil is usually required to landscape a garden, ideally recycling some that was excavated earlier, which may mean having to hire earth-moving equipment.

It's important that ground levels next to the house should be at least a couple of brick courses (150mm) below DPC level. If for any reason this can't be achieved, the next best option is to create a shallow gravel-filled trench around the house. This will allow any damp to evaporate and disperse safely away. Paths and hard surfaces near

the house should slope gently away to disperse rainwater.

As we saw in Chapter 8, the requirements for disabled access in Part M of the Building Regulations may have some influence on the design of your front garden. Pathways approaching the house should facilitate wheelchair access. This requires ramps leading up to external doors at shallow gradients rather than steps, and level thresholds as you enter the house.

Photo: Starline pools

Outbuildings

You may have decided to delay construction of the garage and any outbuildings in a bid to preserve cash flow. Being non-habitable, most are exempt from Building Regulations as long as they're smaller than 30m^2 floor area. However, where they've been built as part of the contracted work they must be 'snagged' at the end of the job, along with the rest of the property, prior to making final payment.

Security

When it comes to the security of your home, there are two main approaches – defence or exposure. A defensive strategy requires high garden walls, security gates, video surveillance, and security shutters on windows. This might be appropriate in very isolated rural areas where 'no one can hear you scream'. But this approach has the disadvantage of announcing to the world that you have something worth stealing, as well as conveying an image of paranoia and tension.

The exposure strategy takes the opposite approach by opening up the areas most likely to be attacked to public view. An open-plan front garden makes much of the house visible to neighbours and passers-by. Used in conjunction with movement-detecting security lights it can be as effective a defence as barbed wire. Trouble is, you're relying on there being a sufficient number of citizens hanging around the immediate area at any one time to deter suspicious characters from trying to break in.

They say there's nothing new under the sun, and sometimes it's old ideas that work best. High iron railings beloved of the Victorians combine both approaches, providing a barrier that prevents easy access whilst exposing the view to the house beyond.

Photo: Demax.co.uk

Photo: Grange Fencing

Completion

The last phase of the project needs to be very carefully managed, so that you stay in control and the builders finish the job properly.

NOTIFY BUILDING CONTROL 7
Final inspection – minimum two days' notice

Completion certificate

The job isn't over until Building Control have carried out their final inspection. This should always be carried out before your contract with the builders has terminated and they've completely vanished. Otherwise any remedial work that Building Control require could become your responsibility to complete. Once the final inspection has been done, before they can issue their completion certificate there are a few outstanding documents they'll need to get hold of, notably the SAP compliance report – see below.

Although not in any way a warranty, Building Control completion certificates are generally recognised as confirming that construction has been carried out to an acceptable basic standard, and mortgage lenders usually want to see this certificate before they'll release the final-stage payment.

But what Building Control consider complete, and what you regard as complete, can be two very different things. For example, they won't be terribly interested in all the cosmetic stuff, like whether the decorations have been done or the skirting boards are secure. Remember, the Building Regulations are only the minimum standard acceptable for health and safety and energy efficiency. They aren't a guarantee that the builders have finished everything they were supposed to have done. That's where snagging comes in, which is pretty much down to you.

Surplus plasterboard must be taken to special recycling centres.

Snagging

The objective of the snagging inspection is to draw up a list of all remaining defects. These are normally very minor things, like paint blemishes, sticking doors and windows, loose joinery, missing mastic to joints, messy tile grouting, and wonky wall sockets – in fact anything that doesn't look too clever. Surveyors have a trained eye and are the best people to do this job. If you do it yourself, you need to methodically inspect each room and component of the building in turn. So it's best to walk round unaccompanied so that you can concentrate. You'll then need to provide your builder with a copy of your snagging list and then take a long, hard walk around the building together, to agree all the points. Later you can check off satisfactorily completed snagging items.

Not surprisingly, snagging is a common bone of contention, but it's pointless to fall out with the contractor having got this far. If the work has been largely to a good standard, a tactful approach can go a long way. But it's no defence for the contractor to say that the defect has been there for weeks and an interim certificate 'passed it'. This doesn't mean the work was approved – just paid for. Any snags that are obvious prior to Practical Completion should be made good before final payment is made. If there are extensive snags, you're within your rights to refuse to accept completion until notified in writing by the contractor that the building has been properly finished.

Where you're directly employing your own tradespeople, you have considerably less leverage unless you thoroughly check their work prior to payment. But most reputable plumbers and electricians will be willing to make a return visit to rectify any significant faults. The best approach is to insist on work being done properly in the first place, and anticipate that you may need to do a small amount of redecoration and minor making good after the initial 'running-in period'.

Test certificates, SAP compliance, and EPCs

Before the job can be considered complete and final payment made you must receive signed commissioning certificates for the boiler/central heating and hot-water systems, to confirm they've been properly installed. Similarly, a test certificate for the electrics (IEE electrical test certificate) must be provided. Also, be sure to obtain operating instructions for all appliances and services, including ovens, hobs, ranges, boilers,

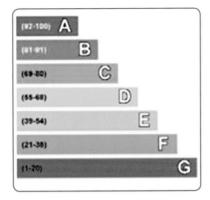

dishwashers, washing machines, alarm systems, and solar panels etc. It's a good idea to retain contact details for the trades who fitted them.

Amongst other things, the Building Control Surveyor conducting the completion inspection will want to be assured that no changes have been made during construction that might detract from the building's thermal performance calculated in the 'Design SAP'. As we saw earlier, a final 'as-built SAP' calculation will now need to be carried out on the finished house. From this, your Energy Performance Certificate (EPC) can be produced and the SAP compliance report filled out.

Practical Completion (PC)

Practical Completion is the date when you can occupy the house without any major inconvenience from building work. This applies where you have an architect or surveyor managing a main contractor. When they formally issue the Practical Completion Certificate, that triggers the release of the last-stage payment due under contract to the builders. If the contractor still needs to complete a few outstanding items, these can be listed along with the PC certificate. The usual safeguard against any minor snagging items not being finished is the retention, which is typically 5% of the contract sum, half of which will be released at Practical Completion. This is also the point when the 'defects liability period' commences, leaving 2.5% to be retained for a further six months – see below.

Final account

After such a long project, totting up exactly how much you owe can frequently become something of a negotiating matter. During the build there may have been some modifications and changes to the design, as well as unexpected extras that caused the original contract value to be amended, and the contract period may have had to be extended.

If you have a supervising architect they will draft a preliminary Final Account for you and the builder to mull over, and this will form the basis of any outstanding final payment due to the contractor. Or the contractor may submit their own version of events. Whether these figures look a little high or not, you'll need to run through them to compare them with your own records. You should especially

All done bar the fencing (and the weather).

check things like change instructions, extras, and prime cost sums. You may want to request copies of suppliers' invoices to help analyse the cost of materials that you're being charged for. If you haven't done so already, endeavour to agree a fair price for any additional work. This process can drag on for several weeks beyond Practical Completion until eventually an agreed Final Account can be prepared and all the loose ends tied up.

It's not unusual for there to be small differences of opinion, often due to fading memories, so some degree of compromise is usually needed on both sides. If the contractor has done a decent job and finished more or less on time it's only fair not to haggle too much over minor items.

Post-completion

This is the joyous occasion that you've been working towards for what must seem like a lifetime. The day you finally move in to your new home you feel a mixture of elation and pride mixed with sheer relief and exhaustion. But even now, the property won't be 100% perfection. There will probably be some minor faults that need to be rectified during the first six months, whilst you still hold the 2.5% retention. 'Latent defects' are problems that appear after a period of occupation, that weren't apparent at handover. Typically this will include things like minor shrinkage cracks to plasterwork and small leaks, which need to be made good with the help of a final snagging list.

At the end of the six months you or your architect should carry out a final inspection, and note any defects that need to be put right. Once rectified, the architect can issue a Final Certificate triggering release of the remaining 2.5% retention money due to the contractor. If you haven't done so already, calculate any outstanding sums due to the builders by deducting all payments made to date from the total in the agreed Final Account. Once paid, that should be the end of the project – apart from claiming lots of lovely money back from the Government.

Making your VAT claim

You can only make one VAT claim to HMRC, and it must be submitted strictly within three months of receiving your Building Control completion certificate. The receipts you send them must be originals, not photocopies, and they must be legible. For a successful application, every receipt needs to be individually listed and grouped by the type of materials. Before sending in your claim, check that you've included all the key information listed in Chapter 2. Once submitted, you can sit back and relax – and look forward to receiving a large cheque within a month.

And finally...

For self-builders, adding the finishing touches can take several more months, and sometimes years, refining the landscaping and perfecting the internal

Photo: oakwrights.co.uk

Main trades needed on site

- **Decorators**: Decorating walls and joinery.
- **Labourers**: Tidying up, cleaning, insulating lofts, landscaping.
- **Tilers**: Wall and floor tiling, external wall tile hanging.
- **Carpenters**: Weatherboarding.
- **Plasterers**: Rendering.

Who does what? See full list at www.Selfbuild-Homes.com.

decor. The good news is that at this stage you'll soon find your pockets are jingling with lots of reclaimed VAT money, and possibly also your final mortgage stage payment. After months of frugal living and watching every penny, there's inevitably a great temptation to spend. But first check that every outstanding bill has been settled.

Now that it's all over, the time has come to liquidate some assets. There's a ready market for second-hand caravans, and you should be able to get back some money for surplus building materials. Try posting an ad online, or approaching a local builder. Large packs can be returned to the store where purchased, although builders' merchants make a deduction for refunds. Oh, and don't forget to adjust your VAT refund claim as necessary!

At this point a celebration is in order. You deserve warm congratulations for seeing the job through. There's no such thing as a stress-free building project, so if you've survived the job reasonably unscathed – well done. Just be sure to give yourself a well-earned break before embarking on your next grand design!

Photo: Grange Fencing

Photo: oakwrights.co.uk

Photo: oakwrights.co.uk

INDEX